Lecture Notes in Control and Information Sciences

Volume 457

About this Series

This series aims to report new developments in the fields of control and information sciences—quickly, informally and at a high level. The type of material considered for publication includes:

1. Preliminary drafts of monographs and advanced textbooks
2. Lectures on a new field, or presenting a new angle on a classical field
3. Research reports
4. Reports of meetings, provided they are
   (a) of exceptional interest and
   (b) devoted to a specific topic. The timeliness of subject material is very important.

More information about this series at http://www.springer.com/series/642

Mohamed Djemai · Michael Defoort
Editors

Hybrid Dynamical Systems

Observation and Control

Springer

Editors
Mohamed Djemai
Michael Defoort
LAMIH, CNRS UMR 8201
University of Valenciennes
and Hainaut-Cambresis
Valenciennes
France

ISSN 0170-8643 ISSN 1610-7411 (electronic)
ISBN 978-3-319-10794-3 ISBN 978-3-319-10795-0 (eBook)
DOI 10.1007/978-3-319-10795-0

Library of Congress Control Number: 2014947659

Springer Cham Heidelberg New York Dordrecht London

Printed on acid-free paper

Springer is part of Springer Science+Business Media (www.springer.com)

To my wife, to Abdelaziz, Yacine and Ayman

Mohamed Djemai

To my wife and my daughter Elena

Michael Defoort

Preface

In any formulation of control, observation, or diagnosis problems, the mathematical model developed in order to establish the control law or to design an observer does not exactly reflect the actual process. These differences may for example be due to unmodeled dynamics, changes in system parameters, or approximations of the complex behavior of the process. Nevertheless, it must ensure that, despite all these uncertainties, the resulting control law achieves the predefined objectives and/or the observer can give the most accurate estimation possible of physical system parameters. This must be done under assumptions clearly defined with working conditions. From a historical perspective, the study of systems has been approached with continuous modeling using differential equations. The impressive progress in recent years, including the use of faster and faster computers, motivates the development of theoretical tools for a qualitative description of the dynamics systems.

However, in many cases, such description does not represent the complexity and richness of the behavior of the systems. Thus, knowledge of continuous dynamics modeling is often necessary to the synthesis of a control law implementable by a controller, or the synthesis of a state observer. The mathematical model resulting from the coupling of differential equations and the automaton modeling the system evolution from what is known as a hybrid automaton leads to hybrid dynamical systems (HDS). HDS are systems involving both continuous dynamics and discrete events. They can operate in several different modes which are described by a dynamical subsystem. The mode changes are governed by some discrete dynamics.

Switching systems (SS) are probably one of the most important class of hybrid systems. They are called variable structures if the laws describing its evolution are still made using differential equations and the system switches between these different laws. A large class of dynamical systems is modeled by a family of continuous subsystems and a logical law orchestrates switching between these subsystems.

Stability of HDS is critical for all applications. Therefore, it has received a lot of attention since the last two decades. Most of the existing literature brings solutions to solve whether the system is stable under arbitrary switching or whether the stability is achieved for a limited class of switching signals (using either common Lyapunov function or multiple Lyapunov approach). The observability analysis and observer design in HDS has received the interest of many researchers.

The problem of observer design is related to the extraction of unknown variables from the knowledge of the measured output. It is well known that, under certain assumptions, the state trajectory and the output response of any dynamical system are uniquely defined once the initial condition and the input are fixed. The problem of observer design usually deals with: (a) estimation of the state, (b) estimation of the external input (which can be either disturbance or faults), and (c) system security analysis (fault detection and isolation (FDI)). Occurrence of faults can be extremely detrimental, not only to the equipment and surroundings but also to the human operator if they are not detected and isolated in time. Hence, the increasing demand for safety and reliability of HDS during the last years has stimulated the attention to FDI techniques.

From the economical point of view, the study of HDS would enable to enlarge the spectrum of applications. Indeed, the hybrid modeling framework covers a large class of systems, which leads, for instance, to their application in a wide range of physical and engineering systems (power converters, control systems, process, computer science, biology, robotics, automotive industry, system temperature control, etc.).

This book is a collection of new trends in *Hybrid Dynamical Systems*. We invited several active researchers in this field to present the recent contributions of their groups.

Chapter 1, by M.D. Di Benedetto, S. Di Gennaro, and A. D'Innocenzo proposes a procedure to check diagnosability for hybrid systems. When analyzing a hybrid system, the dimension of the state space is often so large that formal verification is out of the question. Its analysis can be carried out using abstraction, namely the constructing a system with a smaller state space, preserving the properties to verify in the original system. Making use of a notion of diagnosability for hybrid systems, generalizing the notion of observability, in this chapter it is shown an abstraction procedure translating a hybrid system into a timed automaton, in order to verify observability and diagnosability properties. This procedure is applied to an electromagnetic valve system for camless engines.

In Chap. 2, by Zs. Lendek, P. Raica, J. Lauber, and T.M. Guerra an observer is designed for discrete-time switching nonlinear systems with a Takagi–Sugeno representation. For designing the observers, a switching nonquadratic Lyapunov function is used. Such Lyapunov functions have shown real improvement in the design conditions for discrete-time Takagi–Sugeno models. The Lyapunov function can be defined for each subsystem or just for the moments when switching takes place. In the first case, the results are more general, but also more conservative. The second case represents a significant improvement for periodic models. Thanks to the Lyapunov function used, it is possible to design observers for some switching systems with unobservable subsystems. The developed conditions are formulated as linear or bilinear matrix inequalities.

Chapter 3, by M. Petreczky, presents a survey on realization theory for linear hybrid systems. It deals with the problem of existence and minimality of a linear time-invariant state-space representation of an input–output map. The implications of realization theory for estimation and control of hybrid systems are discussed.

Chapter 4, by M. Djemai, N. Manamanni and J.P. Barbot, deals with observability conditions and state observer design for a class of hybrid systems. First, a high-order sliding mode-based observer is used to estimate the continuous state and to generate a discrete output. Secondly, starting from this discrete output, a discrete state reconstructor is designed. An illustrative example is provided to show the efficiency of the proposed observer.

Chapter 5, by N. Manamanni, M. Djemai and J.P. Barbot, deals with nonlinear observer design for autonomous SS with jumps. The jumps can result from the system dynamics or from the diffeomorphism which makes it possible to lead the system to an observability canonical form. In this chapter, the authors relate the design of a second-order sliding mode-based observer ("Super Twisting Algorithm"). It allows for estimating both continuous and discrete state related to the active dynamic.

In Chap. 6, by H. Rios, J. Davila and L. Fridman, the problem of continuous and discrete state estimation for switched nonlinear systems is solved using high-order sliding-mode techniques. In the first part of this chapter, systems with exogenous switching are studied. The solvability of the observation problem, for continuous and discrete states, is proposed using structural properties of the system. The high-order sliding-mode techniques are introduced to guarantee finite time convergence to zero of the estimation error for the continuous state. The discrete state is reconstructed using the information of the equivalent output injection.

In Chap. 7, by A. Tanwani, H. Shim and D. Liberzon, an observer design for switched linear systems with state resets is proposed based on the geometric conditions for large-time observability. Without assuming the observability of individual subsystems, the basic idea is to combine the maximal information available from each mode to obtain a good estimate of the state after a certain time interval has passed.

Chapter 8, by M. Petreczky, A. Tanwani, and S. Trenn, studies the observability of switched linear systems. This chapter focuses on the recently introduced geometric characterization of observability which assumes knowledge of the switching signal. These geometric conditions depend on computing the exponential of the matrix and require the exact knowledge of switching times. To relieve the computational burden, some relaxed conditions that do not rely on the switching times are given; this also allows for a direct comparison of the different observability notions. Furthermore, the generalization of the geometric approach to linear switched differential algebraic systems is possible and presented as well.

Chapter 9, by L. Hetel and E. Fridman, considers the stabilization problem for switched affine systems with a sampled-data switching law. The switching law is assumed to be a function of the system state at sampling instants. Sampling interval may be subject to variations or uncertainty. Switching law design criteria, taking into account the sampled-data implementation and uncertainties, are provided.

Chapter 10, by M. Farza, M. M'Saad, and K. Busawon, addresses the observer design problem for a class of continuous time dynamical systems with nonuniformly sampled measurements. More specifically, an observer is proposed that runs in continuous-time with an output error correction term that is updated in a mixed continuous discrete fashion. The proposed observer is actually an impulsive system. It can be put under the form of a hybrid system composed of a continuous time high gain observer coupled with an inter-sample output predictor. Simulations results dealing with a flexible joint robot arm are given to highlight the performance of the proposed observer.

Chapter 11, by W. Aggoune, B. Castillo Toledo, and S. Di Gennaro, presents results on self-triggering control for nonlinear systems. Conditions guaranteeing the existence of a self-triggered control strategy stabilizing the closed-loop system are presented both for deterministic and stochastic nonlinear systems. The problems addressed are self-triggered stabilization and safety. In the stochastic case, the state equations are described by an Itô differential equation driven by a Wiener noise, where the input enters either in the deterministic dynamics or in the dynamics affected by the noise. This kind of model embraces a quite large class of systems, of particular interest since in practice.

Chapter 12, by M. Defoort, J. Van Gorp and M. Djemai, presents an interesting benchmark for control, observation, and diagnosis of HDS: the multicellular converter. The first part of this chapter deals with the controller design for switching power converters, which are a particular class of switched systems. Then, it deals with the observer design to solve the capacitor voltages estimation while taking into account the hybrid behavior of the converter. A hybrid observer, based on gathering partial information from individual modes of the switched system, is designed. Some simulations highlight the efficiency of the proposed control and observer schemes for the three-cells converter.

Finally, in Chap. 13, by T.M. Laleg-Kirati, Z. Belkhatir and F.D. Ledezma, another application field of HDS is presented: biological systems. In biological systems, discrete behaviors might originate from unexpected changes in normal performance, e.g., a transition from a healthy to an abnormal condition. Simplifications, model assumptions, and/or modeled (and ignored) nonlinearities can be represented by sudden changes in the state. In this chapter, the authors are interested in modeling the cardiovascular system (CVS). Hybrid properties appear naturally in the CVS thanks to the presence of valves, which depending on their state (close or open) divide the cardiac cycle into four phases. The objective of this chapter is to use the properties of hybrid systems to describe this complex system.

It is our hope that this book will provide a clear and complete picture on the current state of the art of HDS theory.

Valenciennes, France, March 2014

Mohamed Djemai
Michael Defoort

Acknowledgments

This work was supported by the International Campus on Safety and Intermodality in Transportation, the European Community, the Regional Delegation for Research and Technology, the Ministry of Higher Education and Research, the Nord/Pas-de-Calais Region and the National Center for Scientific Research.

Contents

Contributors

Woihida Aggoune École Nationale Supérieure de l'Electronique et de ses Applications, Cergy-Pontoise, France

Jean Pierre Barbot ENSEA, Cergy-Pontoise, France

Zehor Belkhatir Computer, Electrical and Mathematical Science and Engineering (CEMSE) Division, King Abduallah University of Sciences and Technology (KAUST), Thuwal, Kingdom of Saudi Arabia

Krishna Busawon Engineering and Environment, Northumbria University, Newcastle upon Tyne, UK

Bernardino Castillo Toledo Centro de Investigación Y de Estudios Avanzados – CINVESTAV Del IPN, Unidad Guadalajara, Zapopan, Jalisco, Mexico

Alessandro D'Innocenzo Department of Information Engineering, Computer Science and Mathematics, Center of Excellence DEWS, University of L'Aquila, L'Aquila, Italy

Jorge Davila Section of Graduate Studies and Research, ESIME-UPT National Polytechnic Institute, Mexico, D.F., Mexico

Michael Defoort LAMIH, CNRS UMR 8201, University of Valenciennes and Hainaut-Cambresis, Valenciennes, France

Maria Domenica Di Benedetto Department of Information Engineering, Computer Science and Mathematics, Center of Excellence DEWS, University of L'Aquila, L'Aquila, Italy

Stefano Di Gennaro Department of Information Engineering, Computer Science and Mathematics, Center of Excellence DEWS, University of L'Aquila, L'Aquila, Italy

Mohamed Djemai LAMIH, CNRS UMR 8201, University of Valenciennes and Hainaut-Cambresis, Valenciennes, France

Mondher Farza GREYC Laboratory (UMR 6072 CNRS), Université de Caen and ENSICAEN, Caen Cedex, France

Leonid Fridman Division of Electrical Engineering, Department of Control and Robotics, National Autonomous University of Mexico, Mexico, D.F., Mexico

Emilia Fridman Department of Electrical Engineering and Systems, Tel Aviv University, Ramat Aviv, Tel Aviv, Israel

Thierry Marie Guerra LAMIH, CNRS UMR 8201, University of Valenciennes and Hainaut-Cambresis, Valenciennes, France

Laurentiu Hetel Laboratoire d'Automatique Genie Informatique et Signal (CNRS LAGIS UMR 8219), Ecole Centrale de Lille, Villeneuve d'Ascq Cedex, France

Taous Meriem Laleg-Kirati Computer, Electrical and Mathematical Science and Engineering (CEMSE) Division, King Abduallah University of Sciences and Technology (KAUST), Thuwal, Kingdom of Saudi Arabia

Jimmy Lauber University of Valenciennes and Hainaut-Cambresis, Valenciennes Cedex 9, France

Fernando Diaz Ledezma Computer, Electrical and Mathematical Science and Engineering (CEMSE) Division, King Abduallah University of Sciences and Technology (KAUST), Thuwal, Kingdom of Saudi Arabia

Zsófia Lendek Department of Automation, Technical University of Cluj-Napoca, Cluj-Napoca, Romania

Daniel Liberzon Coordinated Science Laboratory, University of Illinois at Urbana-Champaign, Urbana, IL, USA

Mohammed M'Saad GREYC Laboratory (UMR 6072 CNRS), Université de Caen and ENSICAEN, Caen Cedex, France

Noureddine Manamanni CReSTIC EA3804, Université de Reims Champagne Ardenne, Reims, France

Mihaly Petreczky Department of Automatic Control and Computer Science, Ecole des Mines de Douai, Douai, France

Paula Raica Department of Automation, Technical University of Cluj-Napoca, Cluj-Napoca, Romania

Hector Ríos Division of Electrical Engineering, Department of Control and Robotics, National Autonomous University of Mexico, Mexico, D.F., Mexico

Hyungbo Shim ASRI School of Electrical Engineering, Seoul National University, Seoul, South Korea

Aneel Tanwani Department of Mathematics, University of Kaiserslautern, Kaiserslautern, Germany

Stephan Trenn Department of Mathematics, University of Kaiserslautern, Kaiserslautern, Germany

Jeremy Van Gorp CReSTIC, UFR SEN, Université de Reims Champagne Ardenne, Reims, France

Chapter 1
Hybrid Systems and Verification by Abstraction

Maria Domenica Di Benedetto, Stefano Di Gennaro and Alessandro D'Innocenzo

Abstract The increase of functionality offered by today's control systems based on embedded systems requires more effort to verify the controlled system, as a malfunction can yield catastrophic results. These systems are usually hybrid systems, mixing continuous and discrete dynamics. When analyzing a hybrid system, the dimension of the state space is often so large that formal verification is out of the question. Its analysis can be carried out using abstraction, namely constructing a system with a smaller state space, preserving the properties to verify in the original system. Making use of a notion of diagnosability for hybrid systems, generalizing the notion of observability, in this paper it is shown an abstraction procedure translating a hybrid system into a timed automaton, in order to verify observability and diagnosability properties. The subclass of hybrid systems here considered is that of the durational graphs. We propose a procedure to check diagnosability, and show that the verification problem belongs to the complexity class P. This procedure is applied to an electromagnetic valve system for camless engines.

1.1 Introduction

The complexity of embedded control systems, due to the increase of their functionalities, renders more difficult to verify the absence of undesired behaviors in the controlled system. Bugs in these embedded controllers could be a delicate issue, since they may concern critical devices whose fault could yield to hazardous situations. Usually, these systems are characterized by continuous dynamics which, coupled with an embedded controller, become hybrid [5], namely a mix of discrete and continuous dynamics. The main difficulty when dealing with the verification

---

M.D. Di Benedetto · S. Di Gennaro (✉) · A. D'Innocenzo
Department of Information Engineering, Computer Science and Mathematics,
Center of Excellence DEWS, University of L'Aquila, Via G. Gronchi, 67100 L'Aquila, Italy
e-mail: stefano.digennaro@univaq.it

M.D. Di Benedetto
e-mail: mariadomenica.dibenedetto@univaq.it

A. D'Innocenzo
e-mail: alessandro.dinnocenzo@univaq.it

M. Djemai and M. Defoort (eds.), *Hybrid Dynamical Systems*, Lecture Notes in Control and Information Sciences 457, DOI 10.1007/978-3-319-10795-0_1

problem of such hybrid systems is the dimension of the state space, which is so large that formal verification cannot give a complete answer to the problem. A possible alternative is then the construction of systems with smaller state spaces, even finite, which preserve the property that has to be verified in the original system. This technique is usually called abstraction.

Two interesting properties to be verified in such hybrid systems are observability and diagnosability. This latter consists of the fault detection in finite time in a system, making use of its outputs. This paper deals with the automatic verification of these two properties in hybrid systems. To answer questions regarding diagnosability for a system controlled by an embedded controller means to be capable of detecting the occurrence of faults in finite time. The study of this property can also lead to an appropriate design of the embedded controller in order to impose prescribed fault detection times. The diagnosability of hybrid automata is discussed considering only those discrete output symbols which are observable, and which are associated to the discrete transitions. The concept of diagnosability is tightly related to observability, since diagnosability generalizes the observability property. Diagnosability has application in several fields, e.g., the detection of errors in an air traffic management systems [10, 11], in automotive systems [22], in industrial plants, in communication systems [36].

Given a hybrid system modeling, a given plant, and a subset of states associated with faulty behaviors, an important problem is the automatic verification of the diagnosability of the system. For the class of discrete event systems (DES), the diagnosability verification problem was shown to be polynomial [23, 30, 34, 35, 40]. While for DES, where the concept of time flow is not present, a system is defined diagnosable if it is possible to detect a fault after a finite number of transitions. After the fault, for timed automata, a definition of diagnosability linked with time was introduced [37]: a plant is Δ-diagnosable if it is possible to detect the fault after a time delay bound $\Delta \in \mathbb{N}$ after the fault. For timed automata, the diagnosability verification problem was shown to belong to PSPACE. For hybrid systems, the diagnosability verification problem for general hybrid systems, its decidability, and the computational complexity have not been characterized yet. A previous work in this directing is in Fourlas et al. [22], where a notion of diagnosability was proposed for input–output automata, and where some diagnosability conditions were given. A hybrid diagnosis problem was also presented in McIlraith et al. [32], along with qualitative techniques for diagnosis of continuous systems. In Zhao et al. [42], a diagnoser and a mode estimation algorithm for hybrid automata were presented.

This paper deals with the diagnosability verification for generic hybrid automata. Using the abstraction technique, a procedure for constructing a durational graph abstracting the hybrid system is constructed. A durational graph is a particular timed automaton, similar to the durational transition graph defined in Laroussinie et al. [28]. In a durational transition graph, the invariant sets are not defined, and the guards are rectangular sets. Durational graphs are more general than discrete event systems, where diagnosability verification is polynomial, and less general than timed automata, where diagnosability verification belongs to PSPACE. The proposed abstraction procedure is not guaranteed to be polynomial, but in some real applications, nontrivial and interesting in practice, it is polynomial, as in the case of electromagnetic valves. Thus, our abstraction method allows efficient diagnosability verification of hybrid systems for real cases of interest in engineering applications.

A further, and more important, contribution of this work is proving that the Δ-diagnosability verification problem for durational graphs belongs to the complexity class P, as in the case of DESs. The solution to the Δ-diagnosability verification problem given in this paper answers the question if a given system is Δ-diagnosable, and allows also computing the minimum value $\Delta_{\min}$ such that the system is $\Delta_{\min}$-diagnosable. Note that it is extremely important to compute the worst case delay for fault detection. The computation of $\Delta_{\min}$ in polynomial time is a novel and important result. The computation of the minimum value Δ for diagnosability of timed automata, presented in Tripakis [37], is done trying out different values of Δ using a binary search, and verifying the Δ-diagnosability at each step of the search. The same method could be applied to durational graphs, but it is not efficient since it depends on the binary search. On the contrary, the algorithm here presented computes $\Delta_{\min}$ in just one step, and in polynomial time with respect to the cardinality of the state space. The proposed verification algorithm provides a novel, constructive, and nontrivial proof, where the computation of $\Delta_{\min}$ reduces to a reachability problem over a set of durational graphs, constructed in polynomial time, starting from the original durational graph. This explains why tools like KRONOS [41], UPPAAL [29] (for timed automata), and HyTech [24] (for hybrid systems) cannot compute $\Delta_{\min}$, and cannot be compared with the presented algorithm. The verification algorithm can be divided into two parts. The first part handles unobservable symbols and represents the core of the procedure. The second part deals with systems with no unobservable transitions, and makes use of product automata algorithms similarly to Tripakis [37].

The theoretical results are here applied to the diagnosability verification of an electromagnetic valve system for camless engines, an innovative device in automotive applications [12]. The output used for diagnosis is the hitting times of the valve with the electromagnet surface.

The paper is organized as follows. In Sect. 1.2 the basic definitions are introduced. In Sect. 1.3, the definition of diagnosability for hybrid automata is given, which generalizes the notion of discrete state observability of D'Innocenzo et al. [13], and the procedure to construct the durational graph abstracting a hybrid automaton is proposed. In Sect. 1.4, it is proved that the diagnosability verification problem for durational graphs belongs to the complexity class P. In Sect. 1.5, the verification procedure is applied to an electromagnetic valve system for camless engines. Some remarks conclude the paper.

1.2 Basic Definitions

Hybrid systems constitute a particular class of systems, whose behavior is modeled considering both discrete and continuous dynamics. A framework which is largely used is that proposed in Lygeros et al. [31], where the discrete dynamics is modeled making use of automata, while the continuous dynamics is described by classical ODE associate to each state of the automaton. The continuous and the discrete dynamics influence each other, and this interaction is modeled using invariant conditions in each discrete state, guard conditions, and reset conditions in each transition arch of the automaton. In the following the subclass of hybrid automata is considered, characterized by autonomous dynamics.

Definition 1.1 A *hybrid automaton* is a tuple

$$\mathscr{H} = (Q \times X, Q_0 \times X_0, \mathscr{E}, E, \Psi, \eta, \mathbb{I}, G, R)$$

where

- $Q \times X$ is the hybrid state space, where Q is a finite set of discrete states and $X \subseteq \mathbb{R}^n$ is the continuous state space;
- $Q_0 \times X_0 \subseteq Q \times X$ is the set of initial conditions;
- $\mathscr{E}$ associates to each discrete state $q \in Q$ the dynamics $\dot{x} = f_q(x)$, with f_q locally Lipschitz;
- $E \subseteq Q \times Q$ is the set of edges;
- Ψ is the finite set of discrete output symbols ψ, including the null (unobservable) output ε;
- $\eta : E \rightarrow \Psi$ is the output function, that associates to each edge a discrete output symbol;
- $\mathbb{I}$ associates to each discrete state $q \in Q$ an invariant set $\mathbb{I}_q \subseteq X$;
- G associates to each edge $e \in E$ a guard set $G_e \subseteq \mathbb{I}_{s(e)}$, with $s(e)$ is the first component (source) of e;
- R associates to each edge $e \in E$ a reset map $R_e : \mathbb{I}_{s(e)} \rightarrow 2^{\mathbb{I}_{t(e)}}$ of the continuous state x, with $t(e)$ is the second component (target) of e.

It is worth noting that, although the continuous dynamics are deterministic, the discrete state of a hybrid automaton evolves according to the guards, possibly in a nondeterministic way.

It is assumed that the observable output is only given by the discrete output symbols associated to the discrete transitions.

A hybrid time basis $\tau = \{I_k\}_{k \geq 0}$ is a finite or infinite sequence of intervals $I_k = [t_k, t'_k]$ such that

1. I_k is closed if τ is infinite; I_k might be right-open if it is the last interval of a finite sequence τ;
2. $t_k \leq t'_k$ for all k and $t'_{k-1} \leq t_k$ for $k > 0$.

The length of the hybrid time basis is $|\tau|$.

A hybrid execution is a triple $\chi = (\tau, q, x)$, where τ is a hybrid time basis, and q, x describes the evolution of the discrete and continuous state by means of functions $q : \tau \rightarrow Q$ piecewise continuous, and $x : \tau \rightarrow X$. A hybrid automaton is nonblocking, if all hybrid executions are defined for all time instants. A hybrid automaton is non-Zeno, if infinite transitions cannot occur in a finite time interval.

Generally, Zeno executions are due to modeling mistakes or to model limitations. To each execution χ it is possible to associate a timed string $\rho(\chi)$ given by the sequence of pairs $\{(q_k, \delta_k)\}_{k=0}^{|\tau|}$, where $q_k \in Q$ is the discrete state assumed during the time interval I_k, and $\delta_k = t'_k - t_k \in \mathbb{R}_+ \cup \{0, \infty\}$ is the dwell time in q_k. Note that $\rho(\chi)$ is the execution of the discrete state of $\mathscr{H}$ and, although it is a timed string, in the following it will be called an execution. Given an execution $\rho = \{(q_k, \delta_k)\}_{k=0}^{|\rho|}$, $\rho|_k = q_k$ is the discrete state in the time interval I_k of the execution, with $k \in [0, \dots, |\rho|]$, $\rho|_{k_1}^{k_2} = q_{k_1}, \delta_{k_1}, \dots, q_{k_2}, \delta_{k_2}$ is the substring of ρ from index k_1 to k_2, and $\text{Time}(\rho|_{k_1}^{k_2}) = \sum_{j=k_1}^{k_2} \delta_j$ is the time duration of $\rho|_{k_1}^{k_2}$ (if $k_1 = 0, k_2 = |\rho|$, for short one writes $\text{Time}(\rho)$).

The set $\mathscr{L} = \{\rho(\chi) : \chi \in \mathscr{X}\}$, with $\mathscr{X}$ the set of all executions χ of a hybrid automaton $\mathscr{H}$, is called the timed language of executions of the discrete state of $\mathscr{H}$. Given a subset of discrete states $Q^\circ \subseteq Q$, $\mathscr{L}_{Q^\circ} = \{\rho \in \mathscr{L} : |\rho| < \infty, \rho|_{|\rho|} \in Q^\circ\}$ is the language of executions with finite cardinality, such that the last visited discrete state belongs to Q°.

The output string associated to an execution ρ is defined as the string $\delta_0, \eta(e_0)$, $\delta_1, \eta(e_1), \delta_2, \dots$, where $e_0 = (q_0, q_1), e_1 = (q_1, q_2), \dots$ are edges. The observation $P(\rho)$ is obtained from this output string erasing all the ε outputs, and adding the adjacent time intervals.

A timed automaton [4] is a particular hybrid automaton where the continuous dynamics are clocks with unitary slope ($\dot{x} = 1$), and where the guards are rectangular sets. A particular class of timed automata are the durational graphs.

Definition 1.2 A durational graph $\mathscr{G} = (Q, Q_0, E, \Psi, \eta, \mathbb{I}, G)$ is a hybrid automaton $\mathscr{H}$ such that

- The continuous state space is $X = \mathbb{R}_+ \cup \{0\}$;
- The continuous initial condition set is $X_0 = \{0\}$;
- The continuous dynamics are $\dot{x} = 1$, and $\mathbb{I}_q$ is a rectangular set, for all $q \in Q$;
- The set G_e is a rectangular set, and $R_e(x) = \{0\}$, for all $e \in E$.

For a durational graph $\mathscr{G}$, it is possible to define formal languages of executions and observations as for the generic hybrid automaton $\mathscr{H}$.

Given a hybrid automaton $\mathscr{H}$, let $Q_c \subset Q$ be a set of discrete states modeling a failure in $\mathscr{H}$. For convenience, we will call Q_c the faulty set.

Definition 1.3 An execution $\rho \in \mathscr{L}$ is Δ-faulty for $\mathscr{H}$, if there exists a finite index $k_c \in \big[0, |\rho|\big]$, if $|\rho| < \infty$, or $k_c \in \big[0, |\rho|\big)$, if $|\rho| = \infty$, such that $q_k = \rho|_k \notin Q_c$, $\forall k < k_c$, $q_{k_c} = \rho|_{k_c} \in Q_c$, and $\text{Time}(\rho|_{k_c}^{|\rho|}) = \Delta$.

A Δ-faulty execution is hence a trajectory that enters Q_c at a certain time instant, and then continues flowing for a time Δ. Let $\mathscr{F}_\Delta$ be the set of all Δ-faulty executions, and $\mathscr{F} = \bigcup_{\Delta \geq 0} \mathscr{F}_\Delta \subseteq \mathscr{L}$ the set of all faulty executions.

Definition 1.4 A set Q_c is Δ-diagnosable for $\mathscr{H}$ if $P(\rho) \neq P(\rho')$, $\forall \rho \in \bigcup_{\Delta^\circ \geq \Delta} \mathscr{F}_{\Delta^\circ}$, $\forall \rho' \in \mathscr{L} \setminus \bigcup_{\Delta^\circ \geq \Delta} \mathscr{F}_{\Delta^\circ}$.

Hence, a set Q_c is Δ-diagnosable for $\mathscr{H}$ when it is possible to detect, using the observable output, if an execution has visited the faulty set Q_c, with a maximal delay Δ. It is clear that, in practice, one is interested to the case in which Δ is finite. It is also worth noting that if Q_c is Δ-diagnosable, it is possible to detect that a state $q_c \in Q_c$ has been visited but it is not possible to distinguish which one, unless Q_c reduces to a singleton $\{q_c\}$.

1.3 Diagnosability and Timed Automaton Construction

In D'Innocenzo et al. [13], the observability of a set Q_c was introduced. More precisely, a set Q_c is observable for $\mathscr{H}$ if $P(\rho) \neq P(\rho')$, for all $\rho \in \mathscr{L}_{Q_c}$, and for all $\rho' \in \mathscr{L}_{Q\setminus Q_c}$. The property of Δ-diagnosability introduced in Definition 1.4 is more general than the discrete state observability, as stated by the following, easily proved from the definition of Δ-diagnosability, and from the fact that $\mathscr{F}_0 \subseteq \mathscr{L}_{Q_c}$.

Proposition 1.5 *Q_c is observable if and only if Q_c is 0-diagnosable.*

Therefore, the observability of a set Q_c is equivalent to the diagnosability with a time $\Delta = 0$. Therefore, *mudatis mutandis*, it is possible to apply the results of this paper to the case of observability of discrete states.

From Definition 1.4 descends the following property.

Proposition 1.6 *The following hold*

1. *If Q_c is Δ-diagnosable for $\mathscr{H}$, then Q_c is also Δ^*-diagnosable, for all $\Delta^* \geq \Delta$.*
2. *If Q_c is not Δ-diagnosable for $\mathscr{H}$, then Q_c is not Δ^*-diagnosable, for all $\Delta^* \leq \Delta$.*

Proposition 1.6 is very interesting in the view of the computation of the minimum value $\Delta_{\min}$ for which $\mathscr{H}$ is $\Delta_{\min}$-diagnosable.

In the following, an abstraction procedure to verify the Δ-diagnosability is proposed. Given a *nonblocking* hybrid automaton $\mathscr{H}$, this procedure leads to a durational graph $\mathscr{G}$ abstracting $\mathscr{H}$. The important point is that $\mathscr{G}$ has to preserve the diagnosability properties of $\mathscr{H}$.

Let us indicate with $\text{inc}(q) = \{e \in E \mid t(e) = q\}$, the set of the incoming edges of q. The procedure is based on the following relation:

$$\gamma = \Big\{(q, (q, l)) \mid q \in Q, l \in \text{inc}(q) \cup Q_0\Big\} \subseteq Q \times \Big(Q \times (E \cup Q_0)\Big)$$

which relates to each state $q \in Q$ of $\mathscr{H}$ one state of (q, g) of $\mathscr{G}$ for each incoming edge $l \in \text{inc}(q)$, and an additional state if $q \in Q_0$, as illustrated in Fig. 1.1.

The intuition behind the algorithm is that each discrete state is split, according to γ, depending on the number of incoming edges and initial conditions, so that any discrete state of $\mathscr{G}$ has only one incoming edge.

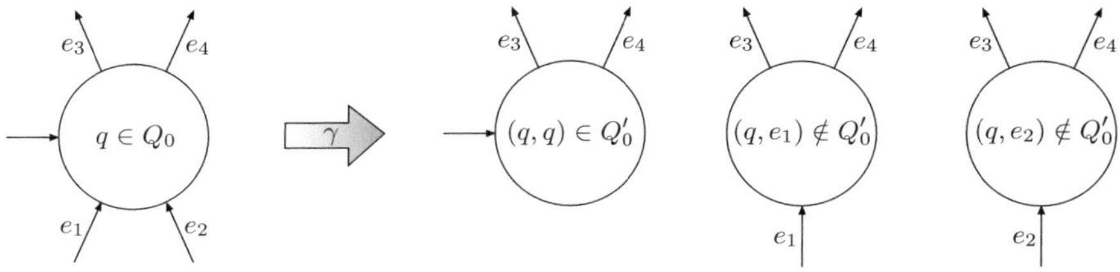

Fig. 1.1 Split induced by the relation γ. The *horizontal arrows* indicate an initial state

Algorithm 1.1 Given a nonblocking hybrid automaton $\mathscr{H} = (Q \times X, Q_0 \times X_0, \mathscr{E}, E, \Psi, \eta, \mathbb{I}, G, R)$, let $\mathscr{G} = (Q', Q'_0, E', \Psi', \eta', \mathbb{I}', G')$ be a durational graph defined as follows

- $Q' = \{(q, l) \mid (q, (q, l)) \in \gamma\}$;
- $Q'_0 = \{(q, q) \in Q' \mid q \in Q_0\}$;
- $E' = \{\big((q_1, l_1), (q_2, l_2)\big) \mid l_1 \in \text{inc}(q_1) \cup Q_0, l_2 = (q_1, q_2)\}$;
- $\eta'(e') = \eta(l_2), \forall e' = \big((q_1, l_1), (q_2, l_2)\big) \in E'$;
- $\mathbb{I}'_{q'} = \{t \in \overline{\mathbb{R}}_+ \mid \exists x_0 \in \mathscr{R}_l,\ x_{f_q}(\tau, x_0) \in \mathbb{I}_q,\ \forall \tau \in [0, t]\}, \forall q' = (q, l) \in Q'$;
- $G'_{e'} = \{t \in \overline{\mathbb{R}}_+ \mid \exists x_0 \in \mathscr{R}_{l_1},\ x_{f_{q_1}}(t, x_0) \in G_{l_2}\}, \forall e' = \big((q_1, l_1), (q_2, l_2)\big) \in E'$;

where $\overline{\mathbb{R}}_+ = \mathbb{R}_+ \cup \{0, \infty\}$, and

$$\mathscr{R}_l = \begin{cases} X_0(q_0) = \{x_0 \in X_0 \mid (x_0, q_0) \in X_0 \times Q_0\} & \text{if } l \in Q_0 \\ \text{Range}(R_e) = \{x \in X \mid \exists x' \in G_e, x \in R_e(x')\} & \text{if } l \in E. \end{cases}$$ ◇

In Algorithm 1.1, $X_0(q_0)$ is the set of initial continuous conditions associated to the initial discrete state $q_0 \in Q_0$, and $\text{Range}(R_e)$ is the range of the reset associated to $e \in E$. Since $\mathscr{H}$ is nonblocking,

$$X_0(q_0) \subseteq \mathbb{I}_{q_0}$$

and

$$\text{Range}(R_e) \subseteq \mathbb{I}_q, \quad \forall q \in Q,\ e \in \text{inc}(q).$$

Moreover, the set $\mathbb{I}'_{q'}$ consists of all time instants t such that there exists an execution of the continuous state, according to the dynamics f_q and with initial condition in $\mathscr{R}_l$ at time 0, that remains in the invariant set $\mathbb{I}_q$ during the time interval $[0, t]$. Notice that the origin $t = 0$ always belongs to $\mathbb{I}'_{q'}$, since we assumed that $\forall q \in Q$, $\forall e \in \text{inc}(q)$, $\text{Range}(R_e) \subseteq \mathbb{I}_q$, and $\forall q \in Q_0$, $X_0(q) \subseteq \mathbb{I}_q$.

Finally, the set $G'_{e'}$ consists of all time instants t such that there exists an execution of the continuous state, according to the dynamics f_{q_1} and with initial condition in $\mathscr{R}_{l_1}$ at time 0, that enables the transition l_2 at time t.

Remark 1.7 The Algorithm for abstraction can also be used to verify more general properties, not easy to check or even undecidable on the general hybrid model, e.g., temporal properties [9]. □

The main difficulty in Algoritm 3 is the computation of the invariants $\mathbb{I}'_{q'}$ and guards $G'_{e'}$ by means of dwell time of the hybrid automaton in each discrete state. If the continuous dynamics are linear, the exact computation is possible when the system has a particular structure [3, 27]. In general, using approaches in the literature, one can compute an overapproximation of

$$\mathbb{I}'_{q'} \supseteq \{t \in \mathbb{R}_+ \cup \{0, \infty\} : \exists x_0 \in \mathscr{R}_l, x_{f_q}(\tau, x_0) \in \mathbb{I}_q, \forall \tau \in [0, t]\} \tag{1.1}$$

$$G'_{e'} \supseteq \{t \in \mathbb{R}_+ \cup \{0\} \mid \exists x_0 \in \mathscr{R}_{l_1}, x_{f_{q_1}}(t, x_0) \in G_{l_2}\}. \tag{1.2}$$

Remark 1.8 There are various computational approaches available for the computation of reach sets for hybrid systems. If the dynamics are *linear*, but the computation of the reach set cannot be solved in closed form, one can obtain approximations of this set making use of various procedures. In Asarin et al. [6], a procedure for automatic verification of safety properties of hybrid systems with linear continuous dynamics, and uncertain bounded input, is proposed. The procedure in Yazarel and Pappas [39] works for high dimensional continuous state spaces, but it is not possible to quantify the overapproximation error. The procedure presented in Chutinan and Krogh [8] can be used to compute, within specific time interval, an overapproximation of the reach set using a sequence of polytopes. By refining this interval, it is possible to determine approximations with arbitrary precision, although this approach suffers from the increase in the computation time. Another procedure, similar to Chutinan and Krogh [8] but sensibly faster [17], is presented in Girard [26], based on the computation of a sequence of zonotopes (i.e., centrally symmetric polytopes). This algorithm can be also extended for the computation of the reach sets of systems with a bounded control input. For the same class of models, in Kurzhanski and Varaiya [19], overapproximating reachable sets with ellipsoidal shapes are proposed. Related results can be found in [18, 20, 21]. If the dynamics are *nonlinear*, the approach developed in Chutinan and Krogh [8] can be used. Moreover, various verification tools for reachability analysis on hybrid systems are d/dt, MATISSE, CheckMate, the Ellipsoidal Toolbox, etc. (see [38] for further details). For the case study of Sect. 1.5, the algorithms proposed in [14, 15] are applied for the abstraction procedure. □

In the following, the properties of $\mathscr{G}$ are studied. If not clear from the context, superscripts are used to refer to $\mathscr{H}$ or to $\mathscr{G}$.

The following statement says that the behavior of $\mathscr{G}$ embeds the behavior of $\mathscr{H}$.

Proposition 1.9 *Given $\mathscr{H}$, $\mathscr{G}$, for each execution $\rho = \{(q_k, \delta_k)\}_{k=0}^{|\rho|} \in \mathscr{L}^{\mathscr{H}}$, there exists an execution $\rho' = \{(q'_k, \Delta'_k)\}_{k=0}^{|\rho'|} \in \mathscr{L}^{\mathscr{G}}$ such that $|\rho| = |\rho'|$, $(q_k, q'_k) \in \gamma$, $\delta_k = \Delta'_k$, $\forall k = 0, \ldots, |\rho|$, and $\eta(q_{k-1}, q_k) = \eta(q'_{k-1}, q'_k)$, $\forall k = 1, \ldots, |\rho|$.*

Proof Consider the execution $\rho = \{(q_k, \delta_k)\}_{k=0}^{|\rho|} \in \mathscr{L}^{\mathscr{H}}$. If $q_0' = (q_0, q_0) \in Q_0'$, by construction of $\mathscr{G}$, $(q_0, q_0') \in \gamma$. If $e_1 = (q_0, q_1) \in E$, there exists $q_1' = (q_1, e_1)$ with $(q_1, q_1') \in \gamma$, such that $e_1' = (q_0', q_1')$ and $\eta'(e_0') = \eta(e_0)$. The invariant $\mathbb{I}'_{q_0'}$ and the guard G'_{e_1} computed as overapproximations (1.1), (1.2), imply that $\delta_0 \in \mathbb{I}'_{q_0'} \cap G'_{e_1}$, and $[0, \delta_0] \subseteq \mathbb{I}'_{q_0'}$. By iteration, one constructs an execution $\rho' = \{(q_k', \Delta_k')\}_{k=0}^{|\rho'|} \in \mathscr{L}^{\mathscr{G}}$ such that $|\rho| = |\rho'|$, $(q_k, q_k') \in \gamma$, $\delta_k = \Delta_k'$, $\forall k = 0, \ldots, |\rho|$, and $\eta(q_{k-1}, q_k) = \eta(q_{k-1}', q_k')$, $\forall k = 1, \ldots, |\rho|$. □

Consider now the following assumption:

Assumption 1.10 *Suppose that the guard and invariant sets of $\mathscr{G}$ are computed exactly, and the reset functions of the system $\mathscr{H}$ are* memoryless *(the system "forgets" its continuous state when a transition occurs)*

$$R_e(x) = \text{Range}(R_e), \qquad \forall e \in E, \quad \forall x \in G_e.$$

If Assumption 1 holds, not only the behavior of $\mathscr{G}$ embeds the behavior of $\mathscr{H}$ but also vice versa.

Proposition 1.11 *Under Assumption 1.10, given $\mathscr{H}$, $\mathscr{G}$, for each execution $\rho' = \{(q_k', \delta_k')\}_{k=0}^{|\rho'|} \in \mathscr{L}^{\mathscr{G}}$, there exists an execution $\rho = \{(q_k, \delta_k)\}_{k=0}^{|\rho|} \in \mathscr{L}^{\mathscr{H}}$ such that $|\rho| = |\rho'|$, $(q_k, q_k') \in \gamma$, $\delta_k = \delta_k'$, $\forall k = 0, \ldots, |\rho|$, and $\eta((q_{k-1}, q_k)) = \eta((q_{k-1}', q_k'))$, $\forall k = 1, \ldots, |\rho|$.*

Proof Consider the execution $\rho' = \{(q_k', \delta_k')\}_{k=0}^{|\rho'|} \in \mathscr{L}^{\mathscr{G}}$. By construction of $\mathscr{G}$, $q_0' = (q_0, q_0)$ for some $q_0 \in Q_0$, with $(q_0, q_0') \in \gamma$. Moreover, $q_1' = (q_1, e)$ for some q_1 successor of q_0 and $e = (q_0, q_1)$, with $(q_1, q_1') \in \gamma$. Then the invariant $\mathbb{I}'_{q_0'}$ and the guard G'_{e_1} computed exactly (as in Algorithm 1.1) imply that there exists $x_0 \in X_0(q_0)$ such that $x_{f_{q0}}(\delta_0', x_0) \in G_e$ and $x_{f_{q0}}(\tau, x_0) \in \mathbb{I}_{q_0}$, $\forall \tau \in [0, \delta_0']$. Under the assumption that the reset functions are memoryless, we construct by iteration an execution $\rho = \{(q_k, \delta_k)\}_{k=0}^{|\rho|} \in \mathscr{L}^{\mathscr{H}}$ such that $|\rho| = |\rho'|$, $(q_k, q_k') \in \gamma$, $\delta_k = \Delta_k'$, $\forall k = 0, \ldots, |\rho|$, and $\eta((q_{k-1}, q_k)) = \eta((q_{k-1}', q_k'))$, $\forall k = 1, \ldots, |\rho|$. □

We now show that the abstraction $\mathscr{G}$ preserves, constructed using $\gamma \subseteq Q^{\mathscr{H}} \times Q^{\mathscr{G}}$ in Algorithm 1.1, diagnosability (and thus observability).

Proposition 1.12 *Given $\mathscr{H}$, $\mathscr{G}$, a faulty set*

$$Q_c^{\mathscr{G}} = \bigcup_{q \in Q_c^{\mathscr{H}}} \{q' \in Q^{\mathscr{G}} \mid (q, q') \in \gamma\}.$$

is Δ-diagnosable for $\mathscr{H}$ if *$Q_c^{\mathscr{G}}$ is Δ-diagnosable for $\mathscr{G}$.*

Proof By contradiction, assume that the hypothesis is true and that $Q_c^{\mathcal{H}}$ is not Δ-diagnosable for $\mathcal{H}$. This implies by Definition 1.4 that there exists $\Delta^* \geq \Delta$, $\rho \in \mathcal{F}_{\Delta^*}^{\mathcal{H}}$ and $\rho' \in \mathcal{L}^{\mathcal{H}} \setminus \mathcal{F}^{\mathcal{H}}$ such that $P(\rho) = P(\rho')$. Let $\rho|_{k_c}$ be the first faulty state of ρ. By Proposition 1.9, there exist $\tilde{\rho}, \tilde{\rho}' \in \mathcal{L}^{\mathcal{G}}$ such that $P(\rho) = P(\tilde{\rho})$ and $P(\rho') = P(\tilde{\rho}')$. Furthermore, $\rho|_{k_c} \in Q_c^{\mathcal{G}}$ and $(\rho|_{k_c}, \tilde{\rho}|_{k_c}) \in \gamma$, thus $\tilde{\rho} \in \mathcal{F}_{\Delta^*}^{\mathcal{G}}$. For the same reasoning, it is clear that $\tilde{\rho}' \in \mathcal{L}^{\mathcal{G}} \setminus \mathcal{F}^{\mathcal{G}}$. This implies that $Q_c^{\mathcal{G}}$ is not Δ-diagnosable for $\mathcal{G}$, that is a contradiction. □

When Assumption 1.10 holds, Proposition 1.11 holds. As a consequence, Proposition 1.12 becomes a necessary and sufficient condition.

Proposition 1.13 *Under Assumption 1.10, given $\mathcal{H}$, $\mathcal{G}$, a faulty set $Q_c^{\mathcal{H}}$ is Δ-diagnosable for $\mathcal{H}$* if and only if *$Q_c^{\mathcal{G}}$ is Δ-diagnosable for $\mathcal{G}$.* □

Proof Straightforward inverting the reasoning in the proof of Proposition 1.12. ■

1.4 Diagnosability Verification

In this section, the diagnosability problem for durational graphs is studied. A verification procedure is proposed and its computational complexity is determined. The verification algorithm consists of two parts. In the first part, the edges associated to an unobservable output symbol are considered, and an algorithm is presented to construct a durational graph without unobservable outputs, preserving diagnosability. In the second part, a verification algorithm for systems which do not generate unobservable outputs is proposed.

Removal of the edges $e \in E$ such that $\eta(e) = \varepsilon$, or ε-edge for short, is discussed in Ozveren and Willsky [33] for discrete event systems to study observability properties, and in Choffrut and Goldwurm [7] for timed automata to preserve simulation relations. In D'Innocenzo et al. [13] a procedure to erase ε-edges from a durational graph is proposed, while preserving observability properties. However, this procedure does not preserve diagnosability since, if a faulty state q_c has at least one outgoing or incoming ε-edge, such state is not observable, while it might be Δ-diagnosable for some $\Delta > 0$. For this reason, in order to preserve Δ-diagnosability, it is necessary to be careful when erasing ε-edges which have a faulty state as endpoint.

Let us consider a durational graph $\mathcal{G} = (Q, \{q_s\}, E, \Psi, \eta, \mathbb{I}, G)$, a destination discrete state $q_d \in Q$, and the set

$$\Lambda(q_d) = \left\{ t \in \mathbb{R}_+ \cup \{0\} \mid \exists\, \rho \in \mathcal{L}_{\{q_d\}}^{\mathcal{G}},\ \mathrm{Time}(\rho) = t \right\}$$

of all time instants t such that there exists an execution of $\mathcal{G}$ with time duration t that terminates in q_d. If there exists an execution $\rho \in \mathcal{L}^{\mathcal{G}}$ such that $\mathrm{Time}(\rho) = \infty$ and $\rho|_i \neq_d, \forall i \geq 0$, (this happens when the system remains confined forever in a cycle of unobservable edges), then $\infty \in \Lambda(q_d)$.

Proposition 1.14 *Given a durational graph $\mathcal{G}$ and a discrete state $q_d \in Q$, then $\Lambda(q_d)$ can be computed in polynomial time. Moreover, $\Lambda(q_d)$ is a rectangular set* if and only if *all paths of $\mathcal{G}$ that connect q_s to q_d do not contain cycles, or there exists at least one path with cycles where at least one edge is associated to a nonsingleton rectangular interval.*

Proof Given $\mathcal{G}$ and q_d, consider a nondeterministic finite automaton [25] $\mathcal{N}$ with set of states Q, initial state $\{q_s\}$ and final state $\{q_d\}$. The alphabet of $\mathcal{N}$ is defined as a finite set of rectangular intervals $\Sigma = \{\sigma_e : e \in E, \sigma_e = G_e \cap \mathbb{I}_{s(e)}\}$. Namely, it is the collection of all the guards G_e intersected with the corresponding invariant $\mathbb{I}_{s(e)}$. The transition relation $\Delta : Q \times \Sigma \to 2^Q$ of $\mathcal{N}$ is defined as $\Delta(q, \sigma_e) = \{q' \in Q \mid e = (q, q')\}$. $\Lambda(q_d)$ can be computed from the regular expression associated to $\mathcal{N}$, replacing

1. alternation $\sigma \mid \sigma'$ with $\sigma \cup \sigma'$ (finite union of rectangular intervals);
2. concatenation $\sigma \cdot \sigma'$ with $\sigma + \sigma'$ (finite sum of rectangular intervals);
3. Kleene star σ^* with $\bigcup_{n\geq 0} n\sigma$ (infinite union of rectangular intervals);

where $\sigma, \sigma' \in \Sigma$. We show that, if the hypothesis of this proposition holds, then $\Lambda(q_d)$ is a rectangular interval. Clearly (1) and (2) always generate rectangular intervals, since a finite union or sum of rectangular intervals is a rectangular interval, and each element of Σ is a rectangular interval. We now prove that (3) generates a rectangular interval if and only if σ^* is not generated by all singleton intervals, that is equivalent to the hypothesis of this proposition. Let $\sigma = [t_1, t_1'] \cup \cdots \cup [t_r, t_r']$, $r \in \mathbb{N}$. Then,

$$\sigma^* = \bigcup_{n\geq 0} n\sigma = \bigcup_{n\geq 0} \left(n[t_1, t_1'] \cup \cdots \cup n[t_r, t_r'] \right).$$

We first prove that there exists a finite value $N_i \in \mathbb{N}$ for each $i \in \{1, \ldots, r\}$, such that

$$\forall n \geq N_i, \quad nt_i < (n+1)t_i \leq nt_i' < (n+1)t_i'.$$

The strict inequalities are clearly true for any value of i, n. Since $t_i' \geq t_i$, $\forall i \in \{1, \ldots, r\}$,

$$(n+1)t_i \leq nt_i' \quad \Leftrightarrow \quad n \geq \frac{t_i}{t_i' - t_i}.$$

Let $N_i = \left\lceil \frac{t_i}{t_i'-t_i} \right\rceil \in \mathbb{N}, \forall i \in \{1, \ldots, r\}$, and define $t_\infty = \min_i\{N_i t_i\} \in \mathbb{R}_+ \cup \{0, \infty\}$. It is clear that t_∞ is finite if and only if $\exists i \mid t_i' > t_i$. It is also clear that for $n \geq N_k$, $k = \text{argmin}\{N_i t_i\}$, all consecutive time intervals $n[t_k, t_k']$ and $(n+1)[t_k, t_k']$ overlap, thus their infinite union $\bigcup_{n\geq N_k} n[t_k, t_k']$ generates the set $[t_\infty, +\infty)$. Let N_∞ be the minimum value of n such that $nt_i \geq t_\infty$, $\forall i \in \{1, \ldots, r\}$. Thus, σ^* can be defined by a finite union of rectangular intervals

$$\sigma^* = \left\{ \bigcup_{n=0}^{N_\infty} n\sigma \right\} \cup [t_\infty, +\infty). \qquad \blacksquare$$

The following algorithm takes as input a durational graph $\tilde{\mathscr{G}}$ and a faulty set $\tilde{Q}_c$ of $\tilde{\mathscr{G}}$, and produces as output a durational graph $\mathscr{G}$ without ε-edges, a faulty set Q_c of $\mathscr{G}$, and a faulty function $\lambda : Q_c \to 2^{\mathbb{R}_+ \cup \{0,\infty\}}$, that associates to each faulty state of $\mathscr{G}$ a rectangular time interval. $\square$

Algorithm 1.2 Given a durational graph $\tilde{\mathscr{G}} = (\tilde{Q}, \tilde{Q}_0, \tilde{E}, \tilde{\Psi}, \tilde{\eta}, \tilde{\mathbb{I}}, \tilde{G})$, and a faulty set $\tilde{Q}_c$, initialize

$$\mathscr{G} = \tilde{\mathscr{G}}, \quad \Psi = \tilde{\Psi} \setminus \{\varepsilon\}, \quad Q_c = \tilde{Q}_c, \quad \lambda(q) = \{0\}, \qquad \forall q \in Q_c.$$

Let $\mathrm{cl}_\varepsilon(q)$ be the classical ε-closure of a discrete state q [25], namely the set of discrete states that can be reached from q via a path of ε-edges. Notice that $q \in \mathrm{cl}_\varepsilon(q)$. For each unvisited state $q_s \in Q$ such that $\exists e \in \mathrm{inc}(q_s), \eta(e) \neq \varepsilon$, and for each $q_d \in Q, \psi \in \Psi$ such that $\exists q^* \in \mathrm{cl}_\varepsilon(q_s) \setminus \{q_s\}, \exists e = (q^*, q_d) \in E, \eta(e) = \psi$, define $Q_c^* = \mathrm{cl}_\varepsilon(q_s) \setminus \{q_s\} \cap Q_c$ and proceed as follows

1. If $Q_c^* \neq \varnothing$, then create a new state q_s^c to Q and Q_c while keeping incoming and outgoing edges of q_s (namely duplicate the state q_s^c). For each $q_c \in Q_c^*$, compute $\Lambda_1(q_c)$ from the durational graph $\mathscr{G}_1$ induced on $\mathscr{G}$ by the set of states $\mathrm{cl}_\varepsilon(q_s)$, the set of edges $\{e \in E \mid \eta(e) = \varepsilon\}$, and with initial state q_s (the durational graph induced on a durational graph $\mathscr{G}$ by the set of states X and the set of edges Y, is a durational graph where the set of states is X, the set of edges is Y, and outputs, guards, invariants, and resets are the same of $\mathscr{G}$ restricted to the sets X, Y). Compute $\Lambda_2(q_d)$ from the durational graph $\mathscr{G}_2$ induced on $\mathscr{G}$ by the set of states $\mathrm{cl}_\varepsilon(q_s) \cup \{q_d\}$ and the set of edges $\{e \in E : \eta(e) = \varepsilon \text{ or } s(e) \in \mathrm{cl}_\varepsilon(q_s) \setminus \{q_s\}, t(e) = q_d, \eta(e) = \psi\}$, with initial state q_c. Set $\lambda(q_s^c) = \lambda(q_s) \cup \bigcup_{q_c \in Q_c^*} \Lambda_1(q_c)$. If $e = (q_s^c, q_d) \notin E$, then add e to E and set $\eta(e) = \psi, G_e = \varnothing$. Set $G_e = G_e \cup \bigcup_{q_c \in Q_c^*} (\Lambda_1(q_c) + \Lambda_2(q_d))$. Mark q_s^c as visited.
2. Compute $\Lambda_3(q_d)$ from the durational graph $\mathscr{G}_3$ induced on $\mathscr{G}$ by the set of states $(\mathrm{cl}_\varepsilon(q_s) \setminus Q_c^*) \cup \{q_d\}$ and the set of edges $\{e \in E \mid \eta(e) = \varepsilon \text{ or } s(e) \in \mathrm{cl}_\varepsilon(q_s) \setminus \{q_s\}, t(e) = q_d, \eta(e) = \psi\}$, with initial state q_s. If $e = (q_s, q_d) \notin E$, then add e to E and set $\eta(e) = \psi, G_e = \varnothing$. Set $G_e = G_e \cup \Lambda_3(q_d)$. Mark q_s as visited. Finally, set $\mathbb{I}(q) = [0, \max_{e \in \mathrm{out}(q)} \{\sup\{G_e\}\}]$ for each $q \in Q$, with $\mathrm{out}(q) = \{e \in E \mid s(e) = q\}$ the set of outgoing edges from q, and erase all states whose incoming edges are all ε-edges, then erase all hanging and unobservable edges.

The idea behind Algorithm 1.2 is to preserve all executions merging the states connected by ε-edges. The main issue is that it is sometimes needed to merge together faulty and not faulty states. In this case, one considers all paths that visit the faulty states, creates a new faulty state q_s^c, and sets $\lambda(q_s^c)$ as the set of all time instants such that a fault can occur starting from q_s, generating only unobservable outputs. The

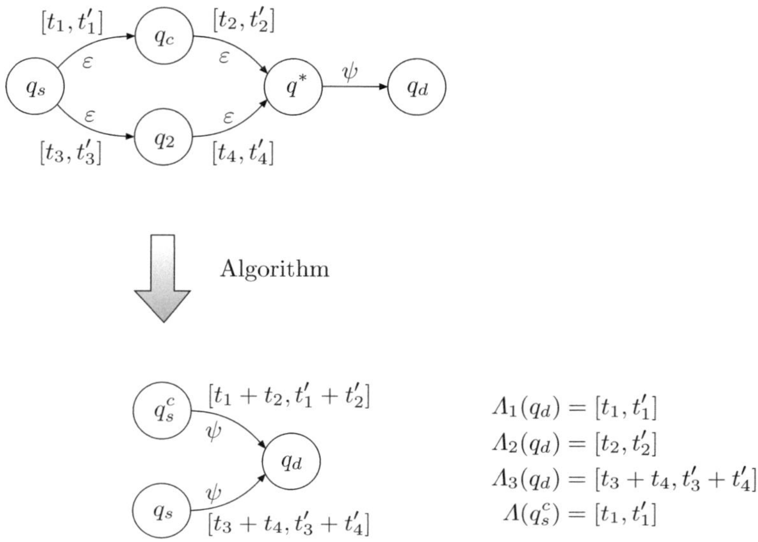

Fig. 1.2 Example of the split procedure operated by Algorithm 1.2

guards of the outgoing edges from q_s are set as the time instant sets such that an observable output is generated, without visiting a faulty state. Moreover, the guards of the outgoing edges from q_s^c are set as the time instant sets such that an observable output is generated, visiting a faulty state. The guards of $\mathscr{G}$ are assumed not singleton sets (actually, it is sufficient to assume that in the cycles of unobservable edges, at least one edge is associated to a guard that is not a singleton set), so that Proposition 1.14 implies that the Algorithm 1.2 constructs a durational graph (Fig. 1.2). In fact, every edge is associated to a guard that is a rectangular set. This is true also if there exist cycles of unobservable edges.

In the following result some properties are given, which hold by construction of Algorithm 1.2.

Proposition 1.15 *Given $\tilde{\mathscr{G}}$, $\mathscr{G}$*

1. *For each execution*

$$\tilde{\rho} \in \mathscr{F}_{\Delta}^{\tilde{\mathscr{G}}}$$

   and for each $\Delta \geq 0$, there exists an execution $\rho \in \mathscr{F}_{\Delta+\lambda^*}^{\mathscr{G}}$ and $\lambda^* \in \lambda(\rho|_{k_c})$, such that $P(\tilde{\rho}) = P(\rho)$;
2. *For each execution $\tilde{\rho} \in \mathscr{L}^{\tilde{\mathscr{G}}} \setminus \mathscr{F}^{\tilde{\mathscr{G}}}$, there exists an execution $\rho \in \mathscr{L}^{\mathscr{G}} \setminus \mathscr{F}^{\mathscr{G}}$ such that $P(\tilde{\rho}) = P(\rho)$;*
3. *For each execution $\rho \in \mathscr{F}_{\Delta}^{\mathscr{G}}$, for each $\lambda^* \in \lambda(\rho|_{k_c})$ and for each $\Delta \geq \lambda^*$, there exists an execution $\tilde{\rho} \in \mathscr{F}_{\Delta-\lambda^*}^{\tilde{\mathscr{G}}}$ such that $P(\rho) = P(\tilde{\rho})$;*

4. *For each execution* $\rho \in \mathcal{L}^{\mathcal{G}} \setminus \mathcal{F}^{\mathcal{G}}$, *there exists an execution* $\rho \in \mathcal{L}^{\tilde{\mathcal{G}}} \setminus \mathcal{F}^{\tilde{\mathcal{G}}}$ *such that* $P(\rho) = P(\tilde{\rho})$.

In the second part of the algorithm, it is possible to assume the absence of ε-edges. The idea for diagnosability verification is to construct a finite set of durational graphs that embed diagnosability properties of $\mathcal{G}$. It is shown how to check if there exists a (minimum) finite Δ such that the starting system $\tilde{\mathcal{G}}$ (with ε-edges) is Δ-diagnosable, by stating conditions on the structure of the constructed durational graphs. Given $\mathcal{G}$, one first defines a constructive procedure of a product durational graph $\mathcal{C}_{q_0',q_0''}$ for each pair $(q_0', q_0'') \in Q_0 \times Q_0$. The following algorithm is similar to the classical product automaton construction: the difference is that one stops the exploration when discovers a faulty state in the first component, and does not explore pairs of states with a faulty state in the second component.

Algorithm 1.3 Given the durational graphs

$$\mathcal{G}_{q_0'} = (Q, \{q_0'\}, E, \Psi, \eta, G, \mathbb{I}), \qquad \mathcal{G}_{q_0''} = (Q, \{q_0''\}, E, \Psi, \eta, G, \mathbb{I})$$

construct the product durational graph

$$\mathcal{C}_{q_0',q_0''} = (\tilde{Q} \subseteq Q \times Q, \{\tilde{q}_0\}, \tilde{E}, \Psi, \tilde{\eta}, \tilde{G}, \tilde{\mathbb{I}})$$

as follows

1. Initialize $\tilde{Q} = \tilde{q}_0 = (q_0', q_0'')$, $\tilde{\mathbb{I}}((q_0', q_0'')) = \mathbb{I}_{q_0'} \cap \mathbb{I}_{q_0''}$, $\tilde{E} = \varnothing$
2. For each unvisited state $(q', q'') \in \tilde{Q}$, do
   - 2.1 For each $e', e'' \in E$ such that $e' = (q', \bar{q}')$, $e'' = (q'', \bar{q}'')$, $\eta(e') = \eta(e'') = \psi \in \Psi$, $\bar{q}'' \notin Q_c$, set $\tilde{Q} = \tilde{Q} \cup (\bar{q}', \bar{q}'')$, $\tilde{\mathbb{I}}((\bar{q}', \bar{q}'')) = \mathbb{I}_{\bar{q}'} \cap \mathbb{I}_{\bar{q}''}$, $e = ((q', q''), (\bar{q}', \bar{q}''))$, $\tilde{E} = \tilde{E} \cup \tilde{e}$, $\tilde{\eta}(\tilde{e}) = \psi$, $\tilde{G}_{\tilde{e}} = G_{e'} \cap G_{e''}$, and then mark (q', q'') as visited;
   - 2.2 If $\bar{q}' \in Q_c$, then mark $(\bar{q}', \bar{q}'')$ as visited.

Given $\mathcal{G}_{q_0'}$, $\mathcal{G}_{q_0''}$, $\mathcal{C}_{q_0',q_0''}$, and any pair (q_c, q) in the state space $\tilde{Q}$ of $\mathcal{C}_{q_0',q_0''}$, then there exist two executions ρ' of $\mathcal{G}_{q_0'}$ and ρ'' of $\mathcal{G}_{q_0''}$ with the same timed observation, where the last visited state of ρ' is q_c, while the last visited state of ρ'' is q. This property is a direct consequence of the classical product automaton construction, and can be formalized by the following proposition:

Proposition 1.16 *Given* $\mathcal{G}$ *and* $\mathcal{C}_{q_0',q_0''}$

1. *For each pair of executions*

$$\rho' = \{(q_k', \delta_k')\}_{k=1}^{|\rho'|} \in \mathcal{F}_0^{\mathcal{G}}, \rho'' = \{(q_k'', \delta_k'')\}_{k=1}^{|\rho''|} \in \mathcal{L}^{\mathcal{G}} \setminus \mathcal{F}^{\mathcal{G}}$$

with $P(\rho') = P(\rho'')$*, there exists an execution* $\rho = \{(q_k, \delta_k)\}_{k=1}^{|\rho|} \in \mathscr{L}^{\mathscr{C}_{q'_1,q''_1}}$ *such that* $(q_k, \delta_k) = ((q'_k, q''_k), \delta'_k = \delta''_k)$*,* $\forall k = 1, \ldots, |\rho|$ *and* $P(\rho) = P(\rho') = P(\rho'')$.

2. *For each execution*

$$\rho = \{(q_k, \delta_k)\}_{k=1\cdots|\rho|} \in \mathscr{L}^{\mathscr{C}_{q'_1,q''_1}}_{\{(q_c,q)\}}$$

$(q_c, q) \in Q_c \times (Q \setminus Q_c), \delta_{|\rho|} = 0$*, there exists a pair of executions* $\rho' = \{(q'_k, \delta'_k)\}_{k=1\ldots|\rho'|} \in \mathscr{F}_0^{\mathscr{T}}$*,* $\rho'' = \{(q''_k, \delta''_k)\}_{k=1\ldots|\rho''|} \in \mathscr{L}^{\mathscr{T}} \setminus \mathscr{F}^{\mathscr{T}}$ *with* $P(\rho') = P(\rho'')$ *such that* $(q_k, \delta_k) = ((q'_k, q''_k), \delta'_k = \delta''_k)$*,* $\forall k = 1 \ldots |\rho|$ *and* $P(\rho) = P(\rho') = P(\rho'')$.

We define the set $Q_0^{\mathscr{C}} = \{(q_c, q) \in Q_c \times (Q \setminus Q_c) : \exists (q'_0, q''_0) \in Q_0 \times Q_0, (q_c, q) \in Q^{\mathscr{C}_{q'_0,q''_0}}\}$. Notice that $Q_0^{\mathscr{C}}$ is the set of pairs $(q_c, q) \in Q_c \times (Q \setminus Q_c)$ such that there exist two executions ρ', ρ'' with the same timed observation, where the last visited state of ρ' is q_c, while the last visited state of ρ'' is q. Consider now, for each pair $(q_c, q) \in Q_0^{\mathscr{C}}$, the product durational graph $\mathscr{D}_{q_c,q}$ is obtained using a version of Algorithm 1.3 where line **2.2** is deleted. By construction, an execution of $\mathscr{D}_{q_c,q}$ models suffixes of two parallel executions of $\mathcal{G}$ having same observation. The first one is a suffix starting from q_c of a faulty execution of $\mathscr{G}$, and the second one is a suffix starting from q of a nonfaulty execution of $\mathscr{G}$. This property is a direct consequence of the classical product automaton construction, and can be formalized by the following proposition.

Proposition 1.17 *Given* $\mathscr{G}$*,* $\mathscr{D}_{q_c,q}$

1. *For each* $\Delta > 0$ *and pair of executions* $\rho' = \{(q'_k, \delta'_k)\}_{k=1}^{|\rho'|} \in \mathscr{F}_\delta{}^{\mathscr{G}}$ *and* $\rho'' = \{(q''_k, \delta''_k)\}_{k=1}^{|\rho''|} \in \mathscr{L}^{\mathscr{G}} \setminus \mathscr{F}^{\mathscr{G}}$ *with* $P(\rho') = P(\rho'')$*,* k_c *the index of the first faulty state of* ρ'*,* $\rho'_{k_c} = q_c \in Q_c$ *and* $\rho''_{k_c} = q \in Q \setminus Q_c$*, there exists an execution* $\rho = \{(q_k, \delta_k)\}_{k=1}^{|\rho|} \in \mathscr{L}^{\mathscr{D}_{q_c,q}}$ *such that* $(q_k, \delta_k) = ((q'_{k+k_c}, q''_{k+k_c}), \delta'_{k+k_c} = \delta''_{k+k_c})$*,* $\forall k = 1, \ldots, |\rho|$*,* $\mathrm{Time}(\rho) = \Delta$*, and* $P(\rho) = P(\rho'|_{k_c}^{|\rho'|}) = P(\rho''|_{k_c}^{|\rho''|})$.
2. *For each execution* $\rho = \{(q_k, \delta_k)\}_{k=1}^{|\rho|} \in \mathscr{L}^{\mathscr{D}_{q_c,q}}$*, there exists a pair of executions* $\rho' = \{(q'_k, \delta'_k)\}_{k=1}^{|\rho'|} \in \mathscr{F}^{\mathscr{G}}_{\mathrm{Time}(\rho)}$*,* $\rho'' = \{(q'_k, \delta''_k)\}_{k=1}^{|\rho''|} \in \mathscr{L}^{\mathscr{G}} \setminus \mathscr{F}^{\mathscr{G}}$ *with* $P(\rho') = P(\rho'')$*,* k_c *the index of the first faulty state of* ρ'*,* $\rho'_{k_c} = q_c$ *and* $\rho''_{k_c} = q \in Q \setminus Q_c$*,* $(q_k, \delta_k) = ((q'_{k+k_c}, q''_{k+k_c}), \delta'_{k+k_c} = \delta''_{k+k_c})$*,* $\forall k = 1, \ldots, |\rho|$ *and* $P(\rho) = P(\rho'|_{k_c}^{|\rho'|}) = P(\rho''|_{k_c}^{|\rho''|})$. ◇

It is now possible to state necessary and sufficient conditions for Δ-diagnosability of the original system $\tilde{\mathscr{G}}$ with ε-edges.

Theorem 1.18 *Given* $\tilde{\mathscr{G}}$*,* Q_c *is* Δ*-diagnosable if and only if*

$$\forall (q_c, q) \in Q_0^{\mathscr{C}}, \quad \forall \rho \in \mathscr{L}^{\mathscr{D}_{q_c,q}}, \quad \mathrm{Time}(\rho) < \Delta + \inf\{\lambda(q_c)\}.$$

Proof (⇒) By contradiction, assume that the hypothesis is true and that there exist $(q_c, q) \in Q_0^{\mathscr{C}}$ and $\rho \in \mathscr{L}^{\mathscr{D}_{q_c,q}}$ such that $\text{Time}(\rho) = \Delta^* \geq \Delta + \inf\{\lambda(q_c)\}$. This implies that, by Propositions 1.16 and 1.17, there exists $\rho' \in \mathscr{F}^{\mathscr{G}}_{\Delta^*}$ with $\rho'_{k_c} = q_c$, and $\rho'' \in \mathscr{L}^{\mathscr{G}} \setminus \mathscr{F}^{\mathscr{G}}$ such that $P(\rho) = P(\rho') = P(\rho'')$. By Proposition 1.15, for all $\lambda^* \in \lambda(q_c)$ there exist $\tilde{\rho}' \in \mathscr{F}^{\tilde{\mathscr{G}}}_{\Delta^*-\lambda^*}$ and $\tilde{\rho}'' \in \mathscr{L}^{\tilde{\mathscr{G}}} \setminus \mathscr{F}^{\tilde{\mathscr{G}}}$, such that $P(\tilde{\rho}') = P(\tilde{\rho}'')$. Let $\lambda^* = \inf\{\lambda(q_c)\}$, then $\Delta^* - \lambda^* \geq \Delta + \inf\{\lambda(q_c)\} - \inf\{\lambda(q_c)\} = \Delta$, that is a contradiction.
(⇐) By contradiction, assume that the hypothesis is true and that Q_c is not Δ-diagnosable for a given $\Delta \geq 0$. This implies that there exist $\Delta^* \geq \Delta$, $\tilde{\rho}' \in \mathscr{F}^{\tilde{\mathscr{G}}}_{\Delta^*}$ and $\tilde{\rho}'' \in \mathscr{L}^{\tilde{\mathscr{G}}} \setminus \mathscr{F}^{\tilde{\mathscr{G}}}$ such that $P(\tilde{\rho}') = P(\tilde{\rho}'')$. By Proposition 1.15, there exist $\rho' \in \mathscr{F}^{\mathscr{G}}_{\Delta^*+\lambda^*}$, $\lambda^* \in \lambda(\rho|_{k_c})$ and $\rho'' \in \mathscr{L}^{\mathscr{G}} \setminus \mathscr{F}^{\mathscr{G}}$, such that $P(\rho') = P(\rho'')$. By Propositions 1.16 and 1.17 there exists $q \in Q \setminus Q_c$ such that $(\rho'_{k_c}, q) \in Q_0^{\mathscr{C}}$. Furthermore, there exists $\rho \in \mathscr{L}^{\mathscr{D}_{\rho'_{k_c},q}}$ such that $\text{Time}(\rho) = \Delta^* + \lambda^* \geq \Delta + \inf\{\lambda(q_c)\}$, that is a contradiction. □

As a first step of the diagnosability verification, one can check if Q_c may be Δ-diagnosable for a finite value of Δ.

Proposition 1.19 *Given $\mathscr{G}$, Q_c is Δ-diagnosable for some finite Δ only if, for all (q_c, q), $\mathscr{D}_{q_c,q}$ has the following properties*

1. *No edges are associated to a guard set $[a, +\infty)$, $a \in \mathbb{R}_+$, and*
2. *Every edge e belonging to a cycle is associated to a guard set $G_e = \{0\}$.*

Proof For each $\mathscr{D}_{q_c,q}$, an execution $\rho \in \mathscr{L}^{\mathscr{D}_{q_c,q}}$ can have an infinite duration $\text{Time}(\rho) = \infty$ only if either a discrete state is visited forever (Condition 1 is not satisfied), or if there exists a cycle of edges that can be crossed in a nonzero amount of time (Condition 2 is not satisfied). □

It is interesting to compute the minimum value $\Delta_{\min}$ for which Q_c is $\Delta_{\min}$-diagnosable. The verification proposed procedure paper does not only allow verifing Δ-diagnosability for a given Δ but, much more important, also directly computing $\Delta_{\min}$ in polynomial time for a given durational graph $\mathscr{G}$.

Theorem 1.20 *Given $\mathscr{G}$ such that Q_c is Δ-diagnosable for some $\Delta < \infty$, the minimum value $\Delta_{\min}$ such that Q_c is $\Delta_{\min}$-diagnosable given by*

$$\Delta_{\min} = \max_{\substack{\rho \in \mathscr{L}^{\mathscr{D}_{q_c,q}} \\ (q_c,q) \in Q_0^{\mathscr{C}}}} \Big\{ \text{Time}(\rho) - \inf\{\lambda(q_c)\} \Big\} \tag{1.3}$$

and can be computed in polynomial time.

Proof Condition (1.3) clearly holds by Theorem 1.21. Computing the maximum duration among all executions for each system $\mathscr{D}_{q_c,q}$ is solvable as follows: let N be the cardinality of the discrete state space of $\mathscr{D}_{q_c,q}$. First construct the set of all paths that contain no cycles: notice that it is bounded by N^2. For each path $q_1, \ldots, q_s$, compute the maximum duration by $\sum_{k=1}^{s-1} \sup\{G_{(q_k,q_{k+1})}\}$. By Proposition 1.19, all cycles (if any) have time duration 0. This implies that the maximum duration among all executions that contain no cycles is also the maximum duration among all executions in $\mathscr{L}^{\mathscr{D}_{q_c,q}}$. □

As a consequence of the proposed verification algorithm, the following new result on verification of diagnosability of durational graphs can be stated.

Theorem 1.21 *The Δ-diagnosability verification problem for the class of durational graphs belongs to the complexity class P.*

1.5 Diagnosability Verification in an Electromagnetic Valve System for Camless Engines

The proposed abstraction and diagnosability verification algorithms are applied to a simple case study, given by an Electromagnetic Valve System for Camless Engines. The mathematical model is at the same time simple enough to apply and test our results, but yet realistic and nontrivial. Camless electromagnetic valves are devices recently considered to decouple the camshaft and the valve lift dynamics, namely to command the opening and closing phases of the intake and exhaust valves. The main advantage of these devices is the possibility of obtaining the optimal engine efficiency in all operating conditions. One of the main open problems is the control of the impact velocities between the valve and the constraints (typically the valve seat), which should be sufficiently low in order to eliminate acoustic noises and avoid damages of the mechanical components. The problem is complicated by the short time (typically $\simeq$ 3–5 ms) available at high-engine speed to make a transition between the two valve's terminal positions, and the constraint in terms of actuator cost and space limitations. These last aspects imply that one typical request is the absence of the valve position sensor.

We consider a simplified model of the electromagnetic valve, represented in Fig. 1.3 (for details see [1, 2, 12] and references therein). It is supposed here to supply only one electromagnet to complete the opening or closing phase. The correct behavior of the valve controlled system can be modeled by the hybrid automaton $\mathscr{H}^1$ shown in Fig. 1.4: q_1 corresponds to the closing phase, q_2 to the valve completely close, q_3 to the opening phase, and q_4 to the valve completely open. The continuous dynamics can be described by the following equations:

$$\dot{x}_v = v_v, \qquad \dot{v}_v = \frac{1}{M}\Big(- kx_v - bv_v + F_m + F_d + F_c\Big)$$

describing the motion of the valve and of the connected anchor, where M is the mass. The valve position x_v ranges from $-\rho$ (open valve) to $+\rho$ (closed valve). Moreover, an elastic force $-kx_v$, due to some springs and a torsion bar, and a viscous friction

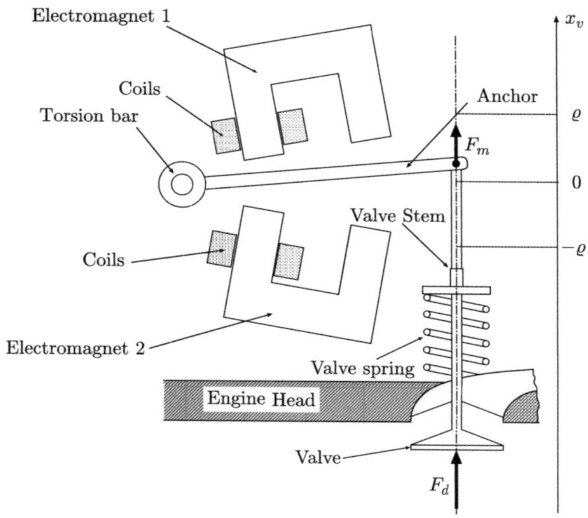

Fig. 1.3 Scheme of an electromagnetic valve system

$-b\dot{x}_v$ act on the valve steam. Finally, F_d is a disturbance whose main contribution is due to the force of the exhaust gases exiting the cylinder, and $F_c(x_v)$ is the constrain force due to the valve seat and electromagnet surfaces, and is always zero except when $x_v = \pm\rho$, when $F_c(\pm\rho) = \pm k\rho - F_m(\pm\rho, \phi_m) - F_d$. The anchor is attracted by the supplied electromagnet to close and to open the valve by means of the electromagnetic force

$$F_m(x_v, \phi_m) = -\frac{1}{2}\mathscr{D}_m(x_v)\phi_m^2, \quad \mathscr{D}_m = a_m e^{-b_m x_v} + c_m$$

where a_m, b_m, c_m are some constants, and ϕ_m is the flux of the supplied electromagnet $m = 1, 2$. The dynamics of ϕ_m is here neglected for simplicity, since it is much faster than the mechanical one. Thus, the squared flux can be considered as the control input of the system, i.e., $u = \phi_m^2$. The signs of the constants are such that $\mathscr{D}_1(x_v) > 0$ for the discrete states q_1, q_2, namely when the valve is closing and the electromagnet 1 is supplied, while $\mathscr{D}_2(x_v) < 0$ for the discrete states q_3, q_4, namely when the valve is opening and the electromagnet 2 is supplied. The discrete dynamics depend on the system state (x_v, v_v) and the control input u, according to the guard sets (defined on arrows in Fig. 1.4) and invariant sets (associated to the discrete states in Fig. 1.4). The reset functions are all identities. We assume w.l.o.g. that the initial hybrid state is $(q_1, (-\rho, 0))$, but our results can be easily extended to more complex sets of initial states. The output of the system is a discrete symbol associated to the edges, i.e., ψ_1 or ψ_2 when the anchor hits, respectively, electromagnet 1 or electromagnet 2. It follows from Di Gennaro et al. [12] that the PD-like control

$$u = \frac{2}{\mathscr{D}_m(x_v)}\Big(p_1(x_v - x_r) + p_2(v_v - v_r) + F_d - kx_r - bv_r - Ma_r\Big)$$

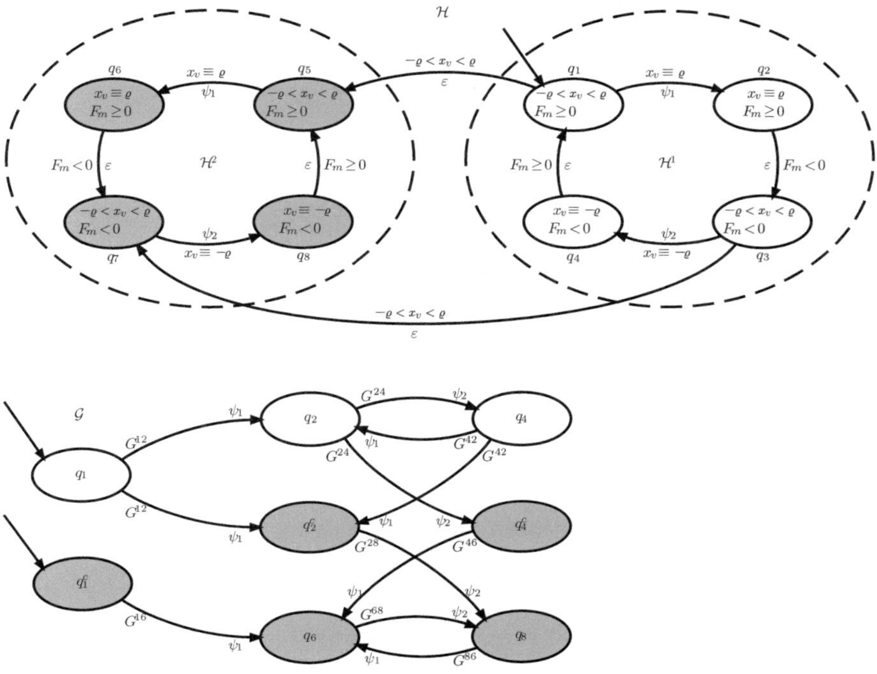

Fig. 1.4 Hybrid model of the electromagnetic valve system $\mathscr{H}$ and abstraction $\mathscr{G}$

$p_1, p_2 > 0$ ensures the correct behavior of the valve. Here $x_r = (-1)^{m+1}\rho$, $m = 1, 2$, $v_r = 0$, $a_r = 0$ are the reference values for appropriately operating the valve.

Setting $e = (x_v - x_r \quad v_v - v_r)^{\mathrm{T}}$, when $F_c = 0$ and using the control above the error dynamics are given by $\dot{e} = A_c e$, where

$$A_c = \begin{pmatrix} 0 & 1 \\ -k_1 & -b_1 \end{pmatrix}, \quad k_1 = \frac{k + p_1}{M} > 0, \quad b_1 = \frac{b + p_2}{M} > 0. \tag{1.4}$$

We want to address here the diagnosability problem due to large parameter variations, which occur in faults of the device we are considering. In fact, the system parameters k, b are subject to abrupt changes due to possible malfunctions. Let k_0, b_0 be their nominal values and $k = k_0 + \Delta k$, $b = b_0 + \Delta b$ be the real ones. The controller parameters (p_1, p_2) must be chosen to satisfy the following constraints for the nominal values $k = k_0$ and $b = b_0$

1. The tracking error goes asymptotically to zero;
2. The norm of the control input is bounded by $u_{\max}$;
3. The seating velocity, i.e., the velocity of the valve when approaching the mechanical constraints, is less than or equal to an appropriate value $v_{\max}$.

Table 1.1 Electromagnetic valve system parameters

$k_0 = 1.17 \times 10^5$ N/m	$b_0 = 6$ Ns/m	$M = 0.1054$ kg
$F_{d,\max} = 3410$ N	$\rho = 4 \times 10^{-3}$ m	$e_{v,\max} = 10$ m/s
$\min_{x_v} \mathscr{D}(x_v) = 1 \times 10^8$	$u_{\max} = 1 \times 10^6$ V	$v_{\max} = 0.05$ m/s

From the first assumption we obtain $p_1 > -k$, $p_2 > -b$. Setting $F_{d,\max} = \max_{t \geq 0} F_d(t)$ and $e_{v,\max}$ the maximum velocity error admissible, from the second we obtain

$$|u| \leq \frac{2}{\min_{x_v} \mathscr{D}_m(x_v)} \Big(p_1 2\rho + p_2 e_{v,\max} + F_{d,\max} + k\rho \Big) \leq u_{\max} \tag{1.5}$$

which can be translated as

$$a_1 p_1 + a_2 p_2 - a_3 \leq 0 \tag{1.6}$$

$a_1 = 1.6 \times 10^{-10}$, $a_2 = 2 \times 10^{-7}$, $a_3 \simeq 1 \times 10^6$, see Table 1.1.

For the third assumption, note that the raising time for the error dynamics is $t_r = \frac{\pi}{\sqrt{4k_1 - b_1^2}}$, where $4k_1 - b_1^2 > 0$ to obtain a fast response, namely

$$p_1 > \frac{1}{4M} (b + p_2)^2 - k. \tag{1.7}$$

Hence, the velocity error has to satisfy $|v_v - v_r|_{t=t_r} = 2\rho e^{-\frac{b_1}{2} t_r} \leq v_{\max}$, thus obtaining

$$p_1 \leq \frac{1 + n^2}{4Mn^2} (b + p_2)^2 - k, \quad n = \frac{2}{\pi} \ln \frac{2\rho}{v_{\max}}. \tag{1.8}$$

A solution to (1.7) and (1.8) exists since $1 + n^2/n^2 > 1$. Conditions (1.6), (1.7) and (1.8) for $k = k_0$, $b = b_0$ define the set of controller parameters ensuring the correct behavior. A pair in this set is for instance $(p_1^*, p_2^*) = (2 \times 10^5, 315)$.

We assume that Δk can vary in the interval $[-k_0, k_0]$, where $-k_0$ corresponds to $k = 0$ (the springs are broken). Moreover, we assume that Δb can vary in the interval $[-0.9b_0, 2b_0]$. In other words, the viscous friction can increase up to 200 % of the nominal value, and can decrease up to 90 % of the nominal value. We define $P = [-k_0, k_0] \times [-0.9b_0, 2b_0]$.

When (k, b) changes, the controller may not ensure the correct valve behavior. In fact, the variations $(\Delta k, \Delta b)$ are allowed to belong to a set $P_{\text{safe}} \subset P$, but when they exit this set a faulty behavior occurs. In order to determine P_{safe}, let us set $p_1 = p_1^*$, $p_2 = p_2^*$ in (1.5), (1.7), (1.8), with $k = k_0 + \Delta k$, $b = b_0 + \Delta b$

$$\Delta k \leq \frac{a_3 - p_1^* a_1 - p_2^* a_2}{a_0}$$
$$\Delta k > \frac{1}{4M}\left(b_0 + \Delta b + p_2^*\right)^2 - p_1^* - k_0$$
$$\Delta k \leq \frac{1+n^2}{4Mn^2}\left(b_0 + \Delta b + p_2^*\right)^2 - p_1^* - k_0$$

with $a_0 = 8 \times 10^{-11}$.

We assume that $(\Delta k, \Delta b)$ may abruptly belong to a faulty value in $P_{\text{faulty}} = P \setminus P_{\text{safe}}$. In that case, the corresponding dynamics of the controlled system switch to the faulty dynamics, modeled in Fig. 1.4 by the hybrid automaton $\mathscr{H}^2$. The dynamics of each discrete state of $\mathscr{H}^2$ is the same as in $\mathscr{H}^1$, except for the value of the parameters (k, b). The sudden change of the system parameters to a faulty value may occur at any time instant from discrete states q_1, q_3, and is associated to an unobservable output. The overall model $\mathscr{H}$ takes into account fault occurrence as illustrated in Fig. 1.4: we assume that the system does not return to a correct behavior once it switches to a faulty behavior.

Although $\mathscr{H}$ is an autonomous hybrid system, it is slightly different from the model in Definition 1.1, since $(\Delta k, \Delta b)$ can be viewed as continuous disturbances that nondeterministically assume values in P_{safe} or P_{faulty}, and determine a correct or faulty behavior. However, since the guards are one-dimensional and the dynamics linear, we can construct our durational graph abstraction by applying the MATLAB® algorithm, developed in D'Innocenzo et al. [14], including the system parameters in the state space and considering P_{safe} and P_{faulty} as sets of initial conditions. Thus, erasing ε-edges by means of Algorithm 1.2, it is possible to construct the durational graph $\mathscr{G}$ shown in Fig. 1.4. In order to determine the guards G^{ij} of $\mathscr{G}$, we compute the minimum and maximum time for the anchor to touch electromagnet 1 starting from electromagnet 2 and vice versa, considering $(\Delta k, \Delta b) \in P_{\text{safe}}$ for the guards of the nonfaulty states and $(\Delta k, \Delta b) \in P_{\text{faulty}}$ for the guards of the faulty states. The invariant sets can be defined by $\mathbb{I}_{q_i} = [0, \max_j\{\sup\{G^{ij}\}\}]$, for each discrete state of $\mathscr{G}$. The construction of $\mathscr{G}$ yields the faulty set $Q_c = \{q_1^c, q_2^c, q_4^c, q_6, q_8\}$ and the faulty function is defined as follows: $\lambda(q_1^c) = G^{12}$, $\lambda(q_2^c) = G^{24}$, $\lambda(q_4^c) = G^{42}$, $\lambda(q_6) = \lambda(q_8) = \{0\}$. It is easy to check that there exists a finite Δ such that $\mathscr{H}$ is Δ-diagnosable only if the following logical formula holds,

$$\begin{gathered}(G^{12} \cap G^{68} = \varnothing) \vee (G^{42} \cap G^{86} = \varnothing) \\ \wedge \\ (G^{12} \cap G^{28} = \varnothing) \vee (G^{12} \cap G^{68} = \varnothing) \vee (G^{42} \cap G^{86} = \varnothing) \\ \wedge \\ (G^{42} \cap G^{46} = \varnothing) \vee (G^{12} \cap G^{68} = \varnothing) \vee (G^{42} \cap G^{86} = \varnothing),\end{gathered}$$

which is true if and only if $G^{12} \cap G^{68} = \varnothing$. Using the MATLAB algorithm developed in D'Innocenzo et al. [14], we compute $G^{12} = [1.81, 2.48]$ ms and

Table 1.2 Summary of notations

$s(e)$	Source state of the edge e
$t(e)$	Target state of the edge e
ε	Unobservable output
$\mathrm{inc}(q)$	Set of incoming edges in the state q
$\mathrm{out}(q)$	Set of outgoing edges from the state q
$\mathrm{cl}_\varepsilon(q)$	ε-closure of the state q
$\mathbb{R}_+$	Positive reals
$\rho\|_i$	Discrete state at index i of the execution associated to ρ
$\rho\|_i^j$	Substring of ρ from index i to index j
$\mathrm{Time}(\rho)$	Time duration of the execution associated to ρ
$\mathcal{L}_{Q^\circ}$	Set of finite executions that terminate in Q°
$P(\rho)$	Observation associated to the string ρ
$\mathcal{F}_\Delta$	Set of Δ faulty executions
$X_0(q_0)$	Set of initial conditions associated to the initial discrete state q_0
$\mathrm{Range}(R_e)$	Range of the reset associated to the edge e
$\Gamma(q)$	Set of time instants t such that there exists an execution of duration t that terminates in the state q

$G^{68} = [1.71, 1.92] \cup [2.30, 3.42]$ ms, thus $G^{12} \cap G^{68} \neq \varnothing$. Namely, if the components of the valve system admit parameters uncertainty within the set P_{faulty}, then P_{faulty} is not Δ-diagnosable for any Δ, and it is not possible to detect faults in finite time. In other words, the diagnosability verification procedure has given a negative answer. However, there may exist a subset $P^*_{\mathrm{faulty}} \subset P_{\mathrm{faulty}}$ which is diagnosable. In this case, one can redesign the control system (e.g., by changing some mechanical characteristic of the valve) so that the parameter uncertainty lies in P^*_{faulty}. A control system designed like this is guaranteed to be diagnosable (Table 1.2).

For this reason, searching for a diagnosable subset of faulty behaviors $P^*_{\mathrm{faulty}} \subset P_{\mathrm{faulty}}$ can be viewed as a support to the design of the control system. In particular, the search for a diagnosable set P^*_{faulty} (e.g., by a trial and error process), may yield to the maximal faulty set (or an approximation of it) such that the system is diagnosable.

For the given valve system, one can define $P^*_{\mathrm{faulty}} = [-1.17 \times 10^5, -0.7 \times 10^5) \times [-5.4, 12]$, as illustrated in Fig. 1.5.

According to the faulty set P^*_{faulty} defined above, one computes $G^{68} = (2.48, 3.42]$ ms. Thus $G^{12} \cap G^{68} = \varnothing$ and P^*_{faulty} is Δ-diagnosable for some finite Δ. Using Theorem 1.20, one obtains that the minimum value $\Delta_{\min}$ such that the set P^*_{faulty} is $\Delta_{\min}$-diagnosable is $\Delta_{\min} = 2\sup\{G^{12}\} = 4.96$ ms. Therefore, one can conclude that if the control system is designed so that parameters uncertainty lies in the set P_{faulty}, then faults cannot be diagnosed in finite time. However, if the control system is designed so that parameters uncertainty lies in the set P^*_{faulty}, then faults can be diagnosed within 4.96 ms.

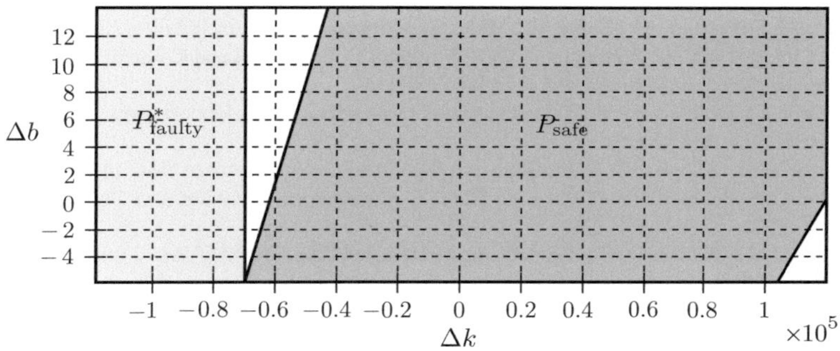

Fig. 1.5 Faulty sets

1.6 Conclusions

In this work, a novel verification procedure is proposed for checking diagnosability for a hybrid automaton whose output is a timed string taking values on a finite set. A definition of Δ-diagnosability is introduced, which generalizes the notion of observability. To verify this property, which is not easy to check and may be even undecidable for a general hybrid model, an abstraction procedure is proposed. The abstracting system belongs to a subclass of timed automata, called durational graph. A novel algorithm is also proposed to construct a durational graph $\mathscr{G}$, from a given hybrid automaton $\mathscr{H}$, and it is showed that it preserves diagnosability. A novel algorithm is proposed to check diagnosability on durational graphs, and to directly compute the minimum value $\Delta_{\min}$ for which a system is $\Delta_{\min}$-diagnosable, and it is proved that the verification problem belongs to the complexity class P. Theoretical results are applied to an electromagnetic valve system for camless engines.

References

1. Acosta Lua, C., Toledo, B.C., Di Benedetto, M.D., Di Gennaro, S.: Feedback regulation of electromagnetic valves for camless engines. In: Proceedings of the European Control Conference—ECC 2007, pp. 4103–4110 (2007)
2. Lua, C.A., Toledo, B.C., Di Benedetto, M.D., Di Gennaro, S.: Output feedback regulation of electromagnetic valves for camless engines. In: Proceedings of the American Control Conference 2007, pp. 2967–2972 (2007)
3. Anai, H., Weispfenning, V.: In: Di Benedetto, M.D., Sangiovanni-Vincentelli, A.L. (eds.) Reach Set computations Using Real Quantifier Elimination, Hybrid Systems: Computation and Control. Lecture Notes in Computer Science, pp. 103–117. Springer, Berlin (2001)
4. Alur, R., Dill, D.L.: A theory of timed automata, l. Theor. Comput. Sci. **126**, 183–235 (1994)
5. Alur, R., Courcoubetis, C., Halbwachs, N., Henzinger, T.A., Ho, P.H., Nicollin, X., Olivero, A., Sifakis, J., Yovine, S.: The algorithmic analysis of hybrid systems. Theor. Comput. Sci. **138**, 3–34 (1995)

6. Asarin, E., Bournez, O., Dang, T., Maler, O.: Approximate Reachability Analysis of Piecewise Linear Dynamical Systems, Hybrid Systems: Computation and Control. Lecture Notes in Computer Science. Springer, Pittsburgh (2000)
7. Choffrut, C., Goldwurm, M.: Timed automata with periodic clock constraints. J. Autom. Lang. Comb. **5**, 371–404 (2000)
8. Chutinan, A., Krogh, B.: Computing polyhedral approximations to flow pipes for dynamic systems. In: Proceedings of the 37th IEEE Conference on Decision and Control, pp. 2089–2094 (1998)
9. Clarke, E.M., Grumberg, O., Peled, D.A.: Model Checking. The MIT Press, Cambridge, Massachusetts (2002)
10. Di Benedetto, M.D., Di Gennaro, S., D'Innocenzo, A.: Error detection within a specific time horizon and application to air traffic management. In: Proceedings of the Joint 44th IEEE Conference on Decision and Control and European Control Conference (CDC-ECC'05), Seville, Spain, pp. 7472–7477 (2005)
11. Di Benedetto, M.D., Di Gennaro, S., D'Innocenzo, A.: Critical States Detection with Bounded Probability of False Alarm and Application to Air Traffic Management. In: Proceedings of the 2nd IFAC Conference on Analysis and Design of Hybrid Systems (ADHS), Alghero, Sardinia, Italy (2006)
12. Di Gennaro, S., Castillo Toledo, B., Di Benedetto, M.D.: Non-linear control of electromagnetic valves for camless engines. Int. J. Control Special Issue Autom. Control **80**(11), 1796–1813 (2007)
13. D'Innocenzo, A., Di Benedetto, M.D., Di Gennaro, S.: In: Hespanha, J., Tiwari, A. (eds.) Observability of Hybrid Automata by Abstraction, Hybrid Systems: Computation and Control. Lecture Notes in Computer Science, pp. 169–183. Springer, Berlin (2006)
14. D'Innocenzo, A., Julius, A.A., Pappas, G.J., Di Benedetto, M.D., Di Gennaro, S.: Verification of temporal properties on hybrid automata by simulation relations. In: Proceedings of the 46th IEEE Conference on Decision and Control, New Orleans (2007)
15. D'Innocenzo, A., Julius, A.A., Di Benedetto, M.D., Pappas, G.J.: Approximate timed abstractions of hybrid automata. In: Proceedings of the 46th IEEE Conference on Decision and Control. New Orleans (2007)
16. Girard, A.: In: Morari, M., Thiele, L. (eds.) Reachability of Uncertain Linear Systems using Zonotopes, Hybrid Systems: Computation and Control. Lecture Notes in Computer Science, pp. 291–305. Springer, Berlin (2005)
17. Guibas, L.J., Nguyen, A., Zhang, L.: Zonotopes as bounding volumes. In: Proceedings of the 14th annual ACM-SIAM symposium on Discrete Algorithms, pp. 803–812, Stanford (2003)
18. Han, Z., Krogh, B.H.: In: Hespanha, J., Tiwari, A. (eds.) Reachability Analysis of Large-Scale Affine Systems Using Low-Dimensional Polytopes, Hybrid Systems: Computation and Control. Lecture Notes in Computer Science, pp. 287–301. Springer, Heidelberg (2006)
19. Kurzhanski, A.B., Varaiya, P.: In: Lynch, N., Krogh, B.H. (eds.) Ellipsoidal Techniques for Reachability Analysis, Hybrid Systems: Computation and Control. Lecture Notes in Computer Science, pp. 202–214. Springer, Heidelberg (2000)
20. Kurzhanski, A.A., Varaiya, P.: Ellipsoidal toolbox. Technical Report, UC, Berkeley (2006)
21. Julius, A.A., Fainekos, G., Anand, M., Lee, I., Pappas, G.J.: Robust Test Generation and Coverage for Hybrid Systems, Hybrid Systems: Computation and Control. Lecture Notes in Computer Science. Springer, Heidelberg (2007)
22. Fourlas, G.K., Kyriakopoulos, K.J., Krikelis, N.J.: Diagnosability of hybrid systems. In: Proceedings of the 10th Mediterranean Conference on Control and Automation - MED2002, pp. 3994–3999, Lisbon (2002)
23. Frank, P.M.: Fault diagnosis in dynamic systems using analytical and knowledge-based redundancy—a survey and some new results. Automatica **26**(3), 459–474 (1990)
24. Henzinger, T.A., Ho, P.-H., Wong-Toi, H.: HyTech: a model checker for hybrid systems. Softw. Tools Technol. Transf. **1**, 110–122 (1997)
25. Hopcroft, J.E., Ullman, J.D.: Introduction to automata theory. Languages and Computation. Addison-Wesley, Reading (1979)

26. Girard, A.: In: Morari, M., Thiele, L. (eds.) Reachability of Uncertain Linear Systems using Zonotopes, Hybrid Systems: Computation and Control. Lecture Notes in Computer Science, pp. 291–305. Springer, New York (2005)
27. Lafferriere, G., Pappas, G.J., Yovine, S.: Symbolic reachability computations for families of linear vector fields. J. Symb. Comput. **32**(3), 231–253 (2001)
28. Laroussinie, F., Markey, N., Schnoebelen, P.: Efficient timed model checking for discrete-time systems. Theor. Comput. Sci. **353**(1–3), 249–271, Elsevier Science Publishers (2006) (http://www.lsv.ens-cachan.fr/Publis/PAPERS/PDF/LMS-TCS05.pdf)
29. Larsen, K.G., Pettersson, P., Yi, W.: Uppaal in a nutshell. Int. J. Softw. Tools Technol. Transf. **1**(1), 134–152 (1997)
30. Lin, F.: Diagnosability of discrete event systems and its applications. J. Discrete Event Dyn. Syst. **4**(1), 197–212 (1994)
31. Lygeros, J., Tomlin, C., Sastry, S.: Controllers for Reachability Specifications for Hybrid Systems. Automatica **35**(3), 349–370 (1999) (Special Issue on Hybrid Systems)
32. McIlraith, S., Biswas, G., Clancy, D., Gupta, V.: In: Lynch, N., Krogh, B. (eds.) Hybrid Systems Diagnosis, in Hybrid Systems: Computation and Control. Lecture Notes in Computer Science, pp. 282–295. Springer, Heidelberg (2000)
33. Ozveren, C.M., Willsky, A.S.: Observability of discrete event dynamic systems. IEEE Trans. Autom. Control **35**(7), 797–806 (1990)
34. Paoli, A., Lafortune, S.: Safe diagnosability for fault tolerant supervision of discrete event systems. Automatica **41**(8), 1335–1347 (2005)
35. Sampath, M., Sengupta, R., Lafortune, S., Sinnamohideen, K., Teneketzis, D.: Diagnosability of discrete-event systems. IEEE Trans. Autom. Control **40**(9), 1555–1575 (1995)
36. Sheth, A., Hartung, C., Han, R.: A decentralized fault diagnosis system for wireless sensor networks. In: Proceedings of the 2nd IEEE International Conference on Mobile Ad-Hoc and Sensor Systems (MASS) 2005, pp. 192–194 (1999)
37. Tripakis, S.: In: Damm, W., Olderog, E.R. (eds.) Fault Diagnosis for Timed Automata. Lecture Notes in Computer Science, pp. 205–221. Springer, Heidelberg (2002)
38. http://wiki.grasp.upenn.edu/hst/index.php?n=Main.HomePage
39. Yazarel, H., Pappas, G.J.: Geometric programming relaxations for linear system reachability. In: Proceedings of the 2004 American Control Conference, Boston, MA (2004)
40. Yoo, T., Lafortune, S.: Polynomial-time verification of diagnosability of partially-observed discrete-event systems. IEEE Trans. Autom. Control **47**(9), 1491–1495 (2002)
41. Yovine, S.: Kronos: a verification tool for real-time systems. Int. J. Softw. Tools Technol. Transf. Springer-Verlag **1**(1), 123–133 (1997)
42. Zhao, F., Koutsoukos, X., Haussecker, H., Reich, J., Cheung, P.: Monitoring and fault diagnosis of hybrid systems. IEEE Trans. Syst. Man Cybern. I—Part B **35**(6), 1225–1240 (2005)

Chapter 2
Observer Design for Discrete-Time Switching Nonlinear Models

Zsófia Lendek, Paula Raica, Jimmy Lauber and Thierry Marie Guerra

Abstract Switched systems are often described by continuous and discrete dynamics as well as their interactions. Although results are available for linear switching systems, for nonlinear switching models few results exist. In this chapter, we consider observer design for discrete-time switching nonlinear systems with a Takagi–Sugeno representation. For designing the observers, a switching nonquadratic Lyapunov function is used. Such Lyapunov functions have shown a real improvement of the design conditions for discrete-time Takagi–Sugeno models. The Lypunov function can be defined for each subsystem or just for the moments when switching takes place. In the first case the results are more general, but also more conservative. The second case represents a significant improvement for periodic models. Thanks to the Lyapunov function used, it is possible to design observers for some switching systems with unobservable subsystems. The developed conditions are formulated as linear or bilinear matrix inequalities. Their advantages and shortcomings are illustrated on numerical examples.

Zs. Lendek (✉) · P. Raica
Department of Automation, Technical University of Cluj-Napoca,
Memorandumului 28, 400114 Cluj-Napoca, Romania
e-mail: zsofia.lendek@aut.utcluj.ro

P. Raica
e-mail: paula.raica@aut.utcluj.ro

J. Lauber · T.M. Guerra
University of Valenciennes and Hainaut-Cambresis,
LAMIH UMR CNRS 8201, Le Mont Houy, 59313 Valenciennes Cedex 9, France
e-mail: jimmy.lauber@univ-valenciennes.fr

T.M. Guerra
e-mail: Thierry.Guerra@univ-valenciennes.fr

M. Djemai and M. Defoort (eds.), *Hybrid Dynamical Systems*, Lecture Notes in Control and Information Sciences 457, DOI 10.1007/978-3-319-10795-0_2

2.1 Introduction

2.1.1 Switching Systems

Switching systems are a class of hybrid systems that switch between a family of modes or subsystems. In the last decades, analysis and synthesis of switching systems has attracted much attention, mostly in the continuous-time case, with linear subsystems.

For instance, linear switching systems where the switching laws can be arbitrarily chosen have been considered in [1] to study the reachable set of such systems. Stabilization and tracking conditions for continuous-time linear switching systems have been developed in [3, 4], delay-dependent stabilization in [33], and observability with unknown input has been investigated in [8]. The results for switching systems have been applied in [26] for the decentralized stabilization of multiagent systems. State feedback controller design for nonlinear switching systems has been presented in [7] and optimal control in [5]. A notable result, although for continuous-time linear switching systems, is the one in [30], which concerns the design of switching sequences for stabilization and proves that it is sufficient for stabilization to employ a periodic switching law.

For the discrete-time case considerably fewer results exist. Most of these concern linear subsystems, such as [31], where stabilization in the presence of input saturation and uncertainties is considered, or [14] which considers the computation of the mode-dependent dwell time. Observability for switching discrete-time linear systems has been investigated in [13], while a linear controller with integral action has been used for the stabilization of switching systems in [7]. Other recent approaches have been reported in [11, 19, 29].

This chapter deals with nonlinear switching models. These models can be found in various domains [41, 47, 58, 60–62], such as automotive, networked control, DC converters, mobile robots, etc. In the case of automotive applications, switching approaches have been used for different parts of the vehicle: engine control, HCCI combustion [40], air path with turbocharger [42–44], clutch actuator control [38].

A very promising recent field of research for switching control concerns systems controlled via network [17, 50]. In this context, stability and stabilization are subject to communication imperfections. Switching methods allow to take into account those constraints.

In power electronics, applications concern power converter structures. Multicellular converters are components which require the control of several switches. A way to consider the different possible modes is to use a switching structure [28].

A particular class of switching models represents those models that switch periodically. Such systems can be found in numerous domains such as automotive, aeronautic, aerospace, and computer control of industrial process. For example, in [10], a periodic dynamic model is used to estimate the air/fuel ratio in each cylinder on an internal combustion engine, Gaiani et al. [25] proposes a periodic model for the rotor blades of helicopter, [56] deals with the problem of an onboard automatic station keeping of a small spacecraft on a specific orbit of reference and proposes a periodic state feedback control law. Other examples are provided in [6] related to computer control and communication systems.

The stability of linear periodic systems is characterized by the monodromy transition matrix and by its eigenvalues, called the characteristic multipliers (often referred to as the poles of the system). If all the characteristic multipliers are in the open unit disc of the complex plane, then the system is asymptotically stable [22]. Concerning the stabilization problem of those models, results are available in [21]. For models including time-varying delays, Stepan and Insperger [51] proposed methods based on Floquet's transformation, which is only applied to autonomous systems, and led to conditions for exponential stability.

Extensions exist to polytopic linear parameter varying periodic models, where the stability analysis is based on the use of quadratic [2, 21] or nonquadratic [12] Lyapunov functions. Results for stabilization has been reported in [32, 34] but observer design has not been considered.

In what follows, we consider observer design for general and periodic switching systems. We represent the switching nonlinear models by switching quasi-linear parameter varying or Takagi–Sugeno (TS) fuzzy systems [52], presented in what follows.

2.1.2 Discrete-Time TS Models

Dynamic systems are modeled in the state space framework, using a state transition model, which describes the evolution of the states over time, and a measurement model, which relates the measurement to the states. Mathematically, we describe such systems, in discrete time, as:

$$\begin{aligned} x_{k+1} &= f(x_k, u_k) \\ y_k &= h(x_k) \end{aligned} \tag{2.1}$$

where f denotes the state transition function, describing the evolution of the states over time, h is the measurement function, relating the measurements to the states, x is the vector of the state variables, u is the vector of the input or control variables, and y denotes the measurement vector.

We represent the system above by Takagi–Sugeno (TS) fuzzy models of the form

$$\begin{aligned} x_{k+1} &= \sum_{i=1}^{r} h_i(z_k)(A_i x_k + B_i u_k) \\ y_k &= \sum_{i=1}^{r} h_i(z_k) C_i x_k \end{aligned} \tag{2.2}$$

where r is the number of local models, A_i, B_i, C_i, are the matrices of the ith local model, z is the vector of the scheduling variables, which may depend on the states, inputs, measurements, or other exogenous variables, and $h_i(z_k), i = 1, 2, \ldots, r$ are normalized membership functions, i.e., $h_i(z_k) \geq 0$ and $\sum_{i=1}^{r} h_i(z_k) = 1, \forall k \in \mathbb{N}$.

Such a model presents several advantages. The TS model is a universal approximator [20], and many nonlinear systems can be exactly represented in a compact set of the state space as TS systems [45]. Moreover, they are convex combination of local affine models, which facilitates stability analysis and controller and observer design for such systems. In addition, many stability and design conditions for TS systems can be formulated as linear matrix inequalities [9, 48, 53, 55], for which efficient algorithms exist.

Once a TS representation of the nonlinear system (2.1) is available, the analysis is performed using Lyapunov's direct method. Most commonly, common quadratic Lyapunov functions have been used and conditions developed independently of the membership functions. In the last years, results have been significantly improved by the use of nonquadratic Lyapunov functions, in particular for the discrete-time case.

Controller and observer design using TS models has gained an increased interest. The most well-known structure used is the so-called PDC (parallel distributed compensation) controller or observer design, where a constant gain corresponds to each rule, and the synthesis is done based on a common quadratic Lyapunov function. With the use of nonquadratic [15, 27, 35] Lyapunov functions, more complex state feedback controllers and observers were developed, and the design conditions became less conservative. These conditions are generally derived in the form of linear matrix inequalities (LMIs).

Switched TS systems have been investigated mainly in the continuous case where the stability is based on the use of a quadratic Lyapunov function [36, 37, 46, 54] or a piecewise one [23, 24]. For discrete-time switching TS models, using nonquadratic Lyapunov functions, few results exist [16, 18].

In this chapter, switching discrete-time TS fuzzy models are considered and observer design conditions are developed. For the ease of the notation, a graph representation of the switching system is employed and to develop the conditions, a nonquadratic switching Lyapunov function is used.

2.2 Switching TS Models

2.2.1 Preliminaries

To design observers for discrete-time switching TS systems, we consider subsystems of the form

$$x_{k+1} = \sum_{i=1}^{r} h_i(z_k)(A_{j,i}x_k + B_{j,i}u_k)$$
$$y_k = \sum_{i=1}^{r} h_i(z_k)C_{j,i}x_k$$

denoted in what follows as

$$\begin{aligned} x_{k+1} &= A_{j,z}x_k + B_{j,z}u_k \\ y_k &= C_{j,z}x_k \end{aligned} \tag{2.3}$$

where j is the number of the current subsystem, $j = 1, 2, \ldots, n_s$, n_s being the number of the subsystems, x denotes the state vector, r is the number of rules, z is the scheduling vector, h_i, $i = 1, 2, \ldots, r$ are normalized membership functions, and $A_{j,i}$, $B_{j,i}$, and $C_{j,i}$, $i = 1, 2, \ldots, r$, $j = 1, 2, \ldots, n_s$, are the local models. Once activated, a subsystem may be active for at least $p_i^m \in \mathbb{N}^+$ and at most $p_i^M \in \mathbb{N}^+$ samples, that are assumed known.

In this chapter, we will make use of the following results:

Lemma 2.1 [49] *Consider a vector* $x \in \mathbb{R}^{n_x}$ *and two matrices* $Q = Q^T \in \mathbb{R}^{n_x \times n_x}$ *and* $R \in \mathbb{R}^{m \times n_x}$ *such that rank*$(R) < n_x$. *The two following expressions are equivalent:*

1. $x^T Qx < 0$, $x \in \{x \in \mathbb{R}^{n_x}, x \neq 0, Rx = 0\}$
2. $\exists M \in \mathbb{R}^{n_x \times m}$ *such that* $Q + MR + R^T M^T < 0$

Analysis and design for TS models often lead to double-sum negativity problems of the form

$$x^T \sum_{i=1}^{r} \sum_{j=1}^{r} h_i(z_k)h_j(z_k)\Gamma_{ij}x < 0 \tag{2.4}$$

where Γ_{ij}, $i, j = 1, 2, \ldots, r$ are matrices of appropriate dimensions.

Lemma 2.2 [59] *The double-sum* (2.4) *is negative, if*

$$\begin{aligned} &\Gamma_{ii} < 0 \\ &\Gamma_{ij} + \Gamma_{ji} < 0, \quad i, j = 1, 2, \ldots, r, \; i < j \end{aligned}$$

Lemma 2.3 [57] *The double-sum* (2.4) *is negative, if*

$$\begin{aligned} &\Gamma_{ii} < 0 \\ &\frac{2}{r-1}\Gamma_{ii} + \Gamma_{ij} + \Gamma_{ji} < 0, \quad i, j = 1, 2, \ldots, r, \; i \neq j \end{aligned}$$

0 and I denote the zero and identity matrices of appropriate dimensions, and a $(*)$ denotes the term induced by symmetry. The subscript $z+m$ (as in $A_{1,z+m}$) stands for the scheduling vector being evaluated at the current sample plus mth instant, i.e., $z(k+m)$.

2.2.2 Graph Representation

For the easier notation, we use a directed graph representation of the switching system (2.3). The associated graph is $\mathscr{G} = \{\mathscr{V}, \mathscr{E}\}$, with $\mathscr{V}$ being the set of vertices representing the subsystems and $\mathscr{E}$ the set of admissible transitions or switches. As such, $(v_i, v_j) \in \mathscr{E}$ if a switch from subsystem i to subsystem j is possible. Note that we assume that self-transitions are also possible: these correspond to the subsystem being active for more than one sample.

A path $\mathscr{P}(v_i, v_j)$ between two vertices v_i and v_j in the graph $\mathscr{G}$ is a sequence of vertices $\mathscr{P}(v_i, v_j) = [v_{p_1}, v_{p_2}, \ldots, v_{p_{n_p}}]$ so that $v_i = v_{p_1}$, $v_j = v_{p_{n_p}}$, and $(v_{p_k}, v_{p_{k+1}}) \in \mathscr{E}$, $k = 1, 2, \ldots, n_p - 1$. A path between two vertices is in general not unique.

A cycle $\mathscr{C} = [c_1, c_2, \ldots, c_{n_c}, c_1]$ is a path having the same initial and final vertex. Two cycles are equivalent if the vertices in one are a cyclic permutation of the vertices in the other. In this chapter, when referring to cycles, we mean elementary cycles. A graph is strongly connected if there is a path between any two vertices in $\mathscr{V}$.

In a weighted graph $\mathscr{G} = \{\mathscr{V}, \mathscr{E}, \mathscr{W}\}$, the weight (adjacency) matrix is defined as $\mathscr{W}(i, j) = w_{i,j}$, with $w_{i,j} \in \mathbb{R}\backslash\{0\}$, if there exists an edge (v_i, v_j), or $\mathscr{W}(i, j) = \infty$, if a switch from subsystem i to subsystem j is not possible. The elements on the diagonal $\mathscr{W}(i, i) = w_{i,i}$ are the weights associated to the vertices.

We define the weight of a path $W(\mathscr{P}(v_i, v_j))$ as the product of all weights of the vertices and edges that appear in the path, i.e.,

$$W([v_{p_1}, v_{p_2}, \ldots, v_{p_{n_p}}]) = \prod_{k=1}^{n_p} w_{p_k, p_k} \cdot \prod_{k=1}^{n_p - 1} w_{p_k, p_{k+1}}$$

The weight of a cycle is similarly defined. A cycle is subunitary, if its weight is less than 1.

A path in a graph associated to a switching system induces a switching law. The length of a path is given by the number of edges it contains. A cycle in a graph associated to a switching system corresponds to a periodic switching law.

The notations above are illustrated on the following example.

Example 2.4 Consider a switching system composed of four subsystems:

$$x_{k+1} = A_{i,z} x_k$$

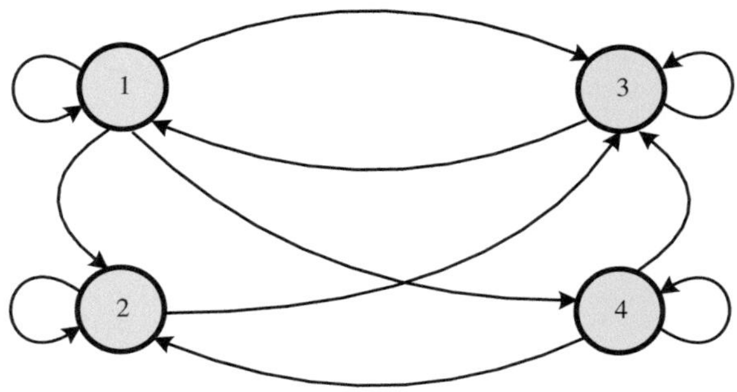

Fig. 2.1 Graph representation of the switching system in Example 2.4

for $i = 1, 2, 3, 4$, and with admissible switches $(1, 2)$, $(1, 3)$, $(1, 4)$, $(2, 3)$, $(3, 1)$, $(4, 2)$, $(4, 3)$. Next to this, each subsystem can be active for more than one sample. The corresponding graph representation is illustrated in Fig. 2.1.

- The graph is $\mathscr{G} = \{\mathscr{V}, \mathscr{E}\}$, with the set of vertices $\mathscr{V} = \{1, 2, 3, 4\}$ and the set of admissible switches:

$$\mathscr{E} = \{(1,1), (1,2), (1,3), (1,4), (2,2), (2,3), (3,1), (3,3), (4,2), (4,3), (4,4)\}$$

- Possible paths between vertices 1 and 3 are $\mathscr{P}(1,3) = [1,\ 3]$, $\mathscr{P}(1,3) = [1,\ 4,\ 3]$ and $\mathscr{P}(1,3) = [1,\ 2,\ 3]$.
- The sequence $[1,\ 3,\ 1]$ is a cycle and is equivalent to $[3,\ 1,\ 3]$.
- The cycle $C_{1,3} = [1,\ 3,\ 1]$ can induce the switching law $[1, 1, 3, 1, \ldots]$ or $[1, 3, 3, 1, \ldots]$ or any switching law of the form: $[\underbrace{1, 1, \ldots, 1}_{p_1}, \underbrace{3, 3, \ldots, 3}_{p_2}, \underbrace{1, 1, \ldots, 1}_{p_1}, \ldots]$.

The associated weight matrix will be constructed based on the possible switches and the number of samples a subsystem is being active. However, to illustrate the definitions, at this point let the associated weight matrix be given by:

$$\mathscr{W} = \begin{pmatrix} 2 & 2 & 0.5 & 2 \\ \infty & 1 & 3 & \infty \\ 0.5 & \infty & 1 & \infty \\ \infty & 1 & 1 & 2 \end{pmatrix}$$

where ∞ corresponds to an inadmissible switch. The graph with the weights given in $\mathscr{W}$ is illustrated in Fig. 2.2.

The weight of the path $\mathscr{P}(1,3) = [1,\ 2,\ 3]$ is $W(\mathscr{P}(1,3)) = w_{11}w_{12}w_{22}w_{23}w_{33} = 12$. The weight of the cycle $\mathscr{C} = [1,\ 3,\ 1]$ is $W(\mathscr{C}) = w_{11}w_{13}w_{33}w_{31} = 0.5 < 1$, so this cycle is subunitary. Since in the graph above there exists a path between any two vertices, the graph is strongly connected.

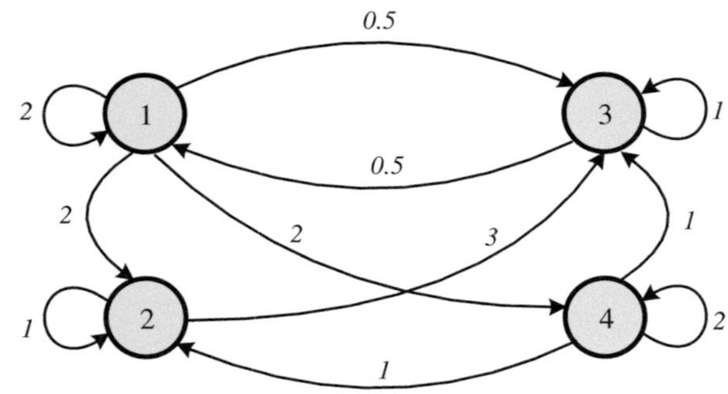

Fig. 2.2 Graph representation of the switching system with weights in Example 2.4

Our goal is to design an observer such that the estimation error dynamics converge to zero. In order to obtain relaxed conditions, we will take into account the switches that are admissible in the system. To see why this is important, consider the following example.

Example 2.5 Consider a TS system with three subsystems, each having two local models, as follows:

$$A_{1,1} = \begin{pmatrix} 0.20 & 0.06 \\ 0.56 & -0.3 \end{pmatrix} \qquad A_{1,2} = \begin{pmatrix} -1.37 & 0.52 \\ 0.96 & -0.24 \end{pmatrix}$$

$$h_{1,1} = \frac{1 - \sin(x_1)}{2} \qquad h_{1,2} = 1 - h_{1,1}$$

$$A_{2,1} = \begin{pmatrix} 0.23 & -0.55 \\ -0.80 & -1.66 \end{pmatrix} \qquad A_{2,2} = \begin{pmatrix} 1.12 & 1.20 \\ 0.49 & -0.29 \end{pmatrix}$$

$$h_{2,1} = \frac{1 - \cos(x_1)}{2} \qquad h_{2,2} = 1 - h_{2,1}$$

$$A_{3,1} = \begin{pmatrix} 3 & 0.5 \\ 0.5 & 1.5 \end{pmatrix} \qquad A_{3,2} = \begin{pmatrix} 2 & 0.1 \\ 0.5 & 2 \end{pmatrix}$$

$$h_{3,1} = \frac{1 - \exp(-x_1^2)}{2} \qquad h_{3,2} = 1 - h_{3,1}$$

One can switch from each subsystem to any other one and any subsystem can be active for any number of samples. However, the local models $A_{1,2}$, $A_{2,1}$, and $A_{2,2}$ are unstable and both local models $A_{3,1}$ and $A_{3,2}$ of the third subsystem are unstable. The associated graph is presented in Fig. 2.3.

Let us see whether this system can be stable based on the switching law that is applied. Due to the instability of the local models, no existing result in the literature can prove the stability of the switching system. Moreover, just switching to one subsystem and keeping it continuously active does not result in a stable system. However, by switching continuously between the first and second subsystem, the states converge to zero. This can be proven by using a periodic Lyapunov function, such as the one proposed in [39]. From the third subsystem, one can switch to the

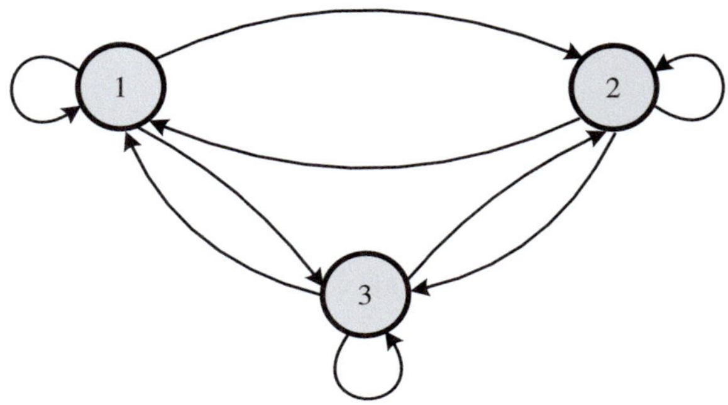

Fig. 2.3 Graph representation of the switching system in Example 2.5

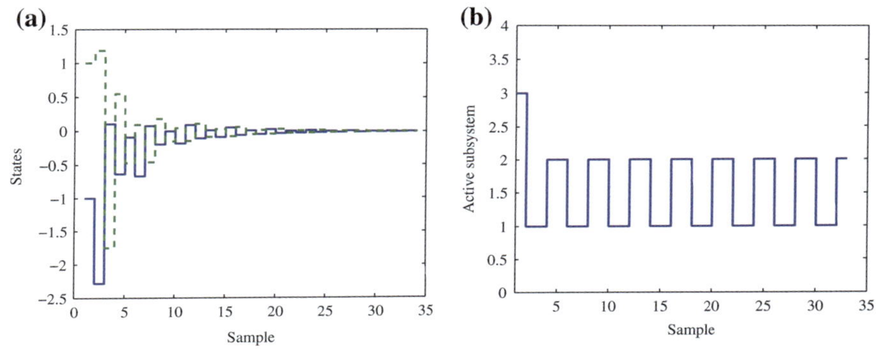

Fig. 2.4 Simulation results for Example 2.5. **a** A trajectory of the switching system. **b** The stabilizing switching law

first or second one and then switch between these two and obtain a stable system. A trajectory that confirms stability of the switching system, starting from $x_0 = [-1,\ 1]^T$ and the corresponding switching law are illustrated in Fig. 2.4a and b, respectively.

As an extension of the stability analysis shown by Example 2.5, by taking into account the switching sequence, it is possible to design more relaxed conditions. Therefore, in what follows, we develop observer design conditions for switching systems.

2.3 Observer Design

We consider observer design for the switching TS system (2.3) of the form (repeated here for convenience):

$$\begin{aligned} x_{k+1} &= A_{j,z} x_k + B_{j,z} u_k \\ y_k &= C_{j,z} x_k \end{aligned}$$

To develop the conditions, we use the switching observer

$$\begin{aligned}\widehat{x}_{k+1} &= A_{j,z}\widehat{x}_k + B_{j,z}u_k + H_{j,z}^{-1}L_{j,z}(y_k - \widehat{y}_k)\\ \widehat{y}_k &= C_{j,z}\widehat{x}_k\end{aligned} \tag{2.5}$$

for the j-th subsystem, the observer switching together with the observed subsystem. The matrices $H_{j,i}$ and $L_{j,i}$, $j = 1, 2 \dots, n_s$, $i = 1, 2, \dots, r$ are to be determined.

The error dynamics $e_k = x_k - \widehat{x}_k$ using this observer, under the assumption that the scheduling variables are available online at sample k, can be written as

$$\begin{aligned}e_{k+1} &= A_{j,z}e_k - H_{j,z}^{-1}L_{j,z}(y_k - \widehat{y}_k)\\ &= (A_{j,z} - H_{j,z}^{-1}L_{j,z}C_{j,z})e_k\end{aligned}$$

which in itself is a switching system.

In what follows, we will consider different possibilities of switching systems and derive observer design conditions that ensure that the estimation error dynamics converges to zero. The conditions will be relaxed depending on the possible switches and the possibility of choosing the switches.

2.3.1 Switching TS Systems

Let us first consider a general switching system, with the possible switches given by the edges in the graph associated to the system. Recall that using the observer (2.5), the error dynamics are given by

$$e_{k+1} = (A_{j,z} - H_{j,z}^{-1}L_{j,z}C_{j,z})e_k \tag{2.6}$$

This is in itself a switching system, with the same switching sequence as the original system and thus with the same graph. Naturally, the simplest way to develop conditions that ensure that the error dynamics is asymptotically stable is to use the common—for all the local models and all the subsystems—Lyapunov function

$$V(e_k) = e_k^T P e_k$$

The difference in the Lyapunov function for two consecutive samples is

$$\begin{aligned}\Delta V &= e_{k+1}^T P e_{k+1} - e_k^T P e_k\\ &= \begin{pmatrix} e_k \\ e_{k+1}\end{pmatrix}^T \begin{pmatrix} -P & 0 \\ 0 & P\end{pmatrix} \begin{pmatrix} e_k \\ e_{k+1}\end{pmatrix}\end{aligned}$$

During the transition from subsystem j to subsystem l, the dynamics of the error system are described by

$$\begin{pmatrix} A_{j,z} - H_{j,z}^{-1} L_{j,z} C_{j,z} & -I \end{pmatrix} \begin{pmatrix} e_k \\ e_{k+1} \end{pmatrix} = 0$$

Using Lemma 2.1, the difference in the Lyapunov function is negative, if there exists M such that

$$\begin{pmatrix} -P & 0 \\ 0 & P \end{pmatrix} + M \begin{pmatrix} A_{j,z} - H_{j,z}^{-1} L_{j,z} C_{j,z} & I \end{pmatrix} + (*) < 0$$

By choosing

$$M = \begin{pmatrix} 0 \\ H_{j,z} \end{pmatrix}$$

we have the conditions

$$\begin{pmatrix} -P & (*) \\ H_{j,z} A_{j,z} - L_{j,z} C_{j,z} & -H_{j,z} - H_{j,z}^T + P \end{pmatrix} < 0 \tag{2.7}$$

and we have

Theorem 2.6 *The error dynamics* (2.6) *is asymptotically stable, if there exist* $P = P^T > 0$, $H_{j,k}$, $j \in \mathscr{V}$, $k = 1, 2, \ldots, r$, *such that* (2.7) *holds for all vertices* $j \in \mathscr{V}$.

Note that the conditions (2.7) are nonlinear, but sufficient LMI conditions can easily be formulated using, e.g., Lemma 2.3, as follows:

Corollary 2.7 *The error dynamics* (2.6) *is asymptotically stable, if there exist* $P = P^T > 0$, $H_{j,k}$, $j \in \mathscr{V}$, $k, l = 1, 2, \ldots, r$, *such that*

$$\begin{aligned} &\Gamma_{i,k}^j < 0 \\ &\frac{2}{r-1}\Gamma_{ii}^j + \Gamma_{ik}^j + \Gamma_{ki}^j < 0 \\ &i, k = 1, 2, \ldots, r \end{aligned}$$

with

$$\Gamma_{ik}^j = \begin{pmatrix} -P & (*) \\ H_{j,k} A_{j,i} - L_{j,k} C_{j,i} & -H_{j,k} + (*) + P \end{pmatrix}$$

Due to the common Lyapunov function for all the local models and all the subsystems, the conditions above are very restrictive. One possibility to reduce the conservativeness is to use a nonquadratic Lyapunov function common for all the subsystems. Thus, one can consider the Lyapunov function

$$V(e_k) = e_k^T P_z e_k \tag{2.8}$$

The difference for two consecutive samples is

$$\begin{aligned}\Delta V &= e_{k+1}^T P_{z+1} e_{k+1} - e_k^T P_z e_k \\ &= \begin{pmatrix} e_k \\ e_{k+1} \end{pmatrix}^T \begin{pmatrix} -P_z & 0 \\ 0 & P_{z+1} \end{pmatrix} \begin{pmatrix} e_k \\ e_{k+1} \end{pmatrix}\end{aligned}$$

and choosing

$$M = \begin{pmatrix} 0 \\ H_{j,z} \end{pmatrix}$$

we have

$$\begin{pmatrix} -P_z & (*) \\ H_{j,z} A_{j,z} - L_{j,z} C_{j,z} & -H_{j,z} - H_{j,z}^T + P_{z+1} \end{pmatrix} < 0 \tag{2.9}$$

However, except for the relaxation brought by the nonquadratic Lyapunov function, this choice does not bring a significant improvement. The next step is choosing a different Lyapunov function for each subsystem, i.e., using

$$V(e_k) = e_k^T P_{j,z} e_k \tag{2.10}$$

for subsystem j.

With this, we obtain

$$\begin{aligned}\Delta V &= e_{k+1}^T P_{l,z+1} e_{k+1} - e_k^T P_{j,z} e_k \\ &= \begin{pmatrix} e_k \\ e_{k+1} \end{pmatrix}^T \begin{pmatrix} -P_{j,z} & 0 \\ 0 & P_{l,z+1} \end{pmatrix} \begin{pmatrix} e_k \\ e_{k+1} \end{pmatrix}\end{aligned}$$

where $[v_j, v_l]$ is an admissible path.

During the transition from j to l, the dynamics of the error system are described by

$$\begin{pmatrix} A_{j,z} - H_{j,z}^{-1} L_{j,z} C_{j,z} & -I \end{pmatrix} \begin{pmatrix} e_k \\ e_{k+1} \end{pmatrix} = 0$$

By choosing in Lemma 2.1

$$M = \begin{pmatrix} 0 \\ H_{j,z} \end{pmatrix}$$

we have

$$\begin{pmatrix} -P_{j,z} & (*) \\ H_{j,z}A_{j,z} - L_{j,z}C_{j,z} & -H_{j,z} - H_{j,z}^T + P_{l,z+1} \end{pmatrix} < 0 \tag{2.11}$$

and the following conditions can be formulated:

Theorem 2.8 *The error dynamics* (2.6) *is asymptotically stable, if there exist* $P_{j,k} = P_{j,k}^T > 0$, $H_{j,k}$, $(v_j, v_l) \in \mathscr{E}$, $k = 1, 2, \ldots, r$, *such that* (2.11) *holds for all admissible edges* $(v_j, v_l) \in \mathscr{E}$.

Although the conditions above are less conservative, it should be noted that they require that the error dynamics of each subsystem is asymptotically stable. However, as it has been shown on Example 2.5, this is not necessary.

Therefore, let us consider a switching Lyapunov function

$$V(e_k) = e_k^T P_{m,j,z} e_k \tag{2.12}$$

defined during the switches, i.e., on the edges of the associated graph $\mathscr{G} = \{\mathscr{V}, \mathscr{E}\}$, with $(v_m, v_j) \in \mathscr{E}$, instead of the subsystems. If a subsystem j may be active for several number of samples, the edge (v_j, v_j) is also considered.

Then, $e_k^T P_{m,j,z} e_k$ is active during the transition from vertex m to vertex j. The difference in the Lyapunov function for two consecutive samples is

$$\begin{aligned} \Delta V &= e_{k+1}^T P_{j,l,z+1} e_{k+1} - e_k^T P_{m,j,z} e_k \\ &= \begin{pmatrix} e_k \\ e_{k+1} \end{pmatrix}^T \begin{pmatrix} -P_{m,j,z} & 0 \\ 0 & P_{j,l,z+1} \end{pmatrix} \begin{pmatrix} e_k \\ e_{k+1} \end{pmatrix} \end{aligned}$$

where $[v_m, v_j, v_l]$ is an admissible path.

During the transition for j to l, the dynamics of the error system are described by

$$\begin{pmatrix} A_{j,z} - H_{j,z}^{-1} L_{j,z} C_{j,z} & -I \end{pmatrix} \begin{pmatrix} e_k \\ e_{k+1} \end{pmatrix} = 0$$

Using Lemma 2.1, the difference in the Lyapunov function is negative, if there exists M such that

$$\begin{pmatrix} -P_{m,j,z} & 0 \\ 0 & P_{j,l,z+1} \end{pmatrix} + M \begin{pmatrix} A_{j,z} - H_{j,z}^{-1} L_{j,z} C_{j,z} & -I \end{pmatrix} + (*) < 0$$

By choosing

$$M = \begin{pmatrix} 0 \\ H_{j,z} \end{pmatrix}$$

we have

$$\begin{pmatrix} -P_{m,j,z} & (*) \\ H_{j,z}A_{j,z} - L_{j,z}C_{j,z} & -H_{j,z} - H_{j,z}^T + P_{j,l,z+1} \end{pmatrix} < 0 \tag{2.13}$$

With this, the following conditions can be formulated:

Theorem 2.9 *The error dynamics* (2.6) *is asymptotically stable, if there exist* $P_{m,j,k} = P_{m,j,k}^T > 0$, $H_{j,k}$, $(v_m, v_j) \in \mathscr{E}$, $(v_j, v_l) \in \mathscr{E}$, $k = 1, 2, \ldots, r$, *such that* (2.13) *holds for all admissible paths* $\mathscr{P}(v_m, v_l) = [v_m, v_j, v_l]$, $v_m \in \mathscr{V}$.

Similarly to the previous cases, the condition (2.13) is nonlinear. However, LMI conditions can be easily formulated using e.g., Lemma 2.3, as follows:

Corollary 2.10 *The error dynamics* (2.6) *is asymptotically stable, if there exist* $P_{m,j,k} = P_{m,j,k}^T > 0$, $H_{j,k}$, $(v_m, v_j) \in \mathscr{E}$, $k, l = 1, 2, \ldots, r$, *such that*

$$\begin{aligned} &\Gamma_{kk}^{m,j,l,\gamma} < 0 \\ &\frac{2}{r-1}\Gamma_{kk}^{m,j,l,\gamma} + \Gamma_{k\beta}^{i,j,l,\gamma} + \Gamma_{\beta k}^{i,j,l,\gamma} < 0 \\ &k, \beta, \gamma = 1, 2, \ldots, r \end{aligned}$$

with

$$\Gamma_{k\beta}^{m,j,l,\gamma} = \begin{pmatrix} -P_{m,j,k} & (*) \\ H_{j,k}A_{j,\beta} - L_{j,k}C_{j,\beta} & -H_{j,k} + (*) + P_{j,l,\gamma} \end{pmatrix}$$

for all admissible paths $\mathscr{P}(v_m, v_l) = [v_m, v_j, v_l]$, $v_j \in \mathscr{V}$.

In what follows, let us illustrate on an example the conditions when using the different Lyapunov functions.

Example 2.11 Consider a TS system with three subsystems, each having two local models, as follows:

$$\begin{aligned} A_{1,1} &= \begin{pmatrix} 0.73 & 0.52 \\ 0.66 & 0.74 \end{pmatrix} & A_{1,2} &= \begin{pmatrix} 0.28 & 0.27 \\ 1.50 & 1.38 \end{pmatrix} \\ C_{1,1} &= \begin{pmatrix} 1 & 0 \end{pmatrix} & C_{1,2} &= \begin{pmatrix} 1 & 0 \end{pmatrix} \\ A_{2,1} &= \begin{pmatrix} 1.05 & 0 \\ 0.5 & 0.1 \end{pmatrix} & A_{2,2} &= A_{2,1} \\ C_{2,1} &= \begin{pmatrix} 0 & 0 \end{pmatrix} & C_{2,2} &= C_{2,1} \\ A_{3,1} &= \begin{pmatrix} 0.86 & 0.72 \\ 0.23 & 0.76 \end{pmatrix} & A_{3,2} &= \begin{pmatrix} 0.25 & 1.51 \\ 1.94 & 1.83 \end{pmatrix} \\ C_{3,1} &= \begin{pmatrix} 1 & 0 \end{pmatrix} & C_{3,2} &= \begin{pmatrix} 0 & 0 \end{pmatrix} \end{aligned}$$

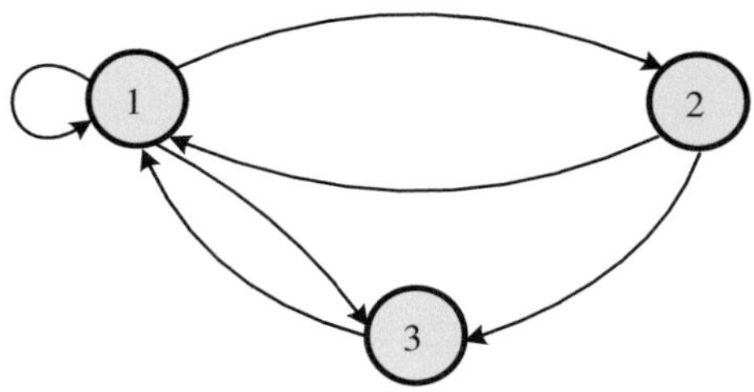

Fig. 2.5 Graph representation of the switching system in Example 2.11

Subsystem 1 may be active for at most five samples, while subsystems 2 and 3 are active only for one sample, i.e., $p_1^m = 1$, $p_1^M = 5$, $p_2^m = p_2^M = p_3^m = p_3^M = 1$. The possible switches are presented in Fig. 2.5. All the local models are unstable. Next to this, the second subsystem and the second local model of the third subsystem are unobservable.

Due to the unobservable and unstable local models, neither the conditions developed using a common quadratic Lyapunov function, nor those based on a common nonquadratic Lyapunov functions are feasible.

Using the conditions of Theorem 2.9, we obtain the gains[1]

$$H_{1,1} = \begin{pmatrix} 0.70 & -0.30 \\ -0.08 & 0.29 \end{pmatrix} \qquad H_{1,2} = \begin{pmatrix} 0.90 & -0.16 \\ -0.33 & 0.19 \end{pmatrix}$$

$$H_{2,1} = \begin{pmatrix} 0.58 & -0.35 \\ -0.10 & 0.51 \end{pmatrix} \qquad H_{2,2} = H_{2,1}$$

$$H_{3,1} = \begin{pmatrix} 0.58 & -0.39 \\ 0.05 & 0.37 \end{pmatrix} \qquad H_{3,2} = \begin{pmatrix} 0.61 & -0.37 \\ -0.37 & 0.24 \end{pmatrix}$$

$$L_{1,1} = \begin{pmatrix} 0.52 \\ 0.37 \end{pmatrix} \qquad L_{1,2} = \begin{pmatrix} 0.01 \\ 0.47 \end{pmatrix}$$

$$L_{3,1} = \begin{pmatrix} 0.28 \\ 0.37 \end{pmatrix} \qquad L_{3,2} = \begin{pmatrix} 0.25 \\ 0.85 \end{pmatrix}$$

Since the second subsystem is not observable, there are no gains $L_{2,1}$ and $L_{2,2}$. Trajectories of the states and the error are presented in Fig. 2.6a and b, respectively. The true initial states were $x_0 = [-1,\ 1]^T$ and the estimated initial states were $\hat{x}_0 = 0$. The corresponding switching law is given in Fig. 2.6c. For testing, the membership functions are assumed to depend on an exogenous measured signal, with h_1 being presented in Fig. 2.6d.

Using the conditions developed with a nonquadratic Lyapunov function for the subsystems, but taking into account the possible switches (notably that subsystems 2 and 3 are active only for one sample), we obtain the gains

[1] Throughout this chapter, computed values are truncated to two decimal places.

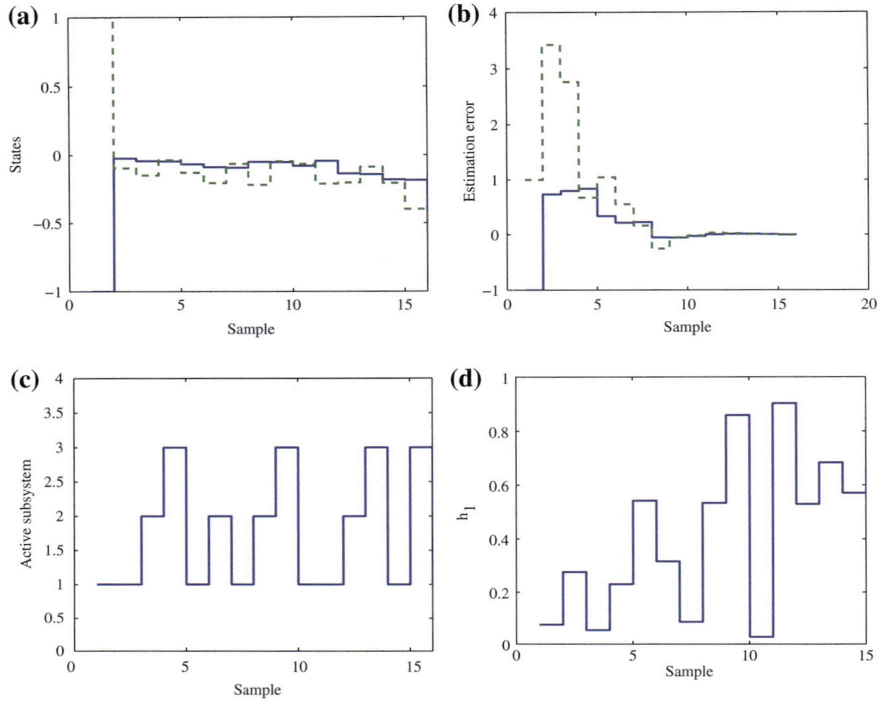

Fig. 2.6 Simulation results for Example 2.11. **a** A trajectory of the switching system. **b** Trajectory of the error. **c** The switching law used. **d** Membership function

$$H_{1,1} = \begin{pmatrix} 0.66 & -0.30 \\ -0.09 & 0.31 \end{pmatrix} \qquad H_{1,2} = \begin{pmatrix} 0.95 & -0.16 \\ -0.36 & 0.20 \end{pmatrix}$$

$$H_{2,1} = \begin{pmatrix} 0.58 & -0.36 \\ -0.01 & 0.36 \end{pmatrix} \qquad H_{2,2} = H_{2,1}$$

$$H_{3,1} = \begin{pmatrix} 0.59 & -0.41 \\ 0.06 & 0.35 \end{pmatrix} \qquad H_{3,2} = \begin{pmatrix} 0.65 & -0.39 \\ -0.38 & 0.25 \end{pmatrix}$$

$$L_{1,1} = \begin{pmatrix} 0.47 \\ 0.42 \end{pmatrix} \qquad L_{1,2} = \begin{pmatrix} 0.08 \\ 0.5 \end{pmatrix}$$

$$L_{3,1} = \begin{pmatrix} 0.28 \\ 0.37 \end{pmatrix} \qquad L_{3,2} = \begin{pmatrix} 0.23 \\ 0.83 \end{pmatrix}$$

which are quite close to those obtained by Theorem 2.9. It should be noted that without taking into account that subsystems 2 and 3 are active only for one sample, the conditions using a nonquadratic Lyapunov function for each subsystem are unfeasible.

In order to reduce the conservativeness by exploiting the knowledge available of the switching sequence, one can also use the observer

$$\begin{aligned} \widehat{x}_{k+1} &= A_{j,z}\widehat{x}_k + B_{j,z}u_k + H_{m,j,z}^{-1}L_{m,j,z}(y_k - \widehat{y}_k) \\ \widehat{y}_k &= C_{j,z}\widehat{x}_k \end{aligned} \tag{2.14}$$

for the j-th subsystem, if the last switch has been from vertex m to vertex j.

The error dynamics $e_k = x_k - \widehat{x}_k$ using this observer, can be written as

$$\begin{aligned} e_{k+1} &= A_{j,z}e_k - H_{m,j,z}^{-1}L_{m,j,z}(y_k - \widehat{y}_k) \\ &= (A_{j,z} - H_{m,j,z}^{-1}L_{m,j,z}C_{j,z})e_k \end{aligned} \tag{2.15}$$

Following the same steps as above, and in Lemma 2.1 choosing

$$M = \begin{pmatrix} 0 \\ H_{m,j,z} \end{pmatrix}$$

we have

Corollary 2.12 *The error dynamics* (2.15) *is asymptotically stable, if there exist* $P_{m,j,k} = P_{m,j,k}^T > 0$, $H_{m,j,k}$, $(v_m, v_j) \in \mathscr{E}$, $(v_j, v_l) \in \mathscr{E}$, $k = 1, 2, \ldots, r$, *such that*

$$\begin{pmatrix} -P_{m,j,z} & (*) \\ H_{m,j,z}A_{j,z} - L_{m,j,z}C_{j,z} & -H_{m,j,z} + (*) + P_{j,l,z+1} \end{pmatrix} < 0 \tag{2.16}$$

for all admissible paths $\mathscr{P}(v_m, v_l) = [v_m, v_j, v_l]$, $v_m \in \mathscr{V}$.

This result can actually be used if the switching sequence is known at least one switch in advance and represents the case when the observer gains are specified for each switch instead of each subsystem.

The conditions of Theorem 2.9 can be further extended to take into account longer sequences of switches. However, this comes with added computational cost and will eventually lead to considering all possible switching trajectories. To avoid this, but still reduce the conservativeness of the approach, consider the α-sample variation of the Lyapunov function. As proven by Kruszewski and Guerra [34], a system is asymptotically stable, if the associated Lyapunov function decreases every α samples, $\alpha \geq 1$, instead of every sample. Thus, for observer design, let us consider the error dynamics (2.6) and the Lyapunov function (2.12), defined on the edges of the associated graph, with $P_{v_i,v_j,z}$ being active during the transition from vertex i to vertex j. The difference in the Lyapunov function for α consecutive samples is

$$\begin{aligned} \Delta V &= e_{k+\alpha}^T P_{v_\alpha,v_{\alpha+1},z+\alpha}e_{k+\alpha} - e_k^T P_{v_0,v_1,z}e_k \\ &= \begin{pmatrix} e_k \\ e_{k+\alpha} \end{pmatrix}^T \begin{pmatrix} -P_{v_0,v_1,z} & 0 \\ 0 & P_{v_\alpha,v_{\alpha+1},z+\alpha} \end{pmatrix} \begin{pmatrix} e_k \\ e_{k+\alpha} \end{pmatrix} \end{aligned}$$

where $[v_0, v_1, \dots, v_{\alpha+1}]$ is an admissible path.

Along the switching sequence $[v_0, v_1, \dots, v_{\alpha+1}]$, the error dynamics are described by

$$\begin{pmatrix} G_1 & -I & 0 & \dots & 0 \\ 0 & G_2 & -I & \dots & 0 \\ \vdots & \vdots & & \vdots & \vdots \\ 0 & 0 & 0 & \dots & -I \end{pmatrix} \begin{pmatrix} e_k \\ e_{k+1} \\ \vdots \\ e_{k+\alpha} \end{pmatrix} = 0$$

with $G_i = A_{v_i,z+i-1} - H^{-1}_{v_i,z+i-1} L_{v_i,z+i-1} C_{v_i,z+i-1}$.

Following the same steps as in the proof of Theorem 2.9, using Lemma 2.1, the difference in the Lyapunov function is negative, if there exists M such that

$$\begin{pmatrix} -P_{v_0,v_1,z} & 0 & \dots & 0 \\ 0 & 0 & \dots & 0 \\ \vdots & \vdots & \dots & \vdots \\ 0 & 0 & \dots & P_{v_\alpha,v_{\alpha+1},z+\alpha} \end{pmatrix} + M \begin{pmatrix} G_1 & -I & 0 & \dots & 0 \\ 0 & G_2 & -I & \dots & 0 \\ \vdots & \vdots & & \vdots & \vdots \\ 0 & 0 & 0 & \dots & -I \end{pmatrix} + (*) < 0$$

with G_i defined as above.

By choosing

$$M = \begin{pmatrix} 0 & 0 & \dots & 0 \\ H_{v_1,z} & 0 & \dots & 0 \\ 0 & H_{v_2,z+1} & \dots & 0 \\ \vdots & \vdots & \dots & \vdots \\ 0 & 0 & \dots & H_{v_\alpha,z+\alpha-1} \end{pmatrix}$$

we have

$$\begin{pmatrix} -P_{v_0,v_1,z} & (*) & \dots & (*) \\ \Omega_1 & -H_{v_1,z} + (*) & \dots & (*) \\ 0 & \Omega_2 & \dots & (*) \\ \vdots & \vdots & \dots & \begin{pmatrix} -H_{v_\alpha,z+\alpha-1} + (*) \\ +P_{v_\alpha,v_{\alpha+1},z+\alpha} \end{pmatrix} \end{pmatrix} < 0 \tag{2.17}$$

with $\Omega_i = H_{v_i,z+i-1} A_{v_i,z+i-1} - L_{v_i,z+i-1} C_{v_i,z+i-1}$, leading to the following theorem:

Theorem 2.13 *The error dynamics* (2.6) *is asymptotically stable, if there exist* $\alpha \in \mathbb{N}^+$, $P_{i,j,k} = P^T_{i,j,k} > 0$, $H_{j,k}$, $(v_i, v_j) \in \mathscr{E}$, $k = 1, 2, \dots, r$, *such that* (2.17) *holds for all admissible paths* $\mathscr{P}(v_0, v_{\alpha+1}) = [v_0, v_1, \dots, v_{\alpha+1}]$.

The conditions above are illustrated on the following example.

Example 2.14 Consider the switching system—actually a periodic switching system—illustrated in Fig. 2.7.

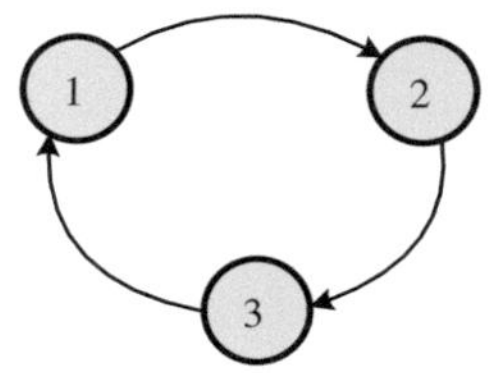

Fig. 2.7 Periodic switching system for Example 2.14

Assuming that each subsystem is active for only one sample, i.e., $p^m = p^M = 1$, the graph is $\mathscr{G} = \{\mathscr{V}, \mathscr{E}\}$, with $\mathscr{V} = \{1, 2, 3\}$ and $\mathscr{E} = \{(1, 2), (2, 3), (3, 1)\}$. Consider the following local models of the TS system above:

$$
\begin{aligned}
A_{1,1} &= \begin{pmatrix} 1.26 & 0.64 \\ 1.92 & 0.61 \end{pmatrix} & A_{1,2} &= \begin{pmatrix} 0.40 & 0.64 \\ 0.09 & 0.10 \end{pmatrix} \\
C_{1,1} &= \begin{pmatrix} 1 & 0 \end{pmatrix} & C_{1,2} &= \begin{pmatrix} 1 & 0 \end{pmatrix} \\
A_{2,1} &= \begin{pmatrix} 1.05 & 0 \\ 0.50 & 0.80 \end{pmatrix} & A_{2,2} &= A_{2,1} \\
C_{2,1} &= \begin{pmatrix} 0 & 0 \end{pmatrix} & C_{2,2} &= \begin{pmatrix} 0 & 0 \end{pmatrix} \\
A_{3,1} &= \begin{pmatrix} 0.40 & 0.45 \\ 0.30 & 0.84 \end{pmatrix} & A_{3,2} &= \begin{pmatrix} 1.67 & 0.43 \\ 1.42 & 0.33 \end{pmatrix} \\
C_{3,1} &= \begin{pmatrix} 0 & 1 \end{pmatrix} & C_{3,2} &= \begin{pmatrix} 0 & 0 \end{pmatrix}
\end{aligned}
$$

The second subsystem is linear, but it is unstable and unobservable. The second local model of the third subsystem is again unstable and unobservable. Due to this, methods available in the literature yield unfeasible LMIs. However, using Lemma 2.2 to formulate LMI conditions for Theorem 2.9, we obtain

$$
\begin{aligned}
H_{1,1} &= \begin{pmatrix} 0.66 & -0.06 \\ 0.03 & 0.56 \end{pmatrix} & H_{1,2} &= \begin{pmatrix} 0.70 & -0.08 \\ 0.06 & 0.55 \end{pmatrix} \\
H_{2,1} &= \begin{pmatrix} 0.46 & -0.09 \\ -0.03 & 0.34 \end{pmatrix} & H_{2,2} &= H_{2,1} \\
H_{3,1} &= \begin{pmatrix} 0.41 & -0.29 \\ -0.33 & 0.58 \end{pmatrix} & H_{3,2} &= \begin{pmatrix} 0.35 & -0.32 \\ -0.38 & 0.54 \end{pmatrix} \\
L_{1,1} &= \begin{pmatrix} 0.98 \\ 1.39 \end{pmatrix} & L_{1,2} &= \begin{pmatrix} 0.57 \\ 0.14 \end{pmatrix} \\
L_{3,1} &= \begin{pmatrix} -0.03 \\ 0.35 \end{pmatrix} & L_{3,2} &= \begin{pmatrix} 0.02 \\ 0.39 \end{pmatrix}
\end{aligned}
$$

Consequently, this observer is able to estimate the states of the switching system above. Since the second subsystem on its own is not observable, there are no observer gains $L_{2,1}$ and $L_{2,2}$. Trajectories of the states and of the error dynamics, together with the membership function h_1, are presented in Fig. 2.5. The true initial states were $x_0 = [-1;\ 1]^T$, while the estimated initial states were $\widehat{x}_0 = [0,\ 0]^T$.

Let us now discuss the conditions developed for the α-sample variation of the Lyapunov function. The previous conditions required that the Lyapunov function

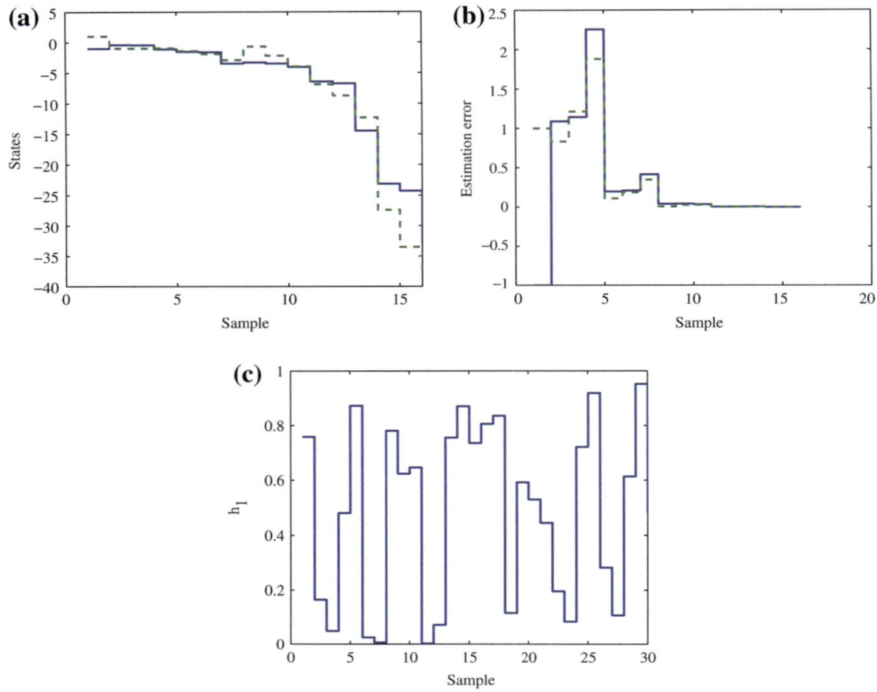

Fig. 2.8 Simulation results for Example 2.14. **a** A trajectory of the switching system. **b** The trajectory of the error. **c** Membership function

decreases with every switch/every sample, i.e., for all $(v_i, v_j) \in \mathscr{E}$. That is, we had the conditions:

$$\begin{pmatrix} -P_{1,2,z} & (*) \\ H_{2,z}A_{2,z} - L_{2,z}C_{2,z} & -H_{2,z} + (*) + P_{2,3,z+1} \end{pmatrix} < 0$$
$$\begin{pmatrix} -P_{2,3,z} & (*) \\ H_{3,z}A_{3,z} - L_{3,z}C_{3,z} & -H_{3,z} + (*) + P_{3,1,z+1} \end{pmatrix} < 0$$
$$\begin{pmatrix} -P_{3,1,z} & (*) \\ H_{1,z}A_{1,z} - L_{1,z}C_{1,z} & -H_{1,z} + (*) + P_{1,2,z+1} \end{pmatrix} < 0$$

A 2-sample variation means that the Lyapunov function has to decrease along paths of length 2, i.e., we have the conditions:

$$\begin{pmatrix} -P_{1,2,z} & (*) & (*) \\ \begin{pmatrix} H_{2,z}A_{2,z} \\ -L_{2,z}C_{2,z} \end{pmatrix} & -H_{2,z} + (*) & (*) \\ 0 & \begin{pmatrix} H_{3,z+1}A_{3,z+1} \\ -L_{3,z+1}C_{3,z+1} \end{pmatrix} & \begin{pmatrix} -H_{3,z+1} + (*) \\ +P_{3,1,z+2} \end{pmatrix} \end{pmatrix} < 0$$

$$\begin{pmatrix} -P_{2,3,z} & (*) & (*) \\ \begin{pmatrix} H_{3,z}A_{3,z} \\ -L_{3,z}C_{3,z} \end{pmatrix} & -H_{3,z} + (*) & (*) \\ 0 & \begin{pmatrix} H_{1,z+1}A_{1,z+1} \\ -L_{1,z+1}C_{1,z+1} \end{pmatrix} & \begin{pmatrix} -H_{1,z+1} + (*) \\ +P_{1,2,z+2} \end{pmatrix} \end{pmatrix} < 0$$

$$\begin{pmatrix} -P_{3,1,z} & (*) & (*) \\ \begin{pmatrix} H_{1,z}A_{1,z} \\ -L_{1,z}C_{1,z} \end{pmatrix} & -H_{1,z} + (*) & (*) \\ 0 & \begin{pmatrix} H_{2,z+1}A_{2,z+1} \\ -L_{2,z+1}C_{2,z+1} \end{pmatrix} & \begin{pmatrix} -H_{2,z+1} + (*) \\ +P_{2,3,z+2} \end{pmatrix} \end{pmatrix} < 0$$

Furthermore, 3-sample variation means that the Lyapunov function has to decrease along paths of length 3, which, in this case, is equivalent to the whole period of switching.

2.3.2 *Periodic TS Systems*

As shown above, in case of periodic systems, a lesser number of edges have to be considered and the conditions become circular. Therefore, in what follows, we consider periodic systems:

$$\begin{aligned} x_{k+1} &= A_{j,z}x_k + B_{j,z}u_k \\ y_k &= C_{j,z}x_k \end{aligned}$$

Since the system is periodic, it is assumed that the subsystems are activated in a sequence $\underbrace{1,\ 1,\dots,\ 1}_{p_1},\ \underbrace{2,\ 2,\dots,\ 2}_{p_2},\ \dots,\ \underbrace{n_s,\ n_s,\dots,\ n_s}_{p_{n_s}},\ \underbrace{1,\ 1,\dots,\ 1}_{p_1}$, etc., where p_i denotes the number of samples for which the ith subsystem is active. In what follows, we will refer to p_i as the period of the ith subsystem.

Recall that using the same switching observer as above, i.e.,

$$\begin{aligned} \widehat{x}_{k+1} &= A_{j,z}\widehat{x}_k + B_{j,z}u_k + H_{j,z}^{-1}L_{j,z}(y_k - \widehat{y}_k) \\ \widehat{y}_k &= C_{j,z}\widehat{x}_k \end{aligned}$$

the error dynamics $e_k = x_k - \widehat{x}_k$ can be written as

$$e_{k+1} = (A_{j,z} - H_{j,z}^{-1}L_{j,z}C_{j,z})e_k \tag{2.18}$$

which, due to the periodicity of the switches, in this case will also be periodic.

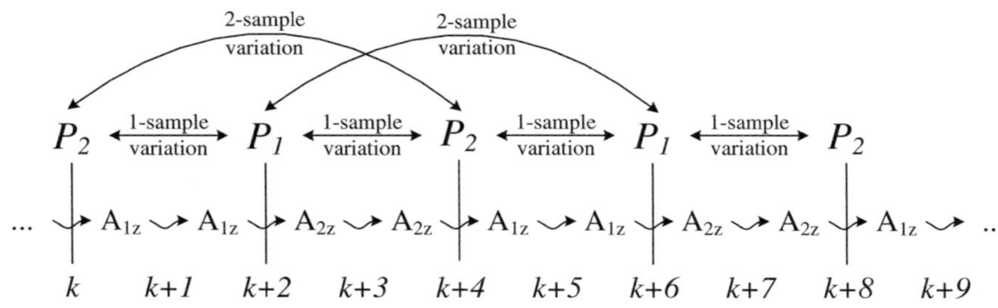

Fig. 2.9 Switches in the system and in the Lyapunov function

Although all the results presented in Sect. 2.3.1 apply for this case, to further reduce the conservativeness of the conditions, we will use a periodic Lyapunov function defined only when a switching takes place. As long as the periodicity of each subsystem is known and constant, the active subsystems' dynamic can easily be written for more than one sample.

To illustrate the periodicity of the system and the definition of the Lyapunov function, consider the following example.

Example 2.15 Consider a periodic TS model consisting of two subsystems, each with period 2, i.e., we have:

$$x_{k+1} = \begin{cases} \sum_{i=1}^{r} h_i(z_k) A_{1i} x_k \text{ if } k = 4m, 4m+1 \\ \sum_{i=1}^{r} h_i(z_k) A_{2i} x_k \text{ if } k = 4m+2, 4m+3 \end{cases} \tag{2.19}$$

The switching in the system and in the Lyapunov function are depicted in Fig. 2.9. As can be seen, the Lyapunov function (with matrices P_1 and P_2) is defined only in the moments when there is a switching in the system: from $A_{1,z}$ to $A_{2,z}$ or from $A_{2,z}$ to $A_{1,z}$, respectively. A 1-sample variation of the Lyapunov function corresponds to the difference between two consecutive values of the Lyapunov function. A 2-sample variation corresponds to the difference after 2 samples of the Lyapunov function, etc.

Therefore, to develop observer design conditions, consider the periodic Lyapunov function defined only in the instants when a switching takes place in the system:

$$V(e_k) = e_k^T P_{j,z} e_k$$

if the active subsystem was j, $j = 1, 2 \ldots, n_s$.

The difference in the Lyapunov function is

$$V(e_{k+p_{\underline{j+1}}}) - V(e_k) = \begin{pmatrix} e_k \\ e_{k+p_{\underline{j+1}}} \end{pmatrix}^T \begin{pmatrix} -P_{j,z} & 0 \\ 0 & P_{\underline{j+1},z+p_{\underline{j+1}}} \end{pmatrix} \begin{pmatrix} e_k \\ e_{k+p_{\underline{j+1}}} \end{pmatrix}$$

where $\underline{j} = j \mod n_s$.

The error dynamics during the $p_{\underline{j+1}}$ samples are

$$\begin{pmatrix} \Upsilon_{j+1,1} & -I & \dots & 0 & 0 \\ 0 & \Upsilon_{j+1,2} & -I & \dots & 0 \\ \vdots & \vdots & \vdots & \vdots & \vdots \\ 0 & 0 & \dots & \Upsilon_{j+1,p_{\underline{j+1}}-1} & -I \end{pmatrix} \begin{pmatrix} e_k \\ e_{k+1} \\ \vdots \\ e_{k+p_{\underline{j+1}}} \end{pmatrix} = 0$$

with

$$\Upsilon_{j+1,i} = A_{\underline{j+1},z+i} - H^{-1}_{\underline{j+1},z+i} L_{\underline{j+1},z+i} C_{\underline{j+1},z+i}$$

for $i = 1,\ 2 \dots,\ p_{\underline{j+1}} - 1$.

Choosing $M = \begin{pmatrix} 0 & 0 & \dots & 0 & 0 \\ H_{\underline{j+1},z+1} & 0 & \dots & 0 & 0 \\ 0 & H_{\underline{j+1},z+2} & \dots & 0 & 0 \\ \vdots & \vdots & \vdots & \vdots & \vdots \\ 0 & 0 & \dots & H_{\underline{j+1},z+p_{\underline{j+1}}-1} & 0 \\ 0 & 0 & \dots & 0 & H_{\underline{j+1},z+p_{\underline{j+1}}} \end{pmatrix}$ and applying Lemma 2.1 leads to

$$\begin{pmatrix} -P_{j,z} & (*) & \dots & (*) & (*) \\ \Omega_{j+1,1} & -H_{\underline{j+1},z+1} - (*) & \dots & (*) & (*) \\ \vdots & \vdots & \vdots & \vdots & \vdots \\ 0 & 0 & \dots & \Omega_{j+1,p_{\underline{j+1}}} & -H_{\underline{j+1},z+p_{\underline{j+1}}} - (*) + P_{j+1,z+p_{\underline{j+1}}} \end{pmatrix} < 0$$

with $\Omega_{j+1,i} = H_{\underline{j+1},z+i} A_{\underline{j+1},z+i-1} - L_{\underline{j+1},z+i-1} C_{\underline{j+1},z+i-1}$ for $i = 1,\ 2,\ \dots,\ p_{\underline{j+1}} - 1$.

Then, the following result can be formulated:

Theorem 2.16 *The error dynamics* (2.18) *is locally asymptotically stable if there exist* $P_{j,i}$, $H_{j,i}$, $j = 1, 2 \dots n_s$, $i = 1, 2, \dots r$ *such that:*

$$\begin{pmatrix} -P_{j,z} & (*) & \dots & (*) & (*) \\ \Omega_{j+1,1} & -H_{\underline{j+1},z+1} - (*) & \dots & (*) & (*) \\ \vdots & \vdots & \vdots & \vdots & \vdots \\ 0 & 0 & \dots & \Omega_{j+1,p_{\underline{j+1}}} & -H_{\underline{j+1},z+p_{\underline{j+1}}} - (*) + P_{j+1,z+p_{\underline{j+1}}} \end{pmatrix} < 0$$

The result above can easily be extended using α-sample variation [35]. Recall that the Lypunov function is only defined in the switching instants, and the α-difference in the Lyapunov function corresponds to α consecutive switches in the system.

Since the Lyapunov function is only defined in the switching instants, the α-difference *in the Lyapunov function* is

$$V(x_{k+t}) - V(x_k) = \begin{pmatrix} e_k \\ e_{k+t} \end{pmatrix}^T \begin{pmatrix} -P_{j,z} & 0 \\ 0 & P_{\underline{j+\alpha},z+t} \end{pmatrix} \begin{pmatrix} e_k \\ x_{k+t} \end{pmatrix}$$

where $t = \sum_{i=1}^{\alpha} p_{\underline{j+1}}$.

The error dynamics during the t samples corresponding to the α switches in the system are

$$\Gamma_{j+1:j+\alpha} \begin{pmatrix} e_k \\ e_{k+1} \\ \vdots \\ e_{k+t} \end{pmatrix} = 0$$

with

$$\Gamma_{j+1:j+\alpha} = \begin{pmatrix} G_{1,0} & -I & \dots & 0 & 0 \\ 0 & G_{1,1} \dots & 0 & 0 & \\ \vdots & \vdots & \vdots & \vdots & \vdots \\ 0 & 0 & \dots & G_{\alpha,t-1} & -I \end{pmatrix}$$

where $G_{i,j} = A_{\underline{j+i},z+j} - H^{-1}_{\underline{j+i},z+j} L_{\underline{j+i},z+j} C_{\underline{j+i},z+j}$, $i = 1, 2, \dots, \alpha$, $j = 1, 2 \dots, t-1$.

Using Lemma 2.1, the difference in the Lyapunov function is negative definite, if there exists M such that

$$\begin{pmatrix} -P_{j,z} & 0 \dots & 0 \\ 0 & 0 \dots & 0 \\ \vdots & \vdots \; \vdots & \vdots \\ 0 & 0 \dots & P_{\underline{j+1},z+t} \end{pmatrix} + M\Gamma_{j+1:j+\alpha} + (*) < 0$$

Choosing $M = \begin{pmatrix} 0 & 0 & \dots & 0 \\ H_{\underline{j+1},z} & 0 & \dots & 0 \\ 0 & H_{\underline{j+1},z+1} & \dots & 0 \\ \vdots & \vdots & \vdots & \vdots \\ 0 & 0 & \dots & H_{\underline{j+\alpha},z+t-1} \end{pmatrix}$ leads directly to

$$\begin{pmatrix} -P_{j,z} & (*) & \dots & (*) & (*) \\ \Omega_{\underline{j+1},z} & -H_{\underline{j+1},z} + (*) & \dots & (*) & (*) \\ \vdots & \vdots & \vdots & \vdots & \vdots \\ 0 & 0 & \dots & \Omega_{\underline{j+\alpha},z+t-1} & \Omega_{j+\alpha,j+\alpha} \end{pmatrix} < 0 \tag{2.20}$$

where $t = \sum_{i=1}^{\alpha} p_{j+1}$, and $\Omega_{j+i,j+k} = H_{\underline{j+i},z+k} A_{\underline{j+i},z+k} - L_{\underline{j+i},z+k} C_{\underline{j+i},z+k}$ and $\Omega_{j+\alpha,j+\alpha} = -H^T_{\underline{j+\alpha},z+t-1} - H^T_{\underline{j+\alpha},z+t-1} + P_{\underline{j+\alpha},z+t-1}$.

Theorem 2.17 *The periodic error dynamics* (2.18) *with periods* $p_1, p_2, \dots, p_{n_s}$ *is asymptotically stable, if there exist* $P_{ji} = P_{ji}^T > 0$, H_{ji}, $j = 1, 2, \dots, n_s$, $i = 1, 2, \dots, r_j$, $l = 1, 2, \dots, \alpha$, *such that* (2.20) *is satisfied.*

Let us illustrate the results for periodic TS systems on the following example.

Example 2.18 Consider the periodic fuzzy system with two subsystems as follows:

$$\begin{aligned} x_{k+1} &= \sum_{i=1}^{2} h_i(z_k) A_{1i} x_k \\ y_k &= \sum_{i=1}^{2} h_i(z_k) C_{1i} x_k \end{aligned}$$

with

$$A_{11} = \begin{pmatrix} -0.44 & -0.26 \\ -0.65 & 0.62 \end{pmatrix} \quad A_{12} = \begin{pmatrix} 1.1 & -0.2 \\ 0.53 & -0.27 \end{pmatrix}$$
$$C_{11} = \begin{pmatrix} 0 & 0 \end{pmatrix} \quad C_{12} = C_{11}$$

$$\begin{aligned} x_{k+1} &= \sum_{i=1}^{2} h_i(z_k) A_{2i} x_k \\ y_k &= \sum_{i=1}^{2} h_i(z_k) C_{2i} x_k \end{aligned}$$

with

$$A_{21} = \begin{pmatrix} 0.02 & 0.6 \\ -0.22 & -0.44 \end{pmatrix} \quad A_{22} = \begin{pmatrix} 0.32 & -0.15 \\ -1 & 0.8 \end{pmatrix}$$
$$C_{21} = \begin{pmatrix} 1 & 0 \end{pmatrix} \quad C_{22} = \begin{pmatrix} 0 & 1 \end{pmatrix}$$

The local models A_{12} and A_{22} are unstable, and unobservable. The periods of the subsystems are $p_1 = 1$ for the first subsystem and $p_2 = 2$ for the second subsystem.

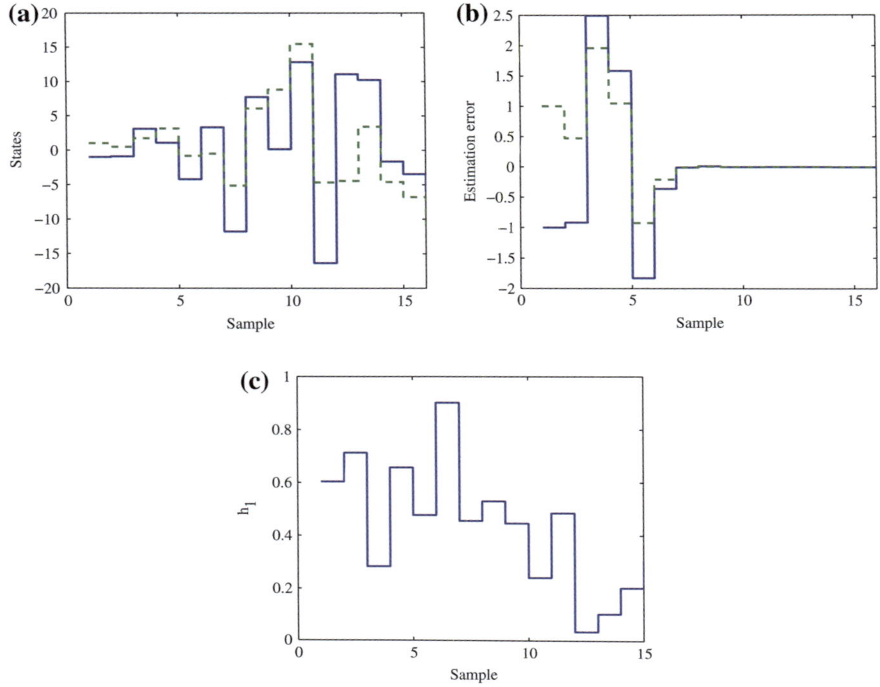

Fig. 2.10 Simulation results for Example 2.18. **a** A trajectory of the switching system. **b** The trajectory of the error. **c** Membership function

Using the conditions of Theorem 2.16 we obtain the observer gains

$$H_{1,1} = \begin{pmatrix} 0.14 & -0.28 \\ -0.23 & 0.54 \end{pmatrix} \qquad H_{1,2} = \begin{pmatrix} 0.15 & -0.26 \\ -0.29 & 0.55 \end{pmatrix}$$
$$H_{2,1} = \begin{pmatrix} 0.28 & -0.24 \\ -0.36 & 0.42 \end{pmatrix} \qquad H_{2,2} = \begin{pmatrix} 0.19 & -0.23 \\ -0.28 & 0.40 \end{pmatrix}$$
$$L_{2,1} = \begin{pmatrix} -0.50 \\ 0.67 \end{pmatrix} \qquad L_{2,2} = \begin{pmatrix} -0.19 \\ 0.40 \end{pmatrix}$$

Consequently, this observer is able to estimate the states of the switching system above. Since the first subsystem on its own is not observable, there are no observer gains $L_{1,1}$ and $L_{1,2}$. Trajectories of the states and of the error dynamics, together with the membership function h_1, are presented in Fig. 2.10. The true initial states were $x_0 = [-1;\ 1]^T$, while the estimated initial states were $\widehat{x}_0 = [0,\ 0]^T$.

For this periodic system, it is not possible to find either a quadratic or a non-quadratic Lyapunov function, common for both subsystems, as the corresponding LMIs are unfeasible.

2.3.3 Controlled Switches

Finally, let us consider the case when an observer has to be designed while the switches may be chosen. Therefore, our goal is to "stabilize" the error dynamics by switching among the subsystems. The switching observer considered,

$$\begin{aligned} \widehat{x}_{k+1} &= A_{j,z}\widehat{x}_k + B_{j,z}u_k + H_{j,z}^{-1}L_{j,z}(y_k - \widehat{y}_k) \\ \widehat{y}_k &= C_{j,z}\widehat{x}_k \end{aligned} \tag{2.21}$$

is the same as before, but next to the matrices $H_{j,i}^{-1}$, $L_{j,i}$, $j = 1, 2 \ldots, n_s$, $i = 1, 2, \ldots, r$, the switching sequence also has to be determined.

The error dynamics is again

$$e_{k+1} = A_{j,z}e_k - H_{j,z}^{-1}L_{j,z}(y_k - \widehat{y}_k) = (A_{j,z} - H_{j,z}^{-1}L_{j,z}C_{j,z})e_k \tag{2.22}$$

Since the switching sequence can be chosen, if there exists a subsystem with an asymptotically stable error dynamics that can be active for an infinite number of samples, the problem can be reformulated as finding a path—a classical graph theoretical problem—from each subsystem to this stable subsystem. If such a path exists, then the switching law is given by this path, and the problem is solved. Therefore, we consider the case when none of the subsystems may be infinitely active. Next to this, for the easier development of the conditions, we assume that the associated graph is strongly connected, i.e., there exists a path between any two vertices.

Recall that we consider the associated directed graph $\mathcal{G} = \{\mathcal{V}, \mathcal{E}\}$, where the vertices $\mathcal{V} = \{v_1, v_2, \ldots, v_{n_s}\}$ correspond to the subsystems, and each edge $e_{i,j} = (v_i, v_j) \in \mathcal{E}$ corresponds to an admissible transition.

In what follows, we build a weight-adjacency matrix, that assigns to each admissible transition, including self-transitions, a weight. By convention, if $(v_i, v_j) \notin \mathcal{E}$, for $i \neq j, i, j = 1, 2, \ldots, n_s$ the corresponding weight $w_{i,j} = \infty$, and if $(v_i, v_i) \notin \mathcal{E}$, $i = 1, 2, \ldots, n_s$, then the corresponding weight $w_{i,i} = 1$. For all other edges, the corresponding weight will be given by an upper bound on the increase of a Lyapunov function associated to the error dynamics.

Consider the Lyapunov function

$$V(e_k) = e_k^T P_{i,z} e_k$$

with $P_{i,z} = P_{i,z}^T > 0$, for the ith subsystem, $i = 1, 2, \ldots, n_s$.

Before constructing the weighting matrix, let us find constants $\delta_{i,j} > 0$ so that $V(e_{k+1}) \leq \delta_{i,j} V(e_k)$, if the transition is (v_i, v_j), i.e., upper bounds on the increase of the Lyapunov function in one sample.

For any $(v_i, v_j) \in \mathcal{E}$ we have

$$
\begin{aligned}
V(e_{k+1}) - \delta_{i,j} V(e_k) &< 0 \\
e_{k+1}^T P_{j,z+1} e_{k+1} - \delta_{i,j} e_k^T P_{i,z} e_k &< 0 \\
\begin{pmatrix} e_k \\ e_{k+1} \end{pmatrix}^T \begin{pmatrix} -\delta_{i,j} P_{i,z} & 0 \\ 0 & P_{j,z+1} \end{pmatrix} \begin{pmatrix} e_k \\ e_{k+1} \end{pmatrix} &< 0
\end{aligned}
$$

At the same time, the system's dynamics can be written as

$$
\begin{pmatrix} A_{i,z} - H_{i,z}^{-1} L_{i,z} C_{i,z} & -I \end{pmatrix} \begin{pmatrix} e_k \\ e_{k+1} \end{pmatrix} = 0
$$

Using Lemma 2.1, we have $V(e_{k+1}) - \delta_{i,j} V(e_k) < 0$ if there exists $\overline{M}_{i,j,z}$ so that

$$
\begin{pmatrix} -\delta_{i,j} P_{i,z} & 0 \\ 0 & P_{j,z+1} \end{pmatrix} + \overline{M}_{i,j,z} \begin{pmatrix} A_{i,z} - H_{i,z}^{-1} L_{i,z} C_{i,z} & -I \end{pmatrix} + (*) < 0
$$

Choosing $\overline{M}_{i,j,z} = \begin{pmatrix} 0 \\ H_{i,j,z} \end{pmatrix}$ we obtain the sufficient conditions

$$
\begin{pmatrix} -\delta_{i,j} P_{i,z} & (*) \\ H_{i,j,z} A_{i,z} - L_{i,z} C_{i,z} & -H_{i,j,z} - H_{i,j,z}^T + P_{j,z+1} \end{pmatrix} < 0 \tag{2.23}
$$

To find all $\delta_{i,j}$, one has to solve (2.23) for all $(v_i, v_j) \in \mathscr{E}$.
Now, define the weight matrix as $\mathscr{W} = [w_{i,j}]$ with

$$
w_{i,j} = \begin{cases} \delta_{i,i}^{p_i^m - 1} & \text{if } \delta_{i,i} > 1 \\ \delta_{i,i}^{p_i^M - 1} & \text{if } \delta_{i,i} < 1 \\ \delta_{i,j} & \text{if } i \neq j \end{cases}
$$

Assume that in the weight matrix constructed above, there exists a subunitary cycle, i.e., there exists $\mathscr{C}_n = \{v_{c1}, v_{c2}, \ldots, v_{cp}, v_{c1}\}$ such that the product of the edges and nodes in this cycle is subunitary, and let this product be denoted by δ_n. For any subsystem i such that $v_i \in \mathscr{C}_n$, we have $V(x_{k+n_c}) < \delta_n V(x_k)$, i.e., after a full cycle, the corresponding Lyapunov function increases at most δ_n times, with $\delta_n < 1$. Consequently, the periodic switching law $\mathscr{C} = [c_1, c_2, \ldots, c_p, c_1]$ stabilizes the periodic error dynamics given by this cycle.

Let us now see the subsystems (if any) that are not in $\mathscr{C}_n$. Since we assumed that $\mathscr{G}$ is strongly connected, for all $v_i \notin \mathscr{C}$, there exists a path $\mathscr{P}(v_i, v_j)$ from the ith subsystem to a subsystem j on the cycle. Consider the switching law $\mathscr{P}(v_i, v_j)\mathscr{C}_n$, i.e., first a switching law that leads to the cycle and then the periodic switching law corresponding to the cycle. Although during the switches corresponding to $\mathscr{P}(v_i, v_j)$ the Lyapunov function might increase, during the periodic switching, it will eventually decrease and converge to zero.

Based on the explanation above, the following result can be stated:

Theorem 2.19 *The error dynamics* (2.22) *is asymptotically stable along a switching law, if its associated graph* $\mathscr{G} = \{\mathscr{V}, \mathscr{E}, \mathscr{W}\}$ *contains a subunitary cycle* $\mathscr{C}_n$. *Furthermore, for the* i*th initial subsystem,* $i = 1, 2, \ldots, n_s$, *the switching law that stabilizes the error dynamics is given by* $\mathscr{P}(v_i, v_j)\mathscr{C}_n$, *where* v_i *denotes the vertex corresponding to the initial subsystems, and* $\mathscr{P}(v_i, v_j)$ *is a path to vertex* v_j, *with* $v_j \in \mathscr{C}_n$.

It should be noted that we assumed that the associated graph is strongly connected, because in this case the existence of a periodic switching law—which is not necessarily stabilizing—is guaranteed. If the graph is not strongly connected, the possibility of developing an observer depends on the starting subsystem. Therefore, in order to design an observer, each strongly connected subgraph has to be analyzed.

2.4 Summary

In this chapter, we considered observer design for switching nonlinear systems represented by TS fuzzy models. We have developed conditions that, when satisfied, guarantee that the estimation error converges to zero. The conditions were derived by taking into account the number of samples a subsystem may be active and the possible switches in the system. We have considered three cases.

The first case is general switching systems, where the switching sequence is not known in advance and it cannot be directly influenced. Due to this assumption, the conditions require that the estimation error dynamics decreases along the trajectory of the subsystems and the switches. This is the worst-case assumption, i.e., all possible combinations of switches between the estimation error subsystems have to be taken into account. The second case, for which the conditions can be relaxed, is when the switching is periodic. Although the conditions developed for the general case apply, due to the reduced number of switches, a smaller number of conditions are necessary. Finally, we considered the case that next to designing the observer gains, the switching sequence can be chosen.

The developed conditions have been extended to the α-sample variation of the Lyapunov function, in order to reduce their conservativeness. Other possibilities to relax the conditions represents the usage of double sums in the Lyapunov function, or using delayed Lyapunov functions or observers.

A shortcoming of the conditions is the computational complexity required to generate the possible combinations of switches and subsystems. In particular for large-scale switching systems, when considering α-sample variation, the number of conditions may be exponential in the number of subsystems, and in consequence, a large number of LMIs that has to be solved.

References

1. Altafini, C.: The reachable set of a linear endogenous switching system. Syst. Control Lett. **47**, 343–353 (2002)
2. Arzelier, D., Peaucelle, D., Farges, C.: Robust analysis and synthesis of linear polytopic discrete-time periodic systems via LMIs. In: Proceedings of the 44th IEEE Conference on Decision and Control, Spain (2005)
3. Baglietto, M., Battistelli, G., Tesi, P.: Stabilization and tracking for switching linear systems under unknown switching sequences. Syst. Control Lett. **62**, 11–21 (2013)
4. Battistelli, G.: On stabilization of switching linear systems. Automatica **49**, 1162–1173 (2013)
5. Bengea, S.C., DeCarlo, R.A.: Optimal control of switching systems. Automatica **41**, 11–27 (2005)
6. Bittanti, S., Colaneri, P.: Invariant representations of discrete-time periodic systems. Automatica **36**, 1777–1793 (2000)
7. Blanchini, F., Miani, S., Savorgnan, C.: Stability results for linear parameter varying and switching systems. Automatica **43**, 1817–1823 (2007)
8. Boukhobza, T., Hamelin, F.: Observability of switching structured linear systems with unknown input. A graph-theoretic approach. Automatica **47**, 395–402 (2011)
9. Boyd, S., El Ghaoui, L., Féron, E., Balakrishnan, V.: Linear Matrix Inequalities in System and Control Theory (Studies in Applied Mathematics). Society for industrial and applied mathematics, Philadelphia, USA (1994)
10. Chauvin, J., Moulin, P., Corde, G., Petit, N., Rouchon, P.: Real-time nonlinear individual cylinder air-fuel ratio observer on a diesel engine test bench. Prague, Czech Republic, In: Preprints of the IFAC World Congress (2005)
11. Chen, Y.-J., Ohtake, H., Wang, W.-J., Wang, H.: Relaxed stabilisation criterion for discrete T-S fuzzy systems by minimum-type piecewise non-quadratic Lyapunov function. IET Control Theor. Appl. **6**(12), 1918–1925 (2012)
12. Daafouz, J., Riedinger, P., Iung, C.: Stability analysis and control synthesis for switched systems: a switched Lyapunov function approach. IEEE Trans. Autom. Control **47**(11), 1883–1887 (2002)
13. De Santis, E.: On location observability notions for switching systems. Syst. Control Lett. **60**, 807–814 (2011)
14. Dehghan, M., Ong, C.-J.: Computations of mode-dependent dwell times for discrete-time switching system. Automatica **49**, 1804–1808 (2013)
15. Ding, B., Sun, H., Yang, P.: Further studies on LMI-based relaxed stabilization conditions for nonlinear systems in Takagi-Sugeno's form. Automatica **42**(3), 503–508 (2006)
16. Dong, J., Yang, G.: H_∞ controller synthesis via switched PDC scheme for discrete-time T-S fuzzy systems. IEEE Trans. Fuzzy Syst. **17**(3), 544–555 (2009)
17. Donkers, M., Heemels, W., van de Wouw, N., Hetel, L.: Stability analysis of networked control systems using a switched linear systems approach. IEEE Trans. Autom. Control **56**(9), 2101–2115 (2011)
18. Doo, J.C., Seung, S.L., PooGyeon, P.: Output-feedback control of discrete-time switching fuzzy system. In: Proceedings of the IEEE International Conference Fuzzy Systems, pp. 441–446. St. Luis, USA (2003)
19. Duan, C., Wu, F.: Switching control synthesis for discrete-time switched linear systems via modified Lyapunov-Metzler inequalities. In: Proceedings of the American Control Conference, pp. 3186–3191. Montreal, Canada (2012)
20. Fantuzzi, C., Rovatti, R.: On the approximation capabilities of the homogeneous Takagi-Sugeno model. In: Proceedings of the Fifth IEEE International Conference on Fuzzy Systems, pp. 1067–1072. USA, New Orleans (1996)
21. Farges, C., Peaucelle, D., Arzelier, D., Daafouz, J.: Robust H_2 performance analysis and synthesis of linear polytopic discrete-time periodic systems via LMIs. Syst. Control Lett. **56**, 159–166 (2007)

22. Farkas, M.: Periodic Motions. Springer, New York (1994)
23. Feng, G.: Controller synthesis of fuzzy dynamic systems based on piecewise Lyapunov functions. IEEE Trans. Fuzzy Syst. **11**(5), 605–612 (2003)
24. Feng, G.: H_∞ controller design of fuzzy dynamic systems based on piecewise Lyapunov functions. IEEE Trans. Syst. Man Cybern. Part B: Cybern. **34**(1), 283–292 (2004)
25. Gaiani, G., Lovera, M., Colaneri, P., Celi, R.: Discrete-time analysis of HHC schemes for helicopter vibration attenuation. In: Proceedings of the IFAC Workshop on Periodic Control Systems, pp. 69–74. Yokohama, Japan (2004)
26. Guan, Y., Ji, Z., Zhang, L., Wang, L.: Decentralized stabilizability of multi-agent systems under fixed and switching topologies. Syst. Control Lett. **62**, 438–446 (2013)
27. Guerra, T.M., Vermeiren, L.: LMI-based relaxed nonquadratic stabilization conditions for nonlinear systems in the Takagi-Sugeno's form. Automatica **40**(5), 823–829 (2004)
28. Hauroigné, P., Riedinger, P., Iung, C.: Observer-based output-feedback of a multicellular converter: control Lyapunov function-sliding mode approach. In: Proceedings of the 51st Conference on Decision and Control, pp. 1727–1732. Maui, Hawaii, USA (2012)
29. Hetel, L., Kruszewski, A., Perruquetti, W., Richard, J.: Discrete-time switched systems, set-theoretic analysis and quasi-quadratic Lyapunov functions. In: Proceedings of the Mediterranean Conference on Control and Automation, pp. 1325–1330. Corfu, Greece (2011)
30. Ji, Z., Wang, L., Guo, X.: Design of switching sequences for controllability realization of switched linear systems. Automatica **43**, 662–668 (2007)
31. Jungers, M., Castelan, E.B., Tarbouriech, S., Daafouz, J.: Finite L_2-induced gain and λ-contractivity of discrete-time switching systems including modal nonlinearities and actuator saturations. Nonlinear Anal. Hybrid Syst. **5**, 289–300 (2011)
32. Kerkeni, H., Guerra, T.M., Lauber, J.: Individual exaust gas mass flow estimation using a periodic observer design. In: Preprints of the IFAC World Congress, Milano, Italy (2011)
33. Kim, S., Campbell, S.A., Liu, X.: Delay independent stability of linear switching systems with time delay. J. Math. Anal. Appl. **339**, 785–801 (2008)
34. Kruszewski, A., Guerra, T.M.: Stabilization of a class of nonlinear model with periodic parameters in the Takagi-Sugeno form. In: Proceedings of the IFAC Workshop Periodic Control Systems, pp. 1–6. Saint Petersburg, Russia (2007)
35. Kruszewski, A., Wang, R., Guerra, T.M.: Nonquadratic stabilization conditions for a class of uncertain nonlinear discrete time TS fuzzy models: a new approach. IEEE Trans. Autom. Control **53**(2), 606–611 (2008)
36. Lam, H.K., Leung, F.H.F., Lee, Y.S.: Design of a switching controller for nonlinear systems with unknown parameters based on a fuzzy logic approach. IEEE Trans. Syst., Man Cybern. Part B **34**(2), 1068–1074 (2004)
37. Lam, H.K., Leung, F.H.F., Tam, P.K.S.: A switching controller for uncertain nonlinear systems. IEEE Control Syst. Mag. **22**(1), 1–14 (2002)
38. Langjord, H., Johansen, T., Hespanha, J.P.: Switched control of an electropneumatic clutch actuator using on/off valves. In: Proceedings of the American Control Conference, pp. 1513–1518. Seattle, Washington, USA (2008)
39. Lendek, Zs, Lauber, J., Guerra, T.M.: Periodic Lyapunov functions for periodic TS systems. Syst. Control Lett. **62**(4), 303–310 (2013)
40. Liao, H.-H., Widd, A., Ravi, N., Jungkunz, A.F., Kang, J.-M., Gerdes, J.C.: Control of recompression HCCI with a three region switching controller. Control Eng. Pract. **21**(2), 135–145 (2013)
41. Moustris, G.P., Tzafestas, S.G.: Switching fuzzy tracking control for mobile robots under curvature constraints. Control Eng. Pract. **19**(1), 45–53 (2011)
42. Nguyen, A., Lauber, J., Dambrine, M.: Robust H_∞ control design for switching uncertain system: Application for turbocharged gasoline air system control. In: Proceedings of the 51st Conference on Decision and Control, pp. 1–6. Maui, Hawaii, USA (2012a)
43. Nguyen, A., Lauber, J., Dambrine, M.: Switching fuzzy control of the air system of a turbocharged SI engine. In: Proceedings of the IEEE International Conference on Fuzzy Systems, pp. 1–6. Brisbane, Australia (2012b)

44. Nguyen, A., Lauber, J., Dambrine, M.: Multi-objective control design for turbocharged spark ignited air system: a switching Takagi-Sugeno model approach. In: Proceedings of the American Control Conference, pp. 1–6, Washington, DC, USA (2013)
45. Ohtake, H., Tanaka, K., Wang, H.: Fuzzy modeling via sector nonlinearity concept. In: Proceedings of the Joint 9th IFSA World Congress and 20th NAFIPS International Conference, vol. 1, pp. 127–132. Vancouver, Canada (2001)
46. Ohtake, H., Tanaka, K., Wang, H.O.: Switching fuzzy controller design based on switching Lyapunov function for a class of nonlinear systems. IEEE Trans. Syst. Man Cybern. Part B **36**(1), 13–23 (2006)
47. Pasamontes, M., Álvarez, J., Guzmán, J., Lemos, J., Berenguel, M.: A switching control strategy applied to a solar collector field. Control Eng. Pract. **19**(2), 135–145 (2011)
48. Scherer, C., Weiland, S.: Linear Matrix Inequalities in Control. Delft University, The Netherlands (2005)
49. Skelton, R.E., Iwasaki, T., Grigoriadis, K.: A unified approach to linear control design. Taylor & Francis, London (1998)
50. Stefan, O., Codrean, A., Dragomir, T.: Stability analysis and control synthesis for a network control system using a nonlinear network transmission model—A switched system approach. In: Proceedings of the IEEE International Conference on Control Applications (CCA), pp. 885–890. Dubrovnik, Croatia (2012)
51. Stepan, G., Insperger, T.: Stability of time-periodic and delayed systems—a route to act and-wait control. Annu. Rev. Control **30**, 159–168 (2006)
52. Takagi, T., Sugeno, M.: Fuzzy identification of systems and its applications to modeling and control. IEEE Trans. Syst. Man Cybern. **15**(1), 116–132 (1985)
53. Tanaka, K., Ikeda, T., Wang, H.: Fuzzy regulators and fuzzy observers: relaxed stability conditions and LMI-based designs. IEEE Trans. Fuzzy Syst. **6**(2), 250–265 (1998)
54. Tanaka, K., Iwasaki, M., Wang, H.O.: Switching control of an R/C hovercraft: stabilization and smooth switching. IEEE Trans. Syst. Man Cybern. Part B **31**(6), 853–863 (2001)
55. Tanaka, K., Wang, H.: Fuzzy regulators and fuzzy observers: a linear matrix inequality approach. In: Proceedings of the 36th IEEE Conference on Decision and Control, vol. 2, pp. 1315–1320. San Diego, CA, USA (1997)
56. Theron, A., Farges, C., Peaucelle, D., Arzelier, D.: Periodic H_2 synthesis for spacecraft in elliptical orbits with atmospheric drag and perturbations. In: Proceedings of American Control Conference, New York, USA (2007)
57. Tuan, H., Apkarian, P., Narikiyo, T., Yamamoto, Y.: Parameterized linear matrix inequality techniques in fuzzy control system design. IEEE Trans. Fuzzy Syst. **9**(2), 324–332 (2001)
58. Venkataramanan, V., Chen, B.M., Lee, T.H., Guo, G.: A new approach to the design of mode switching control in hard disk drive servo systems. Control Eng. Prac. **10**(9), 925–939 (2002)
59. Wang, H., Tanaka, K., Griffin, M.: An approach to fuzzy control of nonlinear systems: stability and design issues. IEEE Trans. Fuzzy Syst. **4**(1), 14–23 (1996)
60. Widyotriatmo, A., Hong, K.-S.: Switching algorithm for robust configuration control of a wheeled vehicle. Control Eng. Prac. **20**(3), 315–325 (2012)
61. Zhao, J., Spong, M.: Hybrid control for global stabilization of the cart-pendulum system. Automatica **37**(12), 1941–1951 (2001)
62. Zwart, H., van Mourik, S., Keesman, K.: Switching control for a class of non-linear systems with an application to post-harvest food storage. Eur. J. Control **16**(5), 567–573 (2010)

Chapter 3
Realization Theory of Linear Hybrid Systems

Mihaly Petreczky

Abstract This chapter presents a survey on realization theory for linear hybrid systems. Recall that for linear systems, realization theory addresses the problem of existence and minimality of a linear time-invariant state-space representation of an input–output map (transfer function). The results of linear realization theory have turned out to be useful for control synthesis, model reduction, and system identification. In this chapter, we address a similar problem for linear hybrid systems and linear switched systems. We will also discuss the implications of realization theory for estimation and control of hybrid systems.

3.1 Introduction

The aim of this chapter is to give an overview of realization theory of linear hybrid and switched systems. Before embarking on technical details, we will recall the origins of realization theory and present the available results in an informal manner.

3.1.1 The Origins of Realization Theory

Realization theory was originally championed by the founder of modern control theory, Kalman. Interest in realization theory was motivated by the following aspects of control engineering:

- What we are truly interested in is the input–output behavior of the physical process: how changes in control inputs influence the outputs. Outputs could either be values we can measure or they can be performance indicators. The latter we cannot necessarily measure directly, but we would like to influence its values by choosing suitable control inputs. In a way, everything beyond the input–output behavior is an artifact, which, on its own, is not interesting for designing a controller, since

M. Petreczky (✉)
Department of Automatic Control and Computer Science, Ecole des Mines de Douai, 941, rue Charles Bourseul, 59508 Douai, France
e-mail: mihaly.petreczky@mines-douai.fr

M. Djemai and M. Defoort (eds.), *Hybrid Dynamical Systems*, Lecture Notes in Control and Information Sciences 457, DOI 10.1007/978-3-319-10795-0_3

the behavior of any controller is determined by the input–output behavior of the underlying physical process.

- Even though we are interested in input–output behavior, for control we need models which are defined by differential and/or difference equations. The latter types of models are known as *state-space representations*. The reason for this is both historic and technological. From a historical perspective, difference/differential equations are objects which have been around for a long time and about which we know a great deal. From a technological perspective, control is carried out by computers with finite memory. In order to control, we should be able to predict how the process will react to a certain choice of control input. In turn, in general, this depends on the past control inputs (and possibly unmeasured disturbances) which the process was subjected to. Differential/difference models have a very important property, namely that they use *finite number* of state variables to encode the effect of past inputs and disturbances. This means that the knowledge of the state variables tells us everything we need to know about the future behavior of the model, and by extension (and a significant leap of faith), about the future behavior of the physical process.
- Models of physical processes, especially those which can be used for control design, are difficult to come by. First principle models are either not available or too complex to be useful for control.

Hence, in order to be successful in control design, we need classes of input–output models and classes of state-space models such that

- We know how to go from state-space models to input–output behaviors and back. This correspondence is not one-to-one: each state-space model usually induces a unique input–output map, but two different state-space models may correspond to the same input–output map.
- We can synthesize controllers for state-space models automatically, but the result should depend only on the input–output map which corresponds to the state-space model, not on the choice of the state-space model, since the latter is an artificially constructed object. Moreover, the resulting controller should only use available measurement and it should achieve the control objectives. Since the controller usually uses the states of the state-space model, this implies that the states of the state-space models should be *observable* (reconstructible from measurements) and they should be *controllable* (we should be able to manipulate them by a careful choice of control inputs).
- We should be able to estimate the parameters of the state-space models from measurement data.

Realization theory aims exactly at resolving these problems. More precisely, realization theory attempts to answer the following question. Fix a class of state-space representations.

- When is it possible to construct a (preferably minimal) state-space representation of the given type which generates the specified input/output behavior? Can we provide an algorithm for constructing a state-space representation from input/output data?

- How to characterize *minimal state-space representations* which generate the specified input/output behavior? By minimal we mean minimal dimensional, i.e., with the least number of state variables. Are minimal state-space representations unique in some sense? Are they *observable* and *controllable*? Can we propose an algorithm for checking minimality?

The standard answers to this are usually as follows: One has to find a class of state-space representations and a class of algebraic structures together with a suitable notion of dimension and basis elements such that the following holds. The algebraic structure (*Hankel structure*) generated by the input–output data has a finite dimension, if and only if the input–output behavior can be generated by a state-space representation from the designated class. The state-space representation then can be constructed from a basis of the Hankel structure. Moreover, minimality is usually equivalent to observability and controllability. In addition, minimal state-space representations realizing the same behavior are isomorphic. Note that for each class of systems the notions of minimality, isomorphism, controllability, observability, Hankel structure has to be defined separately, and the above results have to be proven separately.

For classical *linear systems*, i.e., for state-space representations $\delta x(t) = Ax(t) + Bu(t)$, $y = Cx(t) + Du(t)$, where $\delta x(t) = x(t+1)$ or $\delta x(t) = \dot{x}(t)$; for some matrices A, B, $C$and D, the problems above were resolved as follows. The class of input–output maps considered is the class of *proper rational transfer functions*. It turned out that the input–output behavior of linear time-invariant differential/difference equations can be represented by a proper rational transfer functions, and any proper rational transfer function can be obtained in this way. The corresponding Hankel structure is the classical infinite *Hankel matrix*. The entries of the Hankel matrix, the so-called *Markov parameters*, can be viewed as outcomes of input–output experiments. The fact that an input–output map admits a realization by a linear system is equivalent to the Hankel matrix having a finite rank. Finally, if the rank of the Hankel matrix is finite, then a minimal linear system can be computed from a finite submatrix of the Hankel matrix. The corresponding algorithm is known as *Kalman–Ho realization algorithm*. Moreover, minimal linear state-space representations of a certain input–output map are controllable and observable. Conversely, every controllable and observable state-space representation is minimal. Moreover, all such minimal state-space representations are isomorphic and one can compute them directly from input–output data.

This means that for control, it is enough to work with minimal linear state-space representations, as such representations can indeed be controlled, and their states can be observed, hence controllers can be built. Moreover, due to the fact that minimal representations are isomorphic, the resulting controller will not depend on the choice of the state-space representation. Furthermore, the parameters of such models could be recovered from input–output measurements. This tight relationship between input–output models and state-space representations ensures that several properties such as stability, dissipativity, etc. are in fact properties of the input–output behavior, not of the state-space representation. However, they can be characterized by computationally effective conditions for state-space representations.

In a way, realization theory introduced algebra into the world of Lyapunov functions and optimal control. This combination has resulted in effective control synthesis algorithms. It also provided a tight package of methods ranging from system identification (model building), through model reduction (model simplification) to control algorithms.

Realization theory was developed in 1960s for linear systems [20, 21]. Later realization theory was worked out for nonlinear systems [2, 7, 14–19, 54–56, 60, 62], time-varying systems [22, 59], discrete-event systems [28, 53], and stochastic linear [5, 23, 26, 27] and bilinear systems [8, 10].

3.1.2 Realization Theory of Hybrid Systems

The realization problem for hybrid systems was first formulated in [13], but no solution was provided. In [29, 63] the relationship between input–output equations and the state-space representations was studied. A fairly complete realization theory was developed for linear and bilinear switched and hybrid systems with externally induced switching [30, 31, 33–36, 39, 43–47, 49]. There are some partial results for some classes of hybrid systems with autonomous switching [32, 33, 50] and nonlinear dynamics [42].

In this chapter, we give an overview of realization theory for *linear switched systems (abbreviated by LSS)* and *linear hybrid systems (abbreviated by LHS)*. We focus on these classes of systems due to their simplicity and relevance. A LHS is a hybrid system in the sense of [61] with no guards, whose discrete dynamics is determined by a finite-state automaton, and whose continuous dynamics at each discrete state is governed by a time-invariant linear differential equation. The reset maps are linear, and the discrete events are externally generated inputs. A LSS [25, 57] is a finite collection of time-invariant linear differential equations. Note that LSSs are a subclass of LHSs. During the evolution of an LSS one switches among the various linear differential equations and the trajectory of the LSS is a concatenation of trajectories of the corresponding differential equations. Whenever one switches from one differential equation to another one, the current state of the preceding differential equation becomes the new initial state of the new linear differential equation. In our setting, we assume that the switching signal (i.e., the signal which determines which linear subsystem should be active) is an external input. LSSs can be viewed as a subclass of LHSs, where the discrete dynamics is trivial, the discrete events, discrete outputs, and discrete states coincide and the reset maps are the identity map.

Note that both LHSs and LSS are *state-space representations*, i.e., they have finitely many state variables, knowledge of which is sufficient to predict the future behavior of the system. For this reason, it would perhaps be more logical to call LHSs and LSS linear hybrid and linear switched state-space representations, respectively. However, the latter terminology would be a significant deviation from the established one. Therefore, we prefer to keep using the terms linear hybrid and linear switched systems, with the tacit understanding that these terms refer to classes of state-space representations.

The reason for devoting special attention to LSSs is that while they represent simple subclass of LHSs, most of the techniques developed for LSSs generalize well to LHSs. In fact, many of the algorithmic results we will present only for LSSs, since the corresponding results for LHSs are rather cumbersome, without, however, being conceptually very different. Note, however, that there are certain aspects in which LHSs are truly different from LSSs.

In this chapter, we deal with continuous-time systems only. However, many of the presented results can be extended to discrete-time case, see [39, 45]. If we allow the local subsystems to be bilinear rather than linear, than essentially the same results hold, see [30, 31, 33, 36].

For LHSs, we fix the set of discrete events, the space of continuous inputs, and the spaces of continuous and discrete outputs. For LSSs, we fix the set of continuous inputs and outputs and the set of discrete states. For LHSs, the potential input–output maps are maps which map continuous inputs and timed sequences of discrete events to continuous and discrete outputs. For LSSs, the potential input–output maps are maps which map continuous inputs and timed sequences of discrete states to continuous outputs. Hence, for potential input–output maps of LHSs, the inputs are continuous-valued input signals and timed sequences of discrete events. For the potential input–output maps of LSSs, the inputs are continuous-valued input signals and timed sequences of discrete states (switching signals).

We will say that a LHS or LSS *is a realization* of an input–output map, if for any input, the output generated by this LHS (respectively LSS) equals the value of the input–output map for that input. Note that we consider LHSsand LSSs with a fixed initial state, i.e., the initial state is part of the description of the system. Hence, for any input, the LHS or LSS at hand produce a unique output. If a LHS or LSS is a realization of the input–output map, we will sometimes say that this LHS (respectively, LSS) *realizes* the input–output map. We will be looking for conditions for existence of a LHS or LSS *realizing* a given input–output map.

We will also be interested in defining and characterizing minimality of LHSs and LSSs realizing this input–output map. The main results on realization theory of LHSs and LSSs are as follows.

- **Existence of a realization**: We show that an input–output map has a realization by a LHS or LSSs, if and only if the input–output map satisfies some mild conditions, and its infinite Hankel matrix satisfies certain finiteness conditions. For LSSs, this finiteness condition is that the Hankel matrix has a finite rank. For LHSs, this finiteness condition is that the Hankel matrix has finite rank and that some of its columns form a finite set. If these conditions are satisfied, then a LHS (respectively, LSS) realization of the input–output map can be constructed from a finite submatrix of the Hankel matrix. The corresponding algorithm is similar to the well-known Kalman–Ho algorithm. In order to avoid complicated notation, we will present this algorithm only for LSSs. The case of LHSs is analogous, but slightly more complicated notationally. Note that by Hankel matrix, we mean an infinite matrix, which is analogous to, but different from, the classical Hankel matrix.

- **Observability, reachability**: We define observability and span-reachability for LHSs and we present a characterization of observability and span-reachability via rank conditions. We will also recall the definition of observability and span-reachability of LSSs and present rank conditions for characterizing them. The proposed notions of observability and span-reachability allow for a neat characterization of minimality: a state-space representation is minimal if and only if it is span-reachable and observable. Moreover, minimal state-space representation of the same input–output map turn out to be isomorphic. This is the case both for LHSsand LSSs. Furthermore, any LHS or LSSs can be converted to a minimal one by a suitable algorithm. Again, in order to avoid complicated notation, we present the algorithm only for LSSs. The case of LHSs is analogous, but it is slightly more complicated. Here by minimality we mean minimal state-space dimension. For LSSs, the state-space dimension is defined as the dimension of the common continuous state-space of the linear subsystems. In order to compare dimensions of LSSs, we use the usual ordering of natural numbers. For LHS, the definition of dimension of a state-space representation and minimality is more involved. The reason for this is that the set of discrete states is not fixed and the various linear subsystems live on different state-spaces. For LHSs, we define dimension as a pair of natural numbers: the first component is the number of discrete states, the second component is the sum of dimensions of continuous state-spaces. We define a component-wise partial ordering for these pairs. By a minimal LHS, we mean an LHS whose dimension is minimal among all the LHSs realizing the same input–output map, according this ordering. Since the introduced ordering is a partial one, there exist LHSs whose dimensions are incomparable. Hence, the existence of a minimal LHS is not a trivial fact. In fact, we show that for any input–output map such a minimal LHS exists.

The results mentioned are similar in their spirit to the known results for linear systems. For this reason, we hope that they will be useful for control, model reduction, and system identification of hybrid systems. In fact, in this chapter, we will also present a short overview on the already existing and potential application of realization theory to control theory in a wider sense.

In this chapter, we only present the results, for the proofs we refer to [33, 35, 43, 44, 47]. For LSSs, the proofs rely on reducing the realization problem for LSSs to the realization problem for recognizable formal power series [4, 24, 54]. Similarly, for LHSs the proofs rely on reducing the realization problem for LHSs to that realization problem for recognizable formal power series and Moore automata [9, 12]. Note that recognizable formal power series were used for realization theory of nonlinear systems in the past, see [15, 54] and the references therein.

3.1.3 Outline of the Chapter

In Sect. 3.2, we fix the notation and terminology used in this chapter. In Sect. 3.3, we define the realization problem for LHSs formally and present the necessary background material. In Sect. 3.4, we present realization theory for LSSs. In Sect. 3.5, we present realization theory for general LHSs. In Sect. 3.6, we present an informal overview of potential applications of realization theory.

3.2 Notation

Denote by $\mathbb{N}$ the set of natural numbers including 0. Let $\mathbb{R}_+$ be the *real time-axis*, i.e., $\mathbb{R}_+ = [0, +\infty)$. Denote by $\mathrm{AC}([t_0, t_1], \mathbb{R}^n)$ the set of all absolutely continuous functions of the form $f:[t_0, t_1] \to \mathbb{R}^n$. Denote by $\mathrm{AC}(\mathbb{R}_+, \mathbb{R}^n)$ the set of all absolutely continuous functions of the form $f : \mathbb{R}_+ \to \mathbb{R}^n$. Denote by $\mathrm{PC}(\mathbb{R}_+, \mathbb{R}^m)$ the set of piecewise-continuous functions of the form $f : \mathbb{R}_+ \to \mathbb{R}^n$, i.e., functions $f : \mathbb{R}_+ \to \mathbb{R}^n$ which have finitely many points of discontinuity on each finite interval and at each point of discontinuity, the left- and right-hand side limits exists and are finite. For each $j = 1, \ldots, m$, e_j is the jth unit vector of $\mathbb{R}^m$, i.e., $e_j = (\delta_{1,j}, \ldots, \delta_{n,j})$, $\delta_{i,j}$ is the Kronecker symbol.

The notation described below is standard in automata theory, see [9, 12]. Consider a set X which will be called the *alphabet*. Denote by X^+ the set of finite sequences of elements of X, i.e., each element $w \in X^+$ is of the form $w = a_1a_2 \ldots a_k$ for some $a_1, a_2, \ldots, a_k \in X$ and $k \in \mathbb{N}, k > 0$. Let $\varepsilon \notin X^+$ be a symbol, which we will call the *empty sequence or empty word*. Denote by X^* the set $X^+ \cup \{\varepsilon\}$. The elements of X^* will be referred to as *strings* or *words* over X. If $w \in X^+$, $w = a_1a_2 \ldots a_k$ for some $a_1, a_2, \ldots, a_k \in X$, then a_i is called the ith *letter of* w, for $i = 1, \ldots, k$ and k is called the *length* w. By convention, the length of ε is defined to be zero. The length of a word $w \in X^*$ is denoted by $|w|$. For any two words $w, v \in X^*$, we define the concatenation $wv \in X^*$ of w and v as follows: If $w, v \in X^+$ are of the form $v = v_1v_2 \ldots v_k$, $k > 0$ and $w = w_1w_2 \ldots w_m$, $m > 0$, $v_1, v_2, \ldots, v_k, w_1, w_2, \ldots, w_m \in X$, then define $vw = v_1v_2 \ldots v_kw_1w_2 \ldots w_m$. If $v = \varepsilon$ and $w \in X^*$, then define $vw = w$. Similarly, if $w = \varepsilon$ and $v \in X^*$, then define $vw = v$. With the operation of concatenation, X^* forms a semi-group whose unit element is ε. For $a \in X$ and $k \in \mathbb{N}$, $k > 0$, we denote by a^k the sequence $\overbrace{aa \ldots a}^{k\text{-times}}$; by convention $a^0 = \varepsilon$.

3.3 Linear Hybrid Systems: Definition and Basic Concepts

We start by defining what we mean by linear hybrid systems.

Definition 3.1 (*LHS*) A *linear hybrid system* H (abbreviated as LHS) is a tuple

$$H=(Q, \Gamma, O, \delta, \lambda, \{n_q, A_q, B_q, C_q\}_{q\in Q}, \{M_{q_1,\gamma,q_2}\}_{q_2\in Q,\gamma\in\Gamma,q_1=\delta(q_2,\gamma)}, h_0) \quad (3.1)$$

where

1. Q is a finite set, called *the set of discrete states*,
2. Γ is a finite set, called *the set of discrete events*,
3. O is a finite set, called *the set of discrete outputs*,
4. $\delta: Q \times \Gamma \to Q$ is a function called the *discrete state-transition map*,
5. $\lambda: Q \to O$ is a function called the *discrete readout map*.
6. $\Sigma_q = (A_q, B_q, C_q)$, $q \in Q$ is *the linear system in the discrete state* q and $A_q \in \mathbb{R}^{n_q \times n_q}$, $B_q \in \mathbb{R}^{n_q \times m}$, $C_q \in \mathbb{R}^{p \times n_q}$ are matrices of this linear system.
7. $M_{q_1,\gamma,q_2} \in \mathbb{R}^{n_{q_1} \times n_{q_2}}$ are the matrices for all $q_2 \in Q$, $\gamma \in \Gamma$, $q_1 = \delta(q_2, \gamma)$, which are called *reset maps*.
8. $h_0 = (q_0, x_0)$ is the *initial state*, where $q_0 \in Q$ and $x_0 \in \mathbb{R}^{n_{q_0}}$.

The space $\mathscr{X}_q = \mathbb{R}^{n_q}$, $q \in Q$, $0 < n_q \in \mathbb{N}$, is called the *continuous state associated with the discrete state* q, $\mathbb{R}^m$ is called the *continuous input space*, $\mathbb{R}^p$ is called the *continuous output space*. The *state-space* $\mathscr{H}_H$ *of* H is the set $\mathscr{H}_H = \bigcup_{q \in Q} \{q\} \times \mathscr{X}_q$.

Notice that the linear control systems associated with different states may have different state-spaces, but have the same input and output space. The intuition behind a linear hybrid system is as follows. We associate a linear system

$$\Sigma_q \begin{cases} \dot{x} = A_q x + B_q u \\ y = C_q x \end{cases} \tag{3.2}$$

with each discrete state $q \in Q$. As long as we are in the discrete state q, the state x and the continuous output y develops according to (3.2). A change in the discrete state can occur only if a discrete event $\gamma \in \Gamma$ takes place. If a discrete event γ occurs at time t, then the new discrete state q_+ is determined by applying the discrete state-transition map δ to q, i.e., $q_+ = \delta(q, \gamma)$. The new continuous state $x^+ \in \mathbb{R}^{n_{q+}}$ is computed from the current continuous state $x(t)$ by applying the *reset map* $M_{q_+,\gamma,q}$ to $x(t)$, i.e., $x^+ = M_{q_+,\gamma,q} x(t)$. After the transition, the continuous state x and the continuous output x evolves according to the linear system associated with the new discrete state q_+, started from the initial state x^+. Finally, when in a discrete state $q \in Q$, the system produces a discrete output $o = \lambda(q)$.

Notice that *the discrete events are external inputs. All the continuous subsystems are defined with the same inputs and outputs, but on possibly different state-spaces.* Below we will formalize the intuition described above, by defining input-to-state and input–output maps for LHS. To this end, we need the following.

Definition 3.2 (*Timed sequences*) A *timed sequence of discrete events* is a finite sequence over the set $(\Gamma \times \mathbb{R}_+)$, i.e., it is a sequence of the form

$$w = (\gamma_1, t_1)(\gamma_2, t_2) \cdots (\gamma_k, t_k) \tag{3.3}$$

where $\gamma_1, \gamma_2, \ldots, \gamma_k \in \Gamma$, $k > 0$ are discrete events, and $t_1, t_2, \ldots, t_k \in \mathbb{R}_+$ are time instances. We denote the set of timed sequences of discrete events by Γ^+_{timed}.

By convention, we introduce the symbol ε, which we will call the *empty sequence* and we denote by Γ^*_{timed} the set $\Gamma^+_{\text{timed}} \cup \{\varepsilon\}$.

For each $w \in \Gamma^*_{\text{timed}}$, we define the length of w as follows: If w is of the form (3.3), then its length equals the integer k. If ε, its length is defined as zero. The *length* of w is denoted by $|w|$. The interpretation of a time sequence $w \in \Gamma^*_{\text{timed}}$ above is the following. If $w = \varepsilon$, then this represents the case when no discrete event has yet occurred. If w is of the form (3.3), then w represents the scenario, when the event γ_i took place *after* the event γ_{i-1} and t_i is the *is the time which has passed between the arrival of* γ_{i-1} *and the arrival of* γ_i, i.e., t_i is the difference of the arrival times of γ_i and γ_{i-1}. Hence, $t_i \geq 0$ but we allow $t_i = 0$, i.e., we allow γ_i to arrive instantly after γ_{i-1}. If $i = 1$, then t_1 is simply the time when the first event γ_1 arrived. An input to a LHS is a triple, consisting of a $\mathbb{R}^m$-valued map from $\text{PC}(\mathbb{R}_+, \mathbb{R}^m)$, a timed sequence of discrete events, and a non-negative real indicating the time elapsed since the arrival of the last event.

Definition 3.3 (*Inputs* **U**) Denote by $\mathbf{U} = \text{PC}(\mathbb{R}_+, \mathbb{R}^m) \times \Gamma^*_{\text{timed}} \times \mathbb{R}_+$ the set of inputs of a LHS.

If $(u, w, t) \in \mathbf{U}$, then u represents the continuous-valued input to be fed to the system, w represents the timed-event sequence, and t represents the time which has elapsed since the arrival of the last discrete event described by w. Below we define the notion of input-to-state and input–output maps for LHSs. These functions map elements from **U** to outputs and states, respectively. In the rest of this section, *H denotes a LHS of the form* (3.1).

Definition 3.4 (*Input-to-state map*) The *input-to-state map of H induced by the initial state* $h_I = (q_I, x_I) \in \mathcal{H}_H$ *of H* is the function ξ_{H,h_I}:$\mathbf{U} \to \mathcal{H}$ such that the following holds. For any $w \in \mathbf{U}$, and any $t \in \mathbb{R}_+$, let $\xi_{H,h_I}(u, w, t) \in \mathcal{H}$ be defined recursively on the length of w as follows:

1. $\xi_{H,h_I}(\varepsilon, t) = (q_I, z(t))$ where $z \in \text{AC}([0, t], \mathbb{R}^{n_q})$ is the unique solution (in the sense of Caratheodory) of the differential equation $\dot{z}(t) = A_{q_I}z(t) + B_{q_I}u(t)$, $z(0) = x_I$.
2. Assume that w is of the form (3.3). If $k = 1$, then set $v = \varepsilon$ and if $k > 1$, then set $v = (\gamma, t_1)(\gamma_2, t_2)\cdots(\gamma_{k-1}, t_{k-1})$. Set $T_w = \sum_{i=1}^k t_i$. Assume that $\xi_{H,h_I}(u, v, T_w) = (q, x)$ is defined, then let $\xi_{H,h_I}(u, w, t) = (q^+, z(T_w + t))$, where $q^+ = \delta(q, \gamma_k)$ and $z \in \text{AC}([T_w, T_w + t], \mathbb{R}^{n_q}$ is the unique solution (in the sense of Caratheodory) of the differential equation $\dot{z}(s) = A_{q^+}z(s) + B_{q^+}u(s)$, $z(T_w) = M_{q^+,\gamma_k,q}x$.

Definition 3.5 (*Input–output map*) The *input–output map of the system H induced by a state* $h_I \in \mathcal{H}_H$ *of H* is the function $\upsilon_{H,h_I} : \mathbf{U} \to O \times \mathbb{R}^p$ defined as follows: for all $(u, w, t) \in \mathbf{U}$, if $(q, x) = \xi_{H,h_I}(u, w, t)$, then

$$\upsilon_{H,h_I}(u, w, t) = (\lambda(q), C_q x).$$

The input–output map υ_{H,h_0} induced by the initial state h_0 is called the *input–output map* of H and it is denoted by υ_H.

It is easy to see that the natural candidates for input–output maps of a LHS are functions of the form

$$f : \mathbf{U} \rightarrow O \times \mathbb{R}^p \tag{3.4}$$

In the rest of this section, functions of the form (3.4) will be referred to as *input–output* maps. Next, we define when a LHS H is a realization of an input–output map.

Definition 3.6 (*Realization and input–output equivalence*) The LHS H is a said to be a *realization* of an input–output map f, if the input–output map υ_H of H equals f. Two LHSs H_1 and H_2 are said to be *input–output equivalent*, if their input–output maps are equal, i.e., $\upsilon_{H_1} = \upsilon_{H_2}$.

In the sequel, we will be interested LHSs which are minimal dimensional among all realizations of their input–output maps. Furthermore, we will need the notions of observability, span-reachability, and isomorphism of LHSs.

First, define what we mean by dimension of an LHS.

Definition 3.7 (*Dimension*) For a LHS H, *the dimension* of H is defined as a pair of natural numbers $\dim H = (card(Q), \sum_{q \in Q} n_q)$.

In other words, the first component of $\dim H$ is the cardinality of the discrete state-space, the second component is the sum of dimensions of the continuous state-spaces. For each two pairs of natural numbers (m, n), $(p, q) \in \mathbb{N} \times \mathbb{N}$ define the partial order relation $\leq$ as $(m, n) \leq (p, q)$, if $m \leq p$ and $n \leq q$.

Definition 3.8 (*Minimality*) A LHS H is called a *minimal realization of an input–output map* f, if H is a realization of f and for any LHS $H^{'}$ which is also a realization of f, $\dim H \leq \dim H^{'}$. We call H a minimal, if it is a minimal realization of its own input–output map $f = \upsilon_H$.

Notice that the ordering introduced above is a partial ordering, i.e., there may be two LHSs whose dimensions are not comparable. In the light of this remark, the definition of minimality implicitly requires the dimension of a minimal LHS to be *comparable* to, in fact, to be not greater than, the dimension of any realization of f.

Since not all LHS realizations of f have comparable dimensions, it is not at all clear that one can choose a LHS realization of f whose dimension is minimal. The mathematical proof that such a minimal LHS exists is a problem on its own right.

The definition of dimension above expresses the inherent trade-off between the number of discrete states and dimensionality of each continuous state-space component. The definition of minimality is such that minimal LHS have the smallest possible number of discrete states and the sum of continuous state-space dimensions is the smallest possible as well. However, it can happen that a nonminimal realization has a discrete state, whose linear system has smaller dimension than the linear system of the corresponding discrete state of the minimal LHS.

Another feature of the definition of minimality above is that it allows us to characterize minimality in terms of observability and span-reachability.

Next, we define span-reachability and observability.

Definition 3.9 (*Span-reachability*) For each discrete state $q \in Q$ define the set $R_q(H) = \text{Span}\{x \in \mathbb{R}^{n_q} \mid \exists \underline{w} \in \mathbf{U} : (q, x) = \xi_{H,h_0}(\underline{w})\}$. The LHS H is called *span-reachable* if for all $q \in Q$:

1. there exists $\underline{w} \in \mathbf{U}$ such that $\xi_{H,h_0}(\underline{w}) = (q, x)$ for some $x \in \mathbb{R}^{n_q}$, and
2. $R_q(H) = \mathbb{R}^{n_q}$.

Intuitively, H is span-reachable, if every discrete state can be reached from the initial state and every continuous state can be represented as a linear combination of reachable continuous states (which belong to the same discrete state). Next, we define observability.

Definition 3.10 (*Observability*) The LHS H is called *observable*, if

$$\forall h_1, h_2 \in \mathscr{H}_H : (h_1 \neq h_2 \implies \upsilon_{H,h_1} \neq \upsilon_{H,h_1}).$$

Intuitively, observability means that any two distinct initial states induce two different input–output maps. In other words, for any two distinct initial states there exists a hybrid input $\underline{w}$ such that the corresponding outputs are different.

Finally, we define the notion of LHS (iso) morphisms.

Definition 3.11 (*LHS morphism*) Consider the LHS

$$H^{'} = (Q^{'}, \Gamma, O, \delta^{'}, \lambda^{'}, \{n^{'}_q, A^{'}_q, B^{'}_q, C^{'}_q\}_{q \in Q^{'}}, \{M^{'}_{q_1,\gamma,q_2}\}_{q_2 \in Q^{'},\gamma \in \Gamma, q_1 = \delta(q_2,\gamma)}, h^{'}_0)$$

Let H be of the form (3.1). A function $\mathscr{S} : \mathscr{H}_H \to \mathscr{H}_{H'}$ is called a *LHS morphism* from H to H', denoted by $\mathscr{S} : H \to H'$, if there exists a function $\mathscr{S}_D : Q \to Q^{'}$ and matrices $\mathscr{S}_{C,q} \in \mathbb{R}^{n^{'}_{\mathscr{S}_D(q)} \times n_q}$ such that for all $h = (q, x) \in \mathscr{H}_H$:

1. $\mathscr{S}((q, x)) = (\mathscr{S}_D(q), \mathscr{S}_{C,q}x)$, and
2. for all $\underline{w} \in \mathbf{U}$, $\xi_{H',\mathscr{S}(h)}(\underline{w}) = \mathscr{S}(\xi_{H,h}(\underline{w}))$ and $\upsilon_{H',\mathscr{S}(h)}(\underline{w}) = \upsilon_{H,h}(\underline{w})$, and
3. $h^{'}_0 = \mathscr{S}(h_0)$.

The LHS morphism $\mathscr{S}$ is called and *isomorphism*, if the map $\mathscr{S}_D$ is bijective and the matrices $\mathscr{S}_{C,q}$ are invertible square matrices for all $q \in Q$.

Intuitively, a LHS morphism is a function between the state-spaces of two LHSs which maps state- and output-trajectories of one system to state- and output trajectories of the other system, and it maps initial states to initial states. If $\mathscr{S}$ above is a LHS isomorphism, then the inverse map $\mathscr{S}^{-1} : \mathscr{H}_{H'} \to \mathscr{H}_H$ exists, and it can be viewed as an LHS isomorphism $\mathscr{S}^{-1} : H^{'} \to H$. Note that if $\mathscr{S} : H \to H^{'}$ is a AHLS morphism, then $\mathscr{S}_D$ is a morphism between Moore automata $\mathscr{A} = (Q, \Gamma, O, \delta, \lambda, q_0)$ and $\mathscr{A}^{'} = (Q, \Gamma, O, \delta, \lambda, q_0)$ (see [9, 12, 33, 43, 44, 47] for the definition of

Moore automata and the morphisms between them) and for each $q \in Q$, $\mathscr{S}_{C,q}$ is a system morphism from the linear system (A_q, B_q, C_q) to the linear system $(A'_{\mathscr{S}_D(q)}, B'_{\mathscr{S}_D(q)}, C'_{\mathscr{S}_D(q)})$. In particular, if $\mathscr{S}$ is an LHS isomorphism, then $\mathscr{S}_D$ is a Moore automaton isomorphism between the automata $\mathscr{A}$ and $\mathscr{A}'$, and for each discrete state $q \in Q$, the linear systems (A_q, B_q, C_q) and $(A'_{\mathscr{S}_D(q)}, B'_{\mathscr{S}_D(q)}, C'_{\mathscr{S}_D(q)})$ are related by the algebraic similarity $\mathscr{S}_{C,q}$.

Note that if $\mathscr{S}$ is an LHS morphism, then with the notation of Definition 3.11,

1. for any discrete state $q \in Q$, $\mathscr{S}_{C,q} A_q = A'_{\mathscr{S}_D(q)} \mathscr{S}_{C,q}$, $\mathscr{S}_{C,q} B_q = B'_{\mathscr{S}_D(q)}$, $C_q = C'_{\mathscr{S}_D(q)} \mathscr{S}_{C,q}$ for each discrete event $\gamma \in \Gamma$, and discrete state $q \in Q$, $\mathscr{S}_D(\delta(q, \gamma)) = \delta'(\mathscr{S}_D(q), \gamma)$ and $\lambda(q) = \lambda'(\mathscr{S}_D(q))$,
2. for each discrete event $\gamma \in \Gamma$, and discrete state $q \in Q$, $\mathscr{S}_{C,\delta(q,\gamma)} M_{\delta(q,\gamma),\gamma,q} = M'_{\delta'(\mathscr{S}_D(q),\gamma),\gamma,\mathscr{S}_D(q)} \mathscr{S}_{C,q}$,
3. $\mathscr{S}(h_0) = h'_0$.

The realization problem for linear hybrid systems can be formulated as follows.

Problem 3.12 (*Realization problem*) For a specified input–output map f, find necessary and sufficient conditions for existence of a LHS H which realizes f. Provide a constructive procedure for calculating a LHS realization of f from the values of f. Furthermore, characterize minimal LHS which are realizations of f.

3.4 Special Case: Linear Switched Systems

Before presenting the results on realization theory for LHSsin all their generality, we will consider a special case, that of *linear switched system* (abbreviated LSSs). For linear switched systems, the discrete state dynamics is trivial: the set of discrete states coincides with the set of discrete events and the set of discrete outputs and any discrete state transition is allowed. Moreover, all the linear subsystems are defined on the same state-space and the reset maps equal the identity map.

3.4.1 Definition of Linear Switched Systems

In this section we present the formal definition of linear switched systems and recall a number of relevant definitions. We follow the presentation of [35, 51].

Definition 3.13 (*LSS*) A linear switched system (LSS) is a control system of the form

$$\frac{\mathrm{d}}{\mathrm{d}t} x(t) = A_{\sigma(t)} x(t) + B_{\sigma(t)} u(t), \quad x(t_0) = x_0 \tag{3.5a}$$

$$y(t) = C_{\sigma(t)} x(t) \tag{3.5b}$$

where Q called the set of discrete states, $\sigma \in \mathrm{PC}(\mathbb{R}_+, Q)$ is called the switching signal, $u \in \mathrm{PC}(\mathbb{R}_+, \mathbb{R}^m)$ is called the input, $x \in \mathrm{AC}(\mathbb{R}_+, \mathbb{R}^n)$ is called the state, and $y \in \mathrm{PC}(\mathbb{R}_+, \mathbb{R}^p)$ is called the output. Moreover, $A_q \in \mathbb{R}^{n\times n}$, $B_q \in \mathbb{R}^{n\times m}$, $C_q \in \mathbb{R}^{p\times n}$ are the matrices of the linear system in state $q \in Q$, and x_0 is the initial state. The notation

$$\Sigma = (n, \{(A_q, B_q, C_q) \mid q \in Q\}, x_0) \tag{3.6}$$

is used as a short-hand representation for LSSs of the form (3.5a, 3.5b). The number n is called the *dimension (order) of* Σ and will be denoted by $\dim\Sigma$.

Intuitively, an LSS is just a control system which switches among finitely many linear time-invariant systems. The switching signal is part of the input. Whenever a switch occurs, the continuous state remains the same, only the differential equation governing the state and output evolution changes. That is, whenever we switch to a new linear system, we start the new linear system from the state which is the final state of the previous linear system. For all this to make sense, all the linear systems should have the same input, output, and state-spaces. Finally, we do not have discrete outputs, only continuous ones. In order to formalize the semantics above, we will need the notion of switching sequences, input-to-state and input–output maps.

Definition 3.14 (*Switching sequences*) A *switching sequence* is a sequence of the form

$$w = (q_1, t_1)(q_2, t_2) \cdots (q_k, t_k) \tag{3.7}$$

$q_1, q_2, \ldots, q_k \in Q$ are discrete states and $t_1, t_2, \ldots, t_k \in \mathbb{R}_+$ are *switching times* and $k > 0$. The set of all switching sequences is denoted by Q^+_{timed}.

If w is of the form (3.7) with $k > 0$, then w *determines a switching signal* $\sigma \in \mathrm{PC}([0, \sum_{i=1}^k t_i], Q)$ *which at time* $t \in (\sum_{j=1}^{i-1} t_j, \sum_{j=1}^{i} t_j]$ *takes the value* $\sigma(t) = q_i$, $i = 1, \ldots, k$, $\sigma(0) = q_1$ and for $t > \sum_{j=1}^{k} t_j$, $\sigma(t) = q_k$. The sequence w from (3.7) thus represents the scenario when from 0 to time instance t_1 the active discrete state is q_1, from t_1 to t_2 the active discrete state is q_2, from $t_1 + t_2$ to $t_1 + t_2 + t_3$ the active discrete state is q_3, and so on. That is, t_i indicates the time spent in the discrete state q_i, for all $i = 1, 2, \ldots, k$.

We are now ready to define the input-to-state map. To this end, we introduce the following concept.

Definition 3.15 (*Switched input*) We denote by **SU** the set $\mathrm{PC}(\mathbb{R}_+, \mathbb{R}^m) \times Q^+_{\text{timed}}$.

That is, every element of **SU** represents a pair (u, w), where w is a switching sequence and u is a continuous-valued input signal. This definition merely expresses the fact that we view an LSS as a control system driven by switching signals and continuous-valued inputs. Next, we define the input-to-state map of an LSS.

Definition 3.16 (*State-trajectory*) The *input-to-state map* of Σ induced by a state $x \in \mathbb{R}^n$ is a function

$$X_{\Sigma,x} : \mathbf{SU} \to \mathbb{R}^n$$

such that for all $(u, w) \in \mathbf{SU}$, where w is of the form (3.7), $X_{\Sigma,x_0}(u, w)$ is defined recursively on the length k of w as follows:

- If $k = 1$, then $X_{\Sigma,x}(u, w) = z(t_1)$, where $z \in \mathrm{AC}([0, t_1], \mathbb{R}^n)$ is the unique solution (in Caratheodory sense) $\dot{z}(t) = A_{q_1} z(t) + B_{q_1} u(t)$, $z(0) = x$.
- Assume that $k > 1$ and that for $v = (q_1, t_1) \cdots (q_{k-1}, t_{k-1})$, $X_{\Sigma,x}(u, v)$ is already defined. Then let $X_{\Sigma,x}(u, w) = z(t_k)$, where $z \in \mathrm{AC}([0, t_1], \mathbb{R}^n)$ is the unique solution (in Caratheodory sense) of the differential equation of the differential equation $\dot{z}(t) = A_{q_k} z(t) + B_{q_k} u(t + \sum_{j=1}^{k-1} t_j)$. with $z(0) = X_{\Sigma,x}(u, v)$.

Intuitively, $X_{\Sigma,x}(u, w)$ is the state $x(t_1 + \cdots + t_k)$ reached by (3.5a, 3.5b), if the switching signal corresponding to w is fed to the system, along with the continuous-valued input u. Now we are ready to define the input–output map of Σ.

Definition 3.17 (*Input–output map*) The *input–output map* $Y_{\Sigma,x}$ of Σ induced by the state $x \in \mathbb{R}^n$ is a function

$$Y_{\Sigma,x} : \mathbf{SU} \to \mathbb{R}^p,$$

such that for any $(u, w) \in \mathbf{SU}$, $w \in Q^+_{\mathrm{timed}}$, where w is of the form (3.7),

$$Y_{\Sigma,x}(u, w) = C_{q_k} X_{\Sigma,x}(u, w).$$

We call the input–output map Y_{Σ,x_0} induced by the initial state x_0 of Σ the *input–output map* of Σ, and we denote Y_{Σ,x_0} by Y_Σ.

In other words, the value of $Y_{\Sigma,x}(u, w)$ for $u \in \mathrm{PC}(\mathbb{R}_+, \mathbb{R}^m)$ and $w \in Q^+_{\mathrm{timed}}$ is the output of Σ at time $t_1 + \cdots + t_k$ if Σ is started from x and it is driven by the continuous-valued input u and the switching signal w.

We model the input–output behavior of a system (not necessarily of a finite-dimensional LSS) as a function

$$f : \mathbf{SU} \to \mathbb{R}^p. \tag{3.8}$$

In the rest of this section, functions of the form (3.8) will be called *input–output maps*. Such a function captures the behavior of a black-box, which reacts to piecewise-continuous inputs and switching sequences by generating outputs in $\mathbb{R}^p$. Next, we define what it means that this black-box can be modeled as an LSS, i.e., that an LSS is a realization of f.

Definition 3.18 (*Realization, minimality, equivalence*) The LSS Σ is a *realization of an input–output map f of the form* (3.8), if $Y_\Sigma = f$, i.e., if the input–output map of Σ coincides with f. If Σ is a realization of f, then Σ is *a minimal realization of f*, if for any LSS realization $\hat{\Sigma}$ of f, $\dim \Sigma \leq \dim \hat{\Sigma}$. Two LSSs Σ_1, Σ_2 are said to be *input–output equivalent*, if their input–output maps are equal, i.e., $Y_{\Sigma_1} = Y_{\Sigma_2}$. A LSS Σ is said to be *minimal*, if it is a minimal realization of its own input–output map $f = Y_{\Sigma}$.

Next, we define observability and span-reachability for LSSs. In the subsequent discussion, Σ denotes a LSS of the form (3.5).

Definition 3.19 (*Observability*) An LSS Σ is said to be *observable*, if for any two distinct states $x_1 \neq x_2 \in \mathbb{R}^n$, the input–output maps induced by x_1 and x_2 are different, i.e., if $\forall x_1, x_2 \in \mathbb{R}^n : x_1 \neq x_2 \implies Y_{\Sigma,x_1} \neq Y_{\Sigma,x_2}$.

Let $\mathrm{Reach}_{x_0}(\Sigma) \subseteq \mathbb{R}^n$ denote the reachable set of the LSS Σ from the initial condition x_0, i.e., $\mathrm{Reach}_{x_0}(\Sigma)$ is the range of the map X_{Σ,x_0}.

Definition 3.20 (*Span-Reachability*) The LSS Σ is said to be *reachable*, if $\mathrm{Reach}_{x_0}(\Sigma) = \mathbb{R}^n$. The LSS Σ is *span-reachable* if $\mathbb{R}^n$ is the smallest vector space containing $\mathrm{Reach}_0(\Sigma)$.

We note that span-reachability and reachability are the same in continuous-time, if $x_0 = 0$, [57]. Next, we recall the notion of LSS morphism.

Definition 3.21 (*Isomorphism*) Consider two LSSs

$$\Sigma_1 = (n, \{(A_q, B_q, C_q) \mid q \in Q\}, x_0)$$

and

$$\Sigma_2 = (n^a, Q, \{(A_q^a, B_q^a, C_q^a) \mid q \in Q\}, x_0^a).$$

An *LSS morphism* $\mathscr{S}$ from Σ_1 to Σ_2, denoted by $\mathscr{S}:\Sigma_1 \to \Sigma_2$, is an $n^a \times n$ matrix, such that

$$\forall q \in Q: \quad A_q^a \mathscr{S} = \mathscr{S} A_q, \quad B_q^a = \mathscr{S} B_q, \quad x_0^a = \mathscr{S} x_0, \quad C_q^a \mathscr{S} = C_q.$$

The LSS morphism $\mathscr{S}$ is said to be an *isomorphism*, if the matrix $\mathscr{S}$ is square and invertible.

3.4.2 Linear Switched Systems as a Subclass of LHSs

Roughly speaking, LSSs can be identified with LHSs for which:

1. the discrete dynamics is trivial, i.e., any discrete state transition is allowed and the set of discrete events coincides with the set of discrete states,

2. the continuous subsystems are defined on the same state-, input-, and output-spaces,
3. all the reset maps are identity matrices,
4. the discrete readout map is the identity map.

More precisely, fix the set of discrete states Q and choose a symbol $s_0 \notin Q$. For any LSS $\Sigma = (n, \{(A_q, B_q, C_q) \mid q \in Q\}, x_0)$ define the LHS H_Σ of the form $H_\Sigma = (Q, \Gamma, O, \delta, \lambda, \{n_q, A_q, B_q, C_q\}_{q \in Q}, \{M_{q_1,\gamma,q_2}\}_{q_2 \in Q, \gamma \in \Gamma, q_1 = \delta(q_2,\gamma)}, h_0)$, such that

1. $\Gamma = Q$, $O = S$, $S = \{s_0\} \cup Q$, and for all $s \in S$, $\lambda(s) = s$, $\delta(s, q) = q$, $q \in Q$,
2. $n_{s_0} = n$, $A_{s_0} = 0$, $B_{s_0} = 0$, $C_{s_0} = 0$, $M_{q,q,s_0} = I_n$,
3. for all $q \in Q$, $n_q = n$ and the matrices (A_q, B_q, C_q) of the linear subsystem associated with the discrete state q of H_Σ are the same as the matrices of the linear subsystem associated with the discrete state q of Σ. Furthermore, $M_{q_1,q_1,q_2} = I_n$ for all $q_1, q_2 \in Q$,
4. $h_0 = (s_0, x_0)$

We will call H_Σ the LHS associated with the LSS Σ. The construction of H_Σ roughly follows the intuitive relationship between LSSs and LHSs outlined above. The only additional step is the introduction of the dummy discrete state s_0, which serves the purpose of allowing us to choose the first discrete state. Note that for LSSs, the choice of the discrete state is part of the input. In contrast, for LHSs, the initial discrete state is part of the system's description. By introducing s_0 and setting its continuous dynamics to zero, we can simulate the effect of choosing the discrete state. We can easily express the relationship between state-to-input and input–output maps of Σ and H_Σ. Indeed, consider $(u, w, t) \in \mathbf{U}$ with $\Gamma = Q$. If $w = (q_1, t_0)(q_2, t_1) \cdots (q_k, t_{k-1})$, $q_1, \ldots, q_k \in Q$, $t_0, \ldots, t_k^{-1} \in \mathbb{R}_+$, $k \geq 1$, then for any $q \in Q$, define $v(w, q, t) = (q, t_0)(q_1, t_1) \cdots (q_{k-1}, t_{k-1})(q_k, t) \in Q^+_{\text{timed}}$, and for s_0, define $v(w, s_0, t) = (q_1, t_1) \cdots (q_{k-1}, t_{k-1})(q_k, t)$. If $w = \varepsilon$ and $s \in Q$, then set $v(w, s, t) = (s, t)$. It then follows that

$$
\begin{aligned}
&\forall s \in S, w \in \Gamma^*_{\text{timed}}, w \neq \varepsilon : \\
&\qquad \xi_{H_\Sigma,(s,x)}(u, w, t) = (q_k, X_{\Sigma,x}(u, v(w, s, t)) \text{ and} \\
&\qquad \upsilon_{H_\Sigma,(s,x)}(u, w, t) = (q_k, Y_{\Sigma,x}(u, v(w, s, t)) \\
&\forall s \in Q : \\
&\qquad \xi_{H_\Sigma,(s,x)}(u, \varepsilon, t) = (s, X_{\Sigma,x}(u, v(\varepsilon, s, t)) \text{ and} \\
&\qquad \upsilon_{H_\Sigma,(s,x)}(u, \varepsilon, t) = (s, Y_{\Sigma,x}(u, v(\varepsilon, s, t)) \\
&\xi_{H,(s_0,x)}(u, \varepsilon, t) = (s_0, x) \text{ and } \upsilon_{H,(s_0,x)}(u, \varepsilon, t) = (s_0, 0).
\end{aligned}
\tag{3.9}
$$

Using (3.9), it is easy to see that Σ is span-reachable or observable if and only if H_Σ is span-reachable or respectively observable.

Moreover, there is a one-to-one relationship between LHS morphisms among the LHSs associated with LSSs and LSS morphisms among LSSs. More, precisely, let Σ_1 and Σ_2 be LSSs as in Definition 3.21. Assume that $\mathscr{S}: H_{\Sigma_1} \to H_{\Sigma_2}$ is a LHS morphism. It then follows $\mathscr{S}_D$ is the identity map and $\mathscr{S}_{C,q_1} = \mathscr{S}_{C,q_2}$ for all $q_1, q_2 \in S$. Furthermore, the matrix $\mathscr{S}_{\mathscr{S}} = \mathscr{S}_{C,s_0}$ can be viewed as an LSS morphism $\mathscr{S}_{\mathscr{S}}:\Sigma_1 \to \Sigma_2$. Conversely, if $\mathscr{S}:\Sigma_1 \to \Sigma_2$ is an LSS morphism, then define $\mathscr{S}_{\mathscr{S}}:\mathscr{H}_{H_{\Sigma_1}} \to \mathscr{H}_{H_{\Sigma_2}}$ as $\mathscr{S}_{\mathscr{S}}((s,x)) = (s, \mathscr{S}x)$. It is easy to see that then $\mathscr{S}_{\mathscr{S}}$ is a LHS morphism $\mathscr{S}_{\mathscr{S}}:H_{\Sigma_1} \to H_{\Sigma_2}$. It is easy that the maps $\mathscr{S} \mapsto \mathscr{S}_{\mathscr{S}}$ and $\mathscr{S} \mapsto \mathscr{S}_{\mathscr{S}}$ are each other's inverses and they map LSS (respectively, LHS) isomorphisms to LHS (respectively, LSS) isomorphisms.

Finally, if $f:\mathbf{SU} \to \mathbb{R}^p$ is an input–output map which is realized by an LSS Σ, then H_Σ is a realization of the input–output map $f_H : \mathbf{U} \to S \times \mathbb{R}^p$ defined by $f_H(u, \varepsilon, t) = (s_0, 0)$ and $f_H(u, w, t) = f(u, v(w, s_0, t))$. Moreover, $\dim H_\Sigma = (|Q|+1, (|Q|+1)\dim \Sigma)$.

That is, Definitions 3.16–3.21 can be viewed as special cases of Definitions 3.4–3.11, and thus LSSs can be viewed as a special subclass of LHSs.

Note that the fact that LSSs are subclasses of LHSs does not necessarily imply that realization theory for LHSs implies that of for LSSs. The reason for this is quite obvious: while the necessary conditions remain valid for subclasses of LHSs, the sufficient conditions need not remain valid.

3.4.3 Reachability, Observability, and Minimality

We start by presenting the main results on minimality of LSSs. The following theorem summaries various results on minimality, see [33, 35].

Theorem 3.22 (Minimality, [33, 35]) *An LSS Σ is minimal, if and only if it is span-reachable and observable. If Σ_1 and Σ_2 are two minimal LSSs, and Σ_1 and Σ_2 are input–output equivalent, then Σ_1 and Σ_2 are isomorphic.*

The usefulness of Theorem 3.22 becomes more apparent after presenting an algorithm for minimization of LSSs, i.e., for converting an LSS into a minimal one while preserving its input–output map. This means that as far as the external behavior is concerned, we can always replace an LSS with a minimal one. Moreover, this minimal LSS will have such nice properties as observability and span-reachability. Finally, the fact that minimal and equivalent LSSs are isomorphic is important for system identification: it means that while several LSSs can produce the same observed behavior, as long as we restrict attention to minimal LSSs, all possible models fitting the observed behavior are essentially the same (isomorphic).

In order to formulate the minimization algorithm, we will present a geometric and algebraic characterization of span-reachability and observability. In order to present these conditions, we recall from [51] the definition of the following spaces:

Definition 3.23 ($\mathscr{W}^*$ *and* $\mathscr{V}^*$) Let $\mathscr{V}^* = \mathscr{V}^*(\Sigma)$ be the smallest subspace (with respect to the inclusion) of $\mathbb{R}^n$ which satisfies $A_q\mathscr{V}^* \subseteq \mathscr{V}^*$ and $x_0 \in \mathscr{V}^*$ and $\operatorname{Im} B_q \subseteq \mathscr{V}^*$ for any $q \in Q$. We will call $\mathscr{V}^*(\Sigma)$ the reachable subspace of Σ. Let $\mathscr{W}^* = \mathscr{W}^*(\Sigma)$ be the largest subspace (with respect to inclusion), such that $\mathscr{W}^* \subseteq \ker C_q$ and $A_q\mathscr{W}^* \subseteq \mathscr{W}^*$, for any $q \in Q$. We will call $\mathscr{W}^*$ the unobservable subspace of Σ.

Remark 3.24 (Computing $\mathscr{V}^*, \mathscr{W}^*$) It is not difficult to see that the spaces $\mathscr{V}^*$ and $\mathscr{W}^*$ can be computed as follows. Set V_0 to the space spanned by x_0 and the columns of B_q and define V_k recursively as follows: $V_k = V_0 + \sum_{q \in Q} A_q V_{k-1}$. It then follows that $V_{n-1} = V^*$. Similarly, if we set $W_0 = \bigcap_{q \in Q} \ker C_q$ and $W_k = W_0 \cap \bigcap_{q \in Q} A_q^{-1}(W_{k-1})$, then $W_{n-1} = \mathscr{W}^*$.

We can also define an explicit matrix representation of the spaces $\mathscr{V}^*$ and $\mathscr{W}^*$. The above steps can be implemented, see [3, 33]. To this end, without loss of generality we assume that $Q = \{1, \dots, D\}$ and we introduce the following lexicographic ordering on the sequences from Q^*.

Definition 3.25 (*Lexicographic ordering*) Assume that $Q = \{1, \dots, D\}$. We define a lexicographic ordering $\prec$ on Q^* as follows. For any $v, s \in Q^*$, $v \prec s$, if either $|v| < |s|$ or $0 < |v| = |s|$, $v \neq s$ and for some $l \leq |s|$, $v_l < s_l$ with the usual ordering of integers and $v_i = s_i$ for $i = 1, \dots, l-1$. Here v_i and s_i denote the ith letter of v and s, respectively. Note that $\prec$ is a complete ordering and $Q^* = \{v_1, v_2, \dots\}$ with $v_1 \prec v_2 \prec \dots$. Note that $v_1 = \varepsilon$ and for all $0 < i \in \mathbb{N}$, $q \in Q$, $v_i \prec v_i q$.

Denote by $\mathbf{N}(M)$ the number of sequences from Q^* of length at most M. It then follows that $|v_i| \leq M$ if and only if $i \leq \mathbf{N}(M)$.

The lexicographic ordering defined above will also be used when defining the Hankel matrix of LSSs. Furthermore, we will need the following notation for products of square matrices indexed by sequences from Q^*.

Notation 3.26 (Matrix product) *Consider a collection of matrices* $\{A_q \in \mathbb{R}^{n \times n}\}_{q \in Q}$. *For any* $v \in Q^*$, *define the matrix* $A_v \in \mathbb{R}^{n \times n}$ *as follows. If* $v = \varepsilon$, *then* $A_\varepsilon = I_n$ *is the identity matrix, and if* $v = q_1 \cdots q_k$ *with* $q_1, \dots, q_k \in Q$, $k > 0$, *then*

$$A_v = A_{q_k} A_{q_{k-1}} \cdots A_{q_1}.$$

We are now in position to define the matrix representation of $\mathscr{V}^*$ and $\mathscr{W}^*$.

Definition 3.27 (*Controllability and observability matrix*) Let $\Sigma = (n, \{(A_q, B_q, C_q) \mid q \in Q\}, x_0)$. Define the *controllability matrix* of Σ as

$$\mathscr{R}(\Sigma) = \left[A_{v_1}\widetilde{B},\ A_{v_2}\widetilde{B},\ \dots,\ A_{v_{\mathbf{N}(n-1)}}\widetilde{B}\right] \in \mathbb{R}^{n \times (mD+1)\mathbf{N}(n-1)}$$

with $\widetilde{B} = \left[x_0,\ B_1,\ B_2,\ \dots,\ B_D\right] \in \mathbb{R}^{n \times (Dm+1)}$.

Define the *observability matrix* of Σ as

$$\mathscr{O}(\Sigma) = \left[(\widetilde{C}A_{v_1})^T,\ (\widetilde{C}A_{v_2})^T,\ \dots,\ (\widetilde{C}A_{v_{\mathbf{N}(n-1)}})^T\right]^T \in \mathbb{R}^{pD\mathbf{N}(n-1) \times n}$$

with $\widetilde{C} = \left[C_1^T,\ C_2^T,\ \dots,\ C_D^T\right]^T \in \mathbb{R}^{pD \times n}$.

The relationship between $\mathscr{W}^*$, $\mathscr{V}^*$ and the matrices defined above is as follows.

Proposition 3.28 $\mathscr{V}^* = \operatorname{Im}\mathscr{R}(\Sigma)$ *and* $\mathscr{W}^* = \ker \mathscr{O}(\Sigma)$.

In other words, the controllability matrix can be viewed as a matrix representation of $\mathscr{V}^*$ and the observability matrix can be viewed as a matrix representation of the orthogonal complement of $\mathscr{W}^*$. Theoretically, the controllability and the observability matrices could be used to compute the spaces $\mathscr{W}^*$, $\mathscr{V}^*$, but this approach would not be very practical, as the size of the matrices involved is exponential in the number of continuous states. For this reason, it is more practical to use the ideas of Remark 3.24. We are now ready to state the algebraic and geometric conditions for span-reachability and observability.

Theorem 3.29 [35, 51, 57]. *Let* $\Sigma = (n, \{(A_q, B_q, C_q) \mid q \in Q\}, x_0)$ *be an LSS.*

1. ***Span-Reachability****: The following three statements are equivalent:*

1. Σ *is span-reachable,*
2. $\mathscr{V}^* = \mathbb{R}^n$*,*
3. $\operatorname{rank} \mathscr{R}(\Sigma) = n$.

2. ***Observability****: The following are equivalent:*

1. Σ *is observable,*
2. $\mathscr{W}^* = \{0\}$*,*
3. $\operatorname{rank} \mathscr{O}(\Sigma) = n$.

Remark 3.30 (Minimal LSS may have nonminimal subsystems) Note that observability (span-reachability) of an LSS does not imply observability (reachability) of any of its linear subsystems. In fact, it is easy to construct a counter example [35]: $\Sigma = (n, \{(A_q, B_q, C_q) \mid q \in Q\}, x_0)$, $Q = \{q_1, q_2\}$, $p = m = 1$, $x_0 = (0, 0, 0)^T$ and

$$A_{q_1} = \begin{bmatrix} 0 & 1 & 0 \\ 0 & 0 & 0 \\ 0 & 0 & 1 \end{bmatrix}, \; B_{q_1} = \begin{bmatrix} 0 \\ 1 \\ 0 \end{bmatrix}, \; C_{q_1} = \begin{bmatrix} 1 \\ 1 \\ 0 \end{bmatrix}^T, \; A_{q_2} = \begin{bmatrix} 0 & 0 & 0 \\ 0 & 0 & 0 \\ 0 & 1 & 0 \end{bmatrix},$$

$$B_{q_2} = \begin{bmatrix} 0 \\ 0 \\ 0 \end{bmatrix}, \; C_{q_2} = \begin{bmatrix} 0 \\ 0 \\ 1 \end{bmatrix}^T$$

It is easy to see that Σ is span-reachable and observable, yet none of the subsystems are reachable or observable.

Together with Theorem 3.22, which states that minimal realizations are unique up to isomorphism, this implies that *there exist LSSs which cannot be converted to an equivalent LSS where all (or some) of the linear subsystems are observable (or reachable).*

The proof of Theorem 3.29 can be found in [35, 51, 57]. The intuition behind it is as follows: $\mathscr{V}^*$ is simply the span of all reachable states, and $Y_{\Sigma,x_1} = Y_{\Sigma,x_2} \iff x_1 - x_2 \in \mathscr{W}^*$.

We are now able to recall from [51] the following algorithms for converting a LSS to a reachable, observable, and minimal one, respectively, while preserving input–output behavior. We follow the presentation of [51].

Procedure 3.31 (Reachability reduction) *Assume that* $\dim \mathscr{V}^* = r$ *and choose a basis* $b_1, \dots, b_n$ *of* $\mathbb{R}^n$ *such that* $b_1, \dots, b_r$ *span* $\mathscr{V}^*$. *It is easy to see that in this new basis, the matrices* A_q, B_q, C_q *and the vector* x_0 *can be rewritten as*

$$A_q = \begin{bmatrix} A_q^{\mathrm{R}} & A_q^{'} \\ 0 & A_q^{''} \end{bmatrix}, C_q = \begin{bmatrix} C_q^{\mathrm{R}}, & C_q^{'} \end{bmatrix}, B_q = \begin{bmatrix} B_q^{\mathrm{R}} \\ 0 \end{bmatrix}, x_0 = \begin{bmatrix} x_0^{\mathrm{R}} \\ 0 \end{bmatrix}, \tag{3.10}$$

where $A_q^{\mathrm{R}} \in \mathbb{R}^{r \times r}$, $B_q^{\mathrm{R}} \in \mathbb{R}^{r \times m}$, *and* $C_q^{\mathrm{R}} \in \mathbb{R}^{p \times r}$, $x_0^{\mathrm{R}} \in \mathbb{R}^r$.

Define the LSS $\Sigma^{\mathrm{R}} = (r, Q, \{(A_q^{\mathrm{R}}, B_q^{\mathrm{R}}, C_q^{\mathrm{R}}) \mid q \in Q\}, x_0^{\mathrm{R}})$.

It is easy to see that Σ^R is span-reachable, and it is input-output equivalent to Σ. Intuitively, Σ^{R} is obtained from Σ by restricting the dynamics and the output map of Σ to the subspace $\mathscr{V}^*$.

Procedure 3.32 (Observability reduction) *Assume that* $\dim \mathscr{W}^* = n - o$, *and let* $b_1, \dots, b_n$ *be a basis in* $\mathbb{R}^n$ *such that* $b_{o+1}, \dots, b_n$ *span* $\mathscr{W}^*$. *In this new basis, the matrices* A_q, B_q, C_q *and the vector* x_0 *can be rewritten as*

$$A_q = \begin{bmatrix} A_q^{\mathrm{O}} & 0 \\ A_q^{'} & A_q^{''} \end{bmatrix}, C_q = \begin{bmatrix} C_q^{\mathrm{O}}, & 0 \end{bmatrix}, B_q = \begin{bmatrix} B_q^{\mathrm{O}} \\ B_q^{'} \end{bmatrix}, x_0 = \begin{bmatrix} x_0^{\mathrm{O}} \\ x_0^{'} \end{bmatrix}, \tag{3.11}$$

where $A_q^{\mathrm{O}} \in \mathbb{R}^{o \times o}$, $B_q^{\mathrm{O}} \in \mathbb{R}^{o \times m}$, *and* $C_q^{\mathrm{O}} \in \mathbb{R}^{p \times o}$, $x_0^{\mathrm{O}} \in \mathbb{R}^o$.

Define LSS $\Sigma^{\mathrm{O}} = (o, Q, \{(A_q^{\mathrm{O}}, B_q^{\mathrm{O}}, C_q^{\mathrm{O}}) \mid q \in Q\}, x_0^{\mathrm{O}})$.

It then follows that Σ^{O} is observable and it is input-output equivalent to Σ. If Σ is span-reachable, then so is Σ^{O}. Intuitively, Σ^{O} is obtained from Σ by merging any two states x_1, x_2 of Σ, for which $x_1 - x_2 \in \mathscr{W}^*$.

Finally, by combining Procedures 1 and 2 and using Theorem 3.22, we can formulate the following procedure for minimization of LSSs.

Procedure 3.33 (Minimization) *Transform* Σ *to a reachable LSS* Σ^{R} *by Procedure 1. Subsequently, transform* Σ^{R} *to an observable LSS* $\Sigma^{\mathrm{M}} = (\Sigma^{\mathrm{R}})^{\mathrm{O}}$ *using Procedure 2. Then* Σ^{M} *is a minimal LSS which is input-output equivalent to* Σ.

A more detailed description of the algorithms described in Procedures 1, 2 and 3 can be found in [33].

3.4.4 Existence of a Realization, Kalman–Ho Algorithm

In this section we present conditions on existence of a LSS realization of an input–output map and an algorithm for computing a LSS realization from input–output data. *In this section, f denotes an input–output map of the form* (3.8). Note that at this point we do not assume that f is an input–output map of an LSS. We start with defining the concept of Markov parameters and the Hankel matrix of f. The former is used to define the latter. Similar to the linear case, these concepts are defined using only input–output data. The Hankel matrices are then used for characterizing the existence of a LSS realization and for computing such a realization from input–output data. This is precisely the reason that the Hankel matrix is defined without the assumption that an LSS realization exists. In this section, *we will assume that* $Q = \{1, 2, \dots, D\}$. This can be done without loss of generality.

We start with defining the Markov parameters of f. To this end, we assume that f has a *generalized kernel representation (in the sequel abbreviated as* **GKR**). The definition of a generalized kernel representation is as follows.

Definition 3.34 (*Generalized kernel representation,* [33, 35]) The input–output map f is said to have a *generalized kernel representation* (abbreviated by **GKR**), if for every word $w \in Q^+$, there exist analytic functions

$$K_w^f : \mathbb{R}_+^{|w|} \to \mathbb{R}^p \text{ and } G_w^f : \mathbb{R}_+^{|w|} \to \mathbb{R}^{p\times m},$$

such that the following holds.

1. For all words $w, v \in Q^*$, discrete states $q \in Q$, and time instances $t_1, t_2, \dots, t_{|w|+|v|}, t, \hat{t} \in \mathbb{R}_+$,

$$\begin{aligned} &K^f_{wqqv}(t_1, t_2, \dots, t_{|w|}, t, \hat{t}, t_{|w|+1}, \dots t_{|w|+|v|}) \\ &\quad = K^f_{wqv}(t_1, t_2, \dots t_{|w|}, t + \hat{t}, t_{|w|+1} \dots t_{|w|+|v|}) \\ &G^f_{wqqv}(t_1, t_2, \dots, t_{|w|}, t, \hat{t}, t_{|w|+1}, \dots t_{|w|+|v|}) \\ &\quad = G^f_{wqv}(t_1, t_2, \dots t_{|w|}, t + \hat{t}, t_{|w|+1} \dots t_{|w|+|v|}). \end{aligned}$$

2. For all words $v \in Q^*$, $w \in Q^+$, discrete states $q \in Q$, time instances $t_1, t_2, \dots, t_{|v|+|w|} \in \mathbb{R}_+$,

$$K^f_{vqw}(t_1, t_2, \dots, t_{|v|}, 0, t_{|v|+1}, \dots, t_{|w|+|v|}) = K^f_{vw}(t_1, t_2, \dots, t_{|v|+|w|}).$$

   For each pair of words $v, w \in Q^+$, discrete state $q \in Q$, and time instances $t_1, t_2, \dots, t_{|v|+|w|} \in \mathbb{R}_+$,

$$G^f_{vqw}(t_1, t_2, \dots, t_{|v|}, 0, t_{|v|+1}, \dots, t_{|w|+|v|}) = G^f_{vw}(t_1, t_2, \dots, t_{|v|+|w|}).$$

3. For each switching sequence $w \in Q^+_{\text{timed}}$ of the form (3.7), and input $\mathbb{R}^m$,

$$f(u, w) = K^f_{q_1\cdots q_k}(t_1, \ldots, t_k) + \sum_{j=1}^{k} \int_0^{t_j} G^f_{q_j\cdots q_k}(t_j - s, t_{j+1}, \ldots, t_k) u(s + \sum_{i=1}^{j-1} t_i) \mathrm{d}s. \quad (3.12)$$

If Σ is an LSS of the form (3.5) and it is a realization of f, then

$$K^f_{q_1\ldots q_k}(t_1, \ldots, t_k) = C_q e^{A_{q_k} t_k} \ldots e^{A_{q_1} t_1} x_0$$
$$G^f_{q_j\ldots q_k}(t_j, t_{j+1} \ldots, t_k) = C_{q_k} e^{A_{q_k} t_k} \ldots e^{A_{q_j} t_j} B_{q_j}$$

for any $q_1, \ldots, q_k \in Q$, $t_1, \ldots, t_k \in \mathbb{R}_+$, $j = 1, \ldots, k$, $k > 0$.

The idea behind the definition of Markov parameters is analogous to the one for linear systems. Namely, if f admits a **GKR**, then it is completely determined by the maps K_v and G_v for $v \in Q^+$. In turn, by virtue of their analyticity, K_v and G_v are determined by their high-order derivatives at zero. The derivatives of the maps K_v and G_v can be expressed through the derivatives of f. We define the Markov parameters of f as certain high-order derivatives of f (and hence of K_v and G_v, $v \in Q^+$). In order to formalize the definition of Markov-parameters, we need the following definition.

Notation 3.35 *Consider a sequence* $v = q_1 q_2 \cdots q_k \in Q^+$, $q_1, q_2, \ldots, q_k \in Q$, $k \geq 1$ *and an input* $u \in \mathrm{PC}(\mathbb{R}_+, \mathbb{R}^m)$. *Define the map* $f_{u,v} : \mathbb{R}^k_+ \to \mathbb{R}^p$ *as follows*

$$f_{u,v}(t_1, t_2, \ldots, t_k) = f(u, (q_1, t_1)(q_2, t_2) \cdots (q_k, t_k)).$$

That is, $f_{u,v}$ is obtained from f by fixing the input u and a sequence of discrete states v and varying the switching times only. We denote by $\mathbf{0}$ and by $\mathbf{e}_j$, $j = 1, \ldots, m$ the constant inputs from $\mathrm{PC}(\mathbb{R}_+, \mathbb{R}^m)$, which for each time instance take the value $0 \in \mathbb{R}^m$ and $e_j \in \mathbb{R}^m$, respectively. Recall that e_j is the jth standard basis vector of $\mathbb{R}^m$.

Definition 3.36 (*Markov-parameters*) Define the map $M^f : Q^* \to \mathbb{R}^{pD \times (mD+1)}$ as follows. For all $q, q_0 \in Q$, $j = 1, \ldots, m$, $v \in Q^*$ define the vectors $S^f_q(v), S^f_{q,q_0,j}(v) \in \mathbb{R}^p$ and the matrix $S^f_{q,q_0}(v) \in \mathbb{R}^{p \times m}$ as follows: assume that k is the length of v and define

$$\begin{aligned}
S^f_q(v) &= \frac{\mathrm{d}}{\mathrm{d}t_1}\frac{\mathrm{d}}{\mathrm{d}t_2}\cdots\frac{\mathrm{d}}{\mathrm{d}t_k} f_{0,vq}(t_1, t_2, \ldots, t_k, 0)|_{t_1=\cdots=t_k=0} \\
S^f_{q,q_0,j}(v) &= \frac{\mathrm{d}}{\mathrm{d}t_0}\frac{\mathrm{d}}{\mathrm{d}t_1}\cdots\frac{\mathrm{d}}{\mathrm{d}t_k}(f_{\mathbf{e}_j,q_0 vq}(t_0, t_1, \ldots, t_k, 0) \\
&\quad - f_{\mathbf{0},q_0 vq}(t_0, t_1, \ldots, t_k, 0))|_{t_0=\cdots=t_k=0} \qquad (3.13) \\
S^f_{q,q_0}(v) &= \left[S^f_{q,q_0,1}(v),\ S^f_{q,q_0,2}(v), \ldots,\ S^f_{q,q_0,m}(v)\right]
\end{aligned}$$

Then for any $v \in Q^*$, $M^f(v)$ is defined as

$$M^f(v) = \begin{bmatrix} S_1^f(v), & S_{1,1}^f(v), & \cdots, & S_{1,D}^f(v) \\ S_2^f(v), & S_{2,1}^f(v), & \cdots, & S_{2,D}^f(v) \\ \vdots & \vdots & \cdots & \vdots \\ S_D^f(v), & S_{D,1}^f(v), & \cdots, & S_{D,D}^f(v) \end{bmatrix}. \tag{3.14}$$

The value of M^f are called the *Markov parameters* of f, and $M^f(v)$ is called the Markov parameter of f indexed by $v \in Q^*$.

From [35] it follows that the components of Markov parameters of f can be expressed as derivatives of K_v and G_v as follows

$$S_{q,q_0}^f(v) = \frac{\mathrm{d}}{\mathrm{d}t_1} \cdots \frac{\mathrm{d}}{\mathrm{d}t_k} G_{q_0 v q}^f(0, t_1, \ldots, t_k, 0)|_{t_1 = \cdots = t_k = 0}$$
$$S_q^f(v) = \frac{\mathrm{d}}{\mathrm{d}t_1} \cdots \frac{\mathrm{d}}{\mathrm{d}t_k} K_{vq}^f(t_1, \ldots, t_k, 0)|_{t_1 = \cdots = t_k = 0}$$

Note that the Markov parameters of f determine f uniquely, see [33, 35]. Indeed, if f admits a **GKR**, then $f_{u,v}$, $v \in Q^+$, $u \in \{\mathbf{0}, \mathbf{e}_1, \ldots, \mathbf{e}_m\}$ determines f uniquely. In turn, the Markov parameters determine the Taylor series coefficients of $f_{u,v}$ around 0, for all $v \in Q^+$ and $u \in \{\mathbf{0}, \mathbf{e}_1, \ldots, \mathbf{e}_m\}$. Similarly to the linear case, if f has a realization by an LSS, then the Markov parameters can be written as products of the matrices of an LSS realization.

Lemma 3.37 ([33–35]) *Let Σ be of the form (3.5). Then Σ is a realization of f, if and only if f has a* **GKR**, *and for all $v \in Q^*$,*

$$M^f(v) = \widetilde{C} A_v \left[x_0, \; B_1, \; B_2, \; \ldots, \; B_D\right], \tag{3.15}$$

where $\widetilde{C} = \left[C_1^T, \; C_2^T, \; \ldots, \; C_D^T\right]^T \in \mathbb{R}^{pD \times n}$.

Using the notation of (3.13), we can rewrite (3.15) as $S_q^f(v) = C_q A_v x_0$ and $S_{q,q_0}^f(v) = C_q A_v B_{q_0}$, $\forall q, q_0 \in Q$.

Example 3.38 Consider the linear switched system of the form (3.5), where $Q = \{1, 2\}$, and $x_0 = \left[0, \; 1, \; 1\right]^T$, and

$$A_1 = \begin{bmatrix} 0 & 0 & 0 \\ 0 & 1 & 0 \\ 1 & 0 & 1 \end{bmatrix}, B_1 = \begin{bmatrix} 0 \\ 0 \\ 0 \end{bmatrix}, C_1 = \begin{bmatrix} 0 \\ 0 \\ 1 \end{bmatrix}^T,$$
$$A_2 = \begin{bmatrix} 0 & 1 & 1 \\ 0 & 0 & 0 \\ 0 & 0 & 1 \end{bmatrix}, B_2 = \begin{bmatrix} 0 \\ 1 \\ 0 \end{bmatrix}, C_2 = \begin{bmatrix} 0 \\ 0 \\ 0 \end{bmatrix}^T,$$

Consider the input–output map $f = y_\Sigma$ of Σ. Let us compute the Markov- parameters $M^f(\varepsilon)$, $M^f(1)$ and $M^f(2)$ of f. Since Σ is a realization of f, we can use (3.15)

$$\begin{aligned} M^f(\varepsilon) &= \begin{bmatrix} C_1x_0 & C_1B_1 & C_1B_2 \\ C_2x_0 & C_2B_1 & C_2B_2 \end{bmatrix} = \begin{bmatrix} 1 & 0 & 0 \\ 0 & 0 & 0 \end{bmatrix} \\ M^f(i) &= \begin{bmatrix} C_1A_ix_0 & C_1A_iB_1 & C_1A_iB_2 \\ C_2A_ix_0 & C_2A_iB_1 & C_2A_iB_2 \end{bmatrix} = \begin{bmatrix} 1 & 0 & 0 \\ 0 & 0 & 0 \end{bmatrix}, \end{aligned} \tag{3.16}$$

where $i = 1, 2$.

Next we define the notion of the Hankel matrix of f. Similarly to the linear case, the entries of the Hankel matrix are be formed by the Markov parameters. For the definition of the Hankel matrix of f, we will use lexicographical ordering on the set of sequences Q^*.

Definition 3.39 (*Hankel matrix*) Consider the lexicographic ordering $\prec$ of Q^* from Definition 3.25. Define the *Hankel matrix* H_f of f as the following infinite matrix

$$H_f = \begin{bmatrix} M^f(v_1v_1) & M^f(v_2v_1) & \cdots & M^f(v_kv_1) & \cdots \\ M^f(v_1v_2) & M^f(v_2v_2) & \cdots & M^f(v_kv_2) & \cdots \\ M(v_1v_3) & M^f(v_2v_3) & \cdots & M^f(v_kv_3) & \cdots \\ \vdots & \vdots & \cdots & \vdots & \cdots \end{bmatrix},$$

i.e., the $pD \times (mD + 1)$ block of H_f in the block row i and block column j equals the Markov-parameter $M^f(v_jv_i)$ of f. The rank of H_f, denoted by rankH_f, is the dimension of the linear span of its columns.

In other words, the lth block column of H_f is the sequence of Markov parameters $\{M^f(v_lv_k)\}_{k=1}^{\infty}$. If $|Q| = 1$, then H_f coincides with the Hankel matrix as it was defined for linear systems.

Theorem 3.40 (Existence, [33, 34, 35]) *The input–output map f has a realization by an LSS if and only if f has a* **GKR** *and* rank $H_f < +\infty$. *A minimal realization of f can be constructed from H_f.*

The intuition behind the finite rank condition is the following. If Σ is a realization of f, then by virtue of (3.15), the sequences of Markov parameters $\{M^f(wv)\}_{v\in Q^*}$, $w \in Q^*$ span a space of dimension at most n. Conversely, if rank$H_f = n < +\infty$, then an LSS Σ_f of the form (3.5) can be constructed from the columns of H_f as follows. Choose a finite basis in the column space of H_f. In this basis, let $[x_0, B_1, \ldots, B_D]$ be the coordinates of the first $mD + 1$ columns of H_f, let $[C_1^T, \ldots, C_D^T]^T$ be the matrix of the linear map mapping every column to its first pD rows and let A_q, $q \in Q$, be the matrix of the linear map which maps the column block $\{M^f(wv)\}_{v\in Q^*}$ to $\{M^f(wqv)\}_{v\in Q^*}$ for all $w \in Q^*$. It can be shown that Σ_f is a minimal realization of f, see [33, 35].

In fact, we can formulate a Kalman-Ho-like realization algorithm for LSSs. To this end, for every $M, L \in \mathbb{N}$, we define the following submatrix of the Hankel matrix H_f:

$$H_{f,L,M} = \begin{bmatrix} M^f(v_1v_1) & M^f(v_2v_1) & \cdots & M^f(v_{\mathbf{N}(M)}v_1) \\ M^f(v_1v_2) & M^f(v_2v_2) & \cdots & M^f(v_{\mathbf{N}(M)}v_2) \\ \vdots & \vdots & \cdots & \vdots \\ M^f(v_1v_{\mathbf{N}(L)}) & M^f(v_2v_{\mathbf{N}(L)}) & \cdots & M^f(v_{\mathbf{N}(M)}v_{\mathbf{N}(L)}) \end{bmatrix}. \tag{3.17}$$

Intuitively, $H_{f,L,M}$ is the submatrix of H_f, obtained by keeping the columns of H_f indexed by words of length at most M and keeping the rows indexed by words of length at most L. In contrast to H_f, the matrix $H_{f,L,M}$ is a finite matrix, albeit a very large one: its size is exponential in M and L. We are now read to state the realization algorithm.

Algorithm 3.1 Realization algorithm
Inputs: an integer $N > 0$ and the Hankel matrix $H_{f,N,N+1}$.
Output: LSS Σ_N

1: Compute a decomposition $H_{f,N,N+1} = \mathbf{OR}$, where $\mathbf{O} \in \mathbb{R}^{I_N \times n}$ and $\mathbf{R} \in \mathbb{R}^{n \times J_{N+1}}$ and rank$\mathbf{R} =$ rank$\mathbf{O} = n$, $I_N = \mathbf{N}(N)pD$ and $J_{N+1} = \mathbf{N}(N+1)(mD+1)$.
2: Consider the decomposition

$$\mathbf{R} = \left[\mathbf{C}_{v_1}, \ldots, \mathbf{C}_{v_{\mathbf{N}(N+1)}}\right],$$

such that $\mathbf{C}_{v_i} \in \mathbb{R}^{n\times(Dm+1)}$, $i = 1, 2, \ldots, \mathbf{N}(N+1)$, i.e., $\mathbf{C}_{v_i} \in \mathbb{R}^{n\times(Dm+1)}$, $i = 1, 2, \ldots, \mathbf{N}(N+1)$ are the block columns of $\mathbf{R}$. Define $\overline{\mathbf{R}}$, $\mathbf{R}_q \in \mathbb{R}^{n\times J_N}$, $J_N = \mathbf{N}(N)(mD+1)$, $q \in Q$ as follows

$$\overline{\mathbf{R}} = \left[\mathbf{C}_{v_1}, \ldots, \mathbf{C}_{v_{\mathbf{N}(N)}}\right],$$
$$\mathbf{R}_q = \left[\mathbf{C}_{v_1q}, \ldots, \mathbf{C}_{v_{\mathbf{N}(N)}q}\right].$$

Note that for any $i \in \{1, \ldots, \mathbf{N}(N)\}$ there exists an index $j \in \{2, \ldots, \mathbf{N}(N+1)\}$ such that $v_iq = v_j$. Hence, $\mathbf{R}_q$ is well defined.
3: Construct $\Sigma_N = (n, \{(A_q, B_q, C_q) \mid q \in Q\}, x_0)$ such that

$$\left[x_0, B_1, \ldots, B_D\right] = \text{the first } mD+1 \text{ columns of } \mathbf{R}$$
$$\left[C_1^T, C_2^T, \ldots, C_D^T\right]^T = \text{the first } pD \text{ rows of } \mathbf{O}$$
$$\forall q \in Q : A_q = \mathbf{R}_q\overline{\mathbf{R}}^+,$$

where $\overline{\mathbf{R}}^+$ is the Moore–Penrose pseudoinverse of $\overline{\mathbf{R}}$.
4: Return Σ_N

The following theorem gives conditions under which the state-space representation returned by Algorithm 3.1 is a realization of the map f.

Theorem 3.41 ([33, 46]) *If* $\text{rank}H_{f,N,N} = \text{rank}H_f$, *then Algorithm 3.1 returns a minimal realization of* f. *The condition* $\text{rank}H_{f,N,N} = \text{rank}H_f$ *holds for a given* N, *if there exists a LSS realization* Σ *of* f *such that* $\dim \Sigma \leq N+1$.

Note that $H_{f,N,N}$ can be computed from the responses of f. However, in principle, the computation of $H_{f,N,N}$ requires an exponential number of input/output experiments involving different switching sequences. This is clearly not very practical. It would be more practical to build $H_{f,N,N}$ based on the response of f to a single switching sequence. Preliminary results on the latter approach, for the discrete-time case, can be found in [37].

Remark 3.42 One way to compute the factorization $H_{f,N,N+1} = \mathbf{OR}$ in Algorithm 3.1 is as follows. If $\text{rank}H_{f,N,N+1} = n$ and $H_{f,N,N+1} = USV$ is the SVD decomposition of $H_{f,N,N+1}$ with S being the $n \times n$ diagonal matrix, then define $\mathbf{O} = US^{1/2}$ and $\mathbf{R} = S^{1/2}V$.

3.5 Realization Theory of LHSs

In this section we present an overview of realization theory of LHSs. We start by presenting the main results on minimality, reachability and observability of LHSs. More precisely, we show that minimality is equivalent to span-reachability and observability and any two input–output equivalent minimal LHSs are isomorphic. We also present rank conditions characterizing span-reachability and observability. We conclude this section by presenting conditions for existence of a LHS realization in terms of rank conditions for the Hankel matrices. In order to avoid cumbersome notation, we will not present the corresponding algorithms in detail, the interested reader can consult [33, 43, 44]. Note that the main idea behind these algorithms is similar to the main idea behind the corresponding algorithms for LSSs.

3.5.1 Minimality, Reachability, and Observability of LHSs

We will now turn to presenting the general case of linear hybrid systems. We start by stating the results on minimality.

Theorem 3.43 (Minimality, [33, 43, 47]) *A LHS* H *is minimal, if and only if it is span-reachable and observable. If* H_1 *and* H_2 *are two minimal LHSs, and* H_1 *and* H_2 *are input–output equivalent, then they are isomorphic. Every LHS* H *can be converted to a minimal LHS* H_m, *such that* H_m *and* H *are input–output equivalent.*

Theorem 3.43 implies that as far as the input–output map is concerned, we can restrict attention to minimal LHSs. In fact, we can always convert any LHS to a minimal one. In particular, if an input–output map f has a realization by an LHS, then it also has a realization by a minimal LHS. Moreover, all the minimal realization of the same input–output map are isomorphic. In addition, minimal LHSs are span-reachable and observable.

Similarly to the case of linear time-invariant systems and LSSs, one can formulate rank conditions for observability and span-reachability of LHSs. To this end, we need to introduce additional notation. *In the sequel, without loss of generality, we can assume that* $\Gamma = \{1, \ldots, \mathbf{G}\}$. Furthermore, in the sequel, we will have to encode the case when no discrete event takes place. To this end, we introduce the following notation.

Notation 3.44 (Extended alphabet $\widetilde{\Gamma}$) *Consider the finite set,* $\widetilde{\Gamma} = \Gamma \cup \{e\}$, *where* e *is chosen such that* $e \notin \Gamma$, *i.e.,* e *is not a discrete event.*

Every word $w \in \widetilde{\Gamma}^*$ can uniquely be written as $w = e^{\alpha_1}\gamma_1 e^{\alpha_2}\gamma_2 \cdots e^{\alpha_k}\gamma_k e^{\alpha_{k+1}}$ for some $\gamma_1, \ldots, \gamma_k \in \Gamma$, $\alpha_1, \ldots, \alpha_{k+1} \in \mathbb{N}$, and $k \geq 0$. Recall that e^k denotes the word obtained by repeating the letter e k times and that $e^0 = \varepsilon$. If $w \in \Gamma^*$, i.e., w contains no symbol e, then $\alpha_1 = \cdots = \alpha_{k+1} = 0$. If w contains no element of Γ, then $k = 0$ and $w = e^{\alpha_1}$. In order to get an intuition for the use of words from $\widetilde{\Gamma}$, let us consider the discrete-time situation, i.e., at any sampling time instance, either a discrete event arrives and a change in the discrete state takes place, or no discrete event arrives and the discrete state remains unchanged. Intuitively, one can think of a word $w \in \widetilde{\Gamma}^*$ of the form $w = e^{\alpha_1}\gamma_1 e^{\alpha_2}\gamma_2 \cdots e^{\alpha_k}\gamma_k e^{\alpha_{k+1}}$ as an encoding of the situation when discrete events $\gamma_1, \ldots, \gamma_k$ take place, and $\alpha_1, \ldots, \alpha_{k+1}$ encode the number of steps between the arrival of the discrete events: α_1 is the number of steps before the arrival of the first event γ_1, α_{k+1} represents the number of steps after the arrival of the last discrete event γ_k, and α_i is the number of steps between the arrival of the discrete event γ_{i-1} and γ_i, $i = 2, 3, \ldots, k$.

Notation 3.45 (Discrete state-transition map $\delta: Q \times \widetilde{\Gamma}^* \to Q$) *We extend to discrete-state transition map* δ *to a map* $\widetilde{\delta}: Q \times \widetilde{\Gamma}^* \to Q$ *as follows. For any discrete state* $q \in Q$ *and sequence* $w \in \widetilde{\Gamma}^*$, *define the discrete state* $\widetilde{\delta}(q, w)$ *recursively as follows.*

- *If* $w = \varepsilon$, *then* $\widetilde{\delta}(q, w) = q$.
- *If* $w = v\sigma$ *for some* $\sigma \in \widetilde{\Gamma}$, $v \in \widetilde{\Gamma}^*$, *then*

$$\widetilde{\delta}(q, v\sigma) = \begin{cases} \widetilde{\delta}(q, v) & \text{if } \sigma = e \\ \delta(\widetilde{\delta}(q, v), \gamma) & \text{if } \sigma = \gamma \in \Gamma \end{cases}$$

By abuse of notation, we denote the extension $\widetilde{\delta}$ *of the discrete state-transition map by* δ *as well.*

Notation 3.46 (Product of matrices) *For any* $q \in Q$ *and sequence* $w \in \widetilde{\Gamma}^*$ *define the* $n_{\hat{q}} \times n_q$ *matrix* $\Pi(q, w)$, *where* $\hat{q} = \delta(q, w)$, *recursively as follows.*

- *If* $w = \varepsilon$, *then let* $\Pi(q, w) = I_{n_q}$, *where* I_{n_q} *is the* $n_q \times n_q$ *identity matrix.*

- *If* $w = v\sigma$ *for some* $\sigma \in \widetilde{\Gamma}$, $v \in \widetilde{\Gamma}^*$, *then let* $s = \delta(q, v)$,

$$\Pi(q, v\sigma) = \begin{cases} A_s \Pi(q, v) & \text{if } \sigma = e \\ M_{\delta(s,\gamma),\gamma,s} \Pi(q, v) & \text{if } \sigma = \gamma \in \Gamma \end{cases}$$

Define the $p \times n_q$ *output matrix* $\mathscr{O}(q, w)$ *as follows*

$$\mathscr{O}(q, w) = C_{\delta(q,w)} \Pi(q, w).$$

Intuitively, the matrices $\Pi(q, w)$ and $\mathscr{O}(q, w)$ can be interpreted as follows. If $w = \varepsilon$, then $\Pi(q, w)$ is the identity matrix and $\mathscr{O}(q, w) = C_q$. If $w = e^{\alpha}$ for some $\alpha > 0$, then $\Pi(q, w) = A_q^{\alpha}$ and $\mathscr{O}(q, w) = C_q A_q^{\alpha}$. If $w = e^{\alpha_1}\gamma_1 e^{\alpha_2}\gamma_2 \cdots e^{\alpha_k}\gamma_k e^{\alpha_{k+1}}$ for some $\gamma_1, \ldots, \gamma_k \in \Gamma$, $\alpha_1, \ldots, \alpha_{k+1} \in \mathbb{N}$, and $k > 0$, then

$$\Pi(q, w) = A_{q_k}^{\alpha_{k+1}} M_{q_k,\gamma_k,q_{k-1}} A_{q_{k-1}}^{\alpha_k} M_{q_{k-1},\gamma_{k-1},q_{k-2}} \cdots A_{q_1}^{\alpha_2} M_{q_1,\gamma_1,q_0} A_{q_0}^{\alpha_1},$$
$$\mathscr{O}(q, w) = C_{q_k} A_{q_k}^{\alpha_{k+1}} M_{q_k,\gamma_k,q_{k-1}} A_{q_{k-1}}^{\alpha_k} M_{q_{k-1},\gamma_{k-1},q_{k-2}} \cdots A_{q_1}^{\alpha_2} M_{q_1,\gamma_1,q_0} A_{q_0}^{\alpha_1},$$

where $q_0 = q$ and $q_i = \delta(q_{i-1}, \gamma_i)$ for $i = 1, 2, \ldots, k$.

Next, we introduce the notion of zero-state Markov parameters.

Definition 3.47 (*zero-state Markov parameters*) The *zero-state Markov parameter* of H indexed by discrete state $q \in Q$, is defined as the map $\mathbb{ZM}_q : \widetilde{\Gamma}^* \to \mathbb{R}^{p \times m}$ such that

$$\forall w \in \widetilde{\Gamma}^* : \quad \mathbb{ZM}_q(w) = \begin{cases} \mathscr{O}(\delta(q, v), s) B_{\delta(q,v)}, & \text{if } w = ves, v \in \Gamma^*, s \in \widetilde{\Gamma}^*, \\ 0 & \text{if } w \in \Gamma^* \end{cases}$$

Intuitively, values of the zero-state Markov parameter $\mathbb{ZM}_q$ corresponds to a certain derivative of the continuous output generated from the discrete state q, where the derivatives are taken with respect to the arrival times of discrete events. Note that $\mathbb{ZM}_q(w) = 0$, if w does not contain the symbol e. If w contains the symbol e, then we decompose w as $w = ves$, $v \in \Gamma^*$, i.e., s is the portion of w after the first occurrence of the symbol e. Then $\mathbb{ZM}_q(w)$ equals the matrix product $\mathscr{O}(q_1, s) B_{q_1}$, where $q_1 = \delta(q, v)$ is the discrete state reached by the system from the discrete state q, if the sequence of discrete events v is fed to the system. That is, $w = e^{\alpha}$ for some $\alpha > 0$, then $\mathbb{ZM}_q(w) = C_q A_q^{\alpha-1} B_q$. If $w = e^{\alpha_1}\gamma_1 e^{\alpha_2}\gamma_2 \cdots e^{\alpha_k}\gamma_k e^{\alpha_{k+1}}$ for some $\gamma_1, \ldots, \gamma_k \in \Gamma$, $\alpha_1, \ldots, \alpha_{k+1} \in \mathbb{N}$, and $k > 0$ and $l \in \{1, \ldots, k+1\}$ is such that $\alpha_1 = \cdots = \alpha_{l-1} = 0$ and $\alpha_l > 0$, then

$$\mathbb{ZM}_q(w) = C_{q_k} A_{q_k}^{\alpha_{k+1}} M_{q_k,\gamma_k,q_{k-1}} \cdots M_{q_l,\gamma_l,q_{l-1}} A_{q_{l-1}}^{\alpha_l - 1} B_{q_{l-1}},$$

where $q_0 = q$ and $q_i = \delta(q_{i-1}, \gamma_i)$ for $i = 1, 2, \ldots, k$. That is, the values of $\mathbb{ZM}_q$ contain all the Markov parameters of the linear systems which can be reached from the discrete state q.

Definition 3.48 (*Discrete state indistinguishability relation I_H*) Define the *discrete state indistinguishability relation* $I_H \subseteq Q \times Q$ as follows:

$$\begin{aligned}&(q_1, q_2) \in I_H \iff \\ &\forall v \in \Gamma^* : [\lambda(\delta(q_1, v)) = \lambda(\delta(q_2, v)), \text{ and } \mathbb{ZM}_{\delta(q_1,v)} = \mathbb{ZM}_{\delta(q_2,v)}].\end{aligned} \tag{3.18}$$

Intuitively, $(q_1, q_2) \in I_H$ is equivalent to $\upsilon_{H,(q_1,0)} = \upsilon_{H,(q_2,0)}$. Note that if $\upsilon_{H,(q_1,x_2)} = \upsilon_{H,(q_2,x_2)}$ for some $x_i \in \mathbb{R}^{n_{q_i}}$, $i = 1, 2$, then due to linearity properties of LHSs, $\upsilon_{H,(q_1,0)} = \upsilon_{H,(q_2,0)}$. In fact, I_H is the usual indistinguishability relation for the Moore automaton [9, 12, 47], if the function $\mathbb{ZM}_q$ is considered as an additional discrete output at the discrete state $q \in Q$.

Finally, for any discrete state $q \in Q$ of Σ we define the generalization of unobservability subspace.

Definition 3.49 (*Observability kernel*) For any discrete state $q \in Q$ of Σ define the unobservability subspace $O_{H,q}$ of H as

$$O_{H,q} = \bigcap_{w \in \widetilde{\Gamma}^*} \ker \mathscr{O}(q, w).$$

The space $O_{H,q}$ is a generalization of the unobservability space for linear time-invariant systems. In fact, $O_{H,q}$ is contained in the unobservability space of the linear system (A_q, C_q). However, $O_{H,q}$ also takes into account the output after first, second, etc. discrete-state transition. That is why products of the matrices of the linear subsystems and of the reset maps are considered too. Intuitively, $\forall q \in Q, x_1, x_2 \in \mathbb{R} : (\upsilon_{H,(q,x_1)} = \upsilon_{H,(q,x_2)} \iff x_1 - x_2 \in O_{H,q})$.

We can then state the following theorem.

Theorem 3.50 (Observability, [33]) *The LHS H is observable, if and only if $I_H = \{(q, q) \mid q \in Q\}$ and for each $q \in Q$, $O_{H,q} = \{0\}$.*

The condition $I_H = \{(q, q) \mid q \in Q\}$ is equivalent to saying that $\upsilon_{H,(q_1,x_2)} = \upsilon_{H,(q_2,x_2)}$ implies that $q_1 = q_2$ for all $q_1, q_2 \in Q$, $x_i \in \mathbb{R}^{n_{q_i}}$, $i = 1, 2$. That is, this condition ensures the observability of the discrete state. Condition $\forall q \in Q : O_{H,q} = \{0\}$ is equivalent to $\forall q \in Q, x_1, x_2 \in \mathbb{R} : (\upsilon_{H,(q,x_1)} = \upsilon_{H,(q,x_2)} \implies x_1 = x_2)$. For a detailed proof see [33, 47]. The conditions of Theorem 3.50 can be checked numerically, for the relevant details we refer to [33, 44].

Span-reachability can also be characterized by a rank condition. To this end, for every discrete state $q \in Q$, we define the reachable subspace of a LHS.

Definition 3.51 (*Continuous reachable subspace*) For every $q \in Q$ define the reachable subspace $R_{H,q}$ of H as follows: $R_{H,q}$ is the linear span of the columns of the matrices of the form $\Pi(q_0, w)x_0$, $q = \delta(q_0, w)$, $w \in \widetilde{\Gamma}^*$ or $\Pi(q_1, w)B_{q_1}$, $q = \delta(q_1, w)$, $w \in \widetilde{\Gamma}^*$ and $q_1 = \delta(q_0, v)$ for some $v \in \Gamma^*$.

Note that $R_{H,q}$ contains the controllability subspace of the linear system associated with the discrete state q. However, in addition it contains the images of the controllability subspaces of all the linear systems which are associated with some discrete state occurring on any sequence of discrete state transitions from the initial discrete state to q.

Definition 3.52 (*Discrete reachable subset*) The discrete reachable subset $R_{H,\text{disc}}$ is the set of all discrete states of H which are reachable from q_0, i.e.,

$$R_{H,\text{disc}} = \{q \in Q \mid \exists v \in \Gamma^* : q = \delta(q_0, v)\}.$$

With the definitions above now we are ready to present the main result.

Theorem 3.53 (Span-reachability, [33]) *An LHS H is span-reachable, if and only if* $R_{H,\text{disc}} = Q$ *and for every* $q \in Q$, $\dim R_{H,q} = n_q$.

Again, the conditions of Theorem 3.53 can be checked numerically, see [33, 44] for details.

Similarly to the case of LSSs, it is possible to formulate algorithms which transform an LHS to an observable and span-reachable, i.e., minimal, one, while preserving its input–output map. The details of the minimization algorithm are a little bit more involved than in the case of LSSs, see [33, 44] for details. The main idea behind these algorithms that the linear spaces $R_{H,q}, q \in Q$, and $O_{H,q}, q \in Q$, can be represented as images and, respectively, kernels of finite matrices, in a way which is very similar to what was presented for LSSs. In fact, these algorithms can be derived from the corresponding algorithms for representations of recognizable formal power series [4, 54], both for LSSs and for LHSs.

Note that if all the linear subsystems of a LHS H of the form (3.1) are observable and the Moore automaton part $(Q, \Gamma, O, \delta, \lambda, q_0)$ is observable (see [47] for the definition of observability of Moore automata), then the LHS is observable. Similarly, if all the linear subsystems are reachable and the Moore automaton $(Q, \Gamma, O, \delta, \lambda, q_0)$ is reachable (see again [47] for the definition), then the LHS is span-reachable. The converse is not true, for a counterexample see [47]. In fact, due to the uniqueness of input–output equivalent minimal LHSs, the counterexample of [47] shows that there exist input–output maps which can be realized by LHSs, but for which there exist no LHS realization for which all of the linear subsystems are minimal.

3.5.2 Existence of a Realization

In this section we sketch the conditions for existence of a realization by LHSsfor an input–output map f of the form (3.4). Recall that in this chapter we model potential input–output behaviors of LHSsas functions of the form (3.4). To begin with, we recall that in order for f to be realizable by a LHS, f has to admit a *hybrid kernel representation*. In order to present the definition of hybrid kernel representation, we need the following terminology.

Definition 3.54 (*Discrete and continuous-valued components*) For each input–output map f of the form (3.4), denote by f_C the $\mathbb{R}^p$-valued part, and by f_D the O-valued part of the map f. That is, $f(w) = (f_D(w), f_C(w)) \in O \times \mathbb{R}^p$ for all $w \in \mathbf{U}$.

Informally, f has a hybrid kernel representation if,

(a) f_D depends only on the relative order of discrete events.
(b) f_C is continuous and affine in continuous inputs, moreover for constant continuous inputs, f_C is analytic in the arrival time of the discrete events.

A formal theorem relating hybrid kernel representations with conditions (a) and (b) is presented in [33] Sect. 7.1, Theorem 30.

Definition 3.55 (*Hybrid kernel representation* (**HKR**)) An input–output map f has a *hybrid kernel representation* (abbreviated by **HKR**), if

1. The function f_D depends only on the discrete events, i.e., for any two timed sequences of events, $w_1, w_2 \in \Gamma^*_{\text{timed}}$, $w_1 = (\gamma_1, t_1)(\gamma_2, t_2)\cdots(\gamma_k, t_k)$ and $w_2 = (\gamma_1, \tau_1)(\gamma_2, \tau_2)\cdots(\gamma_k, \tau_k)$, $k \geq 0$, which differ only in the time instances $t_1, \dots, t_k \in \mathbb{R}_+$, $\tau_1, \dots, \tau_k \in \mathbb{R}_+$, for any $t, \tau \in \mathbb{R}_+$, inputs $u_1, u_2 \in \mathrm{PC}(\mathbb{R}_+, \mathbb{R}^m)$, it holds that $f_D(u_1, w_1, t) = f_D(u_2, w_2, \tau)$.
2. For each sequence of events $v \in \Gamma^*$, there exist analytic functions $K^f_v : \mathbb{R}^{|v|+1} \to \mathbb{R}^p$ and $G^f_{v,j} : \mathbb{R}^j \to \mathbb{R}^{p\times m}$ where $j = 1, 2, \dots, |v| + 1$, such that the following holds. For any $u \in \mathrm{PC}(\mathbb{R}_+, \mathbb{R}^m)$ and $t \in \mathbb{R}_+$,

$$f_C(u, \varepsilon, t) = K^f_\varepsilon(t) + \int_0^t G^f_{\varepsilon,1}(t-s)u(s)\mathrm{d}s \tag{3.19}$$

For all $v \in \Gamma^+$ of the form $v = \gamma_1\gamma_2\cdots\gamma_k$, $\gamma_1, \gamma_2, \dots, \gamma_k \in \Gamma$, $k > 0$, for all $t_1, t_2, \dots, t_{k+1} \in \mathbb{R}_+$ and $u \in \mathrm{PC}(\mathbb{R}_+, \mathbb{R}^m)$,

$$\begin{aligned} &f_C(u, (\gamma_1, t_1)(\gamma_2, t_2)\cdots(\gamma_k, t_k), t_{k+1}) \qquad (3.20)\\ &= K^f_v(t_1, \dots, t_{k+1}) + \sum_{i=0}^{k}\int_0^{t_{i+1}} G^f_{v,k+1-i}(t_{i+1}-s, t_{i+2}, \dots, t_{k+1})u(s+\sum_{j=1}^{i} t_j)\mathrm{d}s \end{aligned}$$

Notation 3.56 (f_D **interpreted as a map** $f_D : \Gamma^* \to O$) *In the sequel, if an input–output map f has a **HKR**, then its discrete component f_D will be identified with the map $\tilde{f}_D : \Gamma^* \to O$, defined as follows: $\tilde{f}_D(\varepsilon) = f_D(u, \varepsilon, t)$, and for $\gamma_1, \gamma_2, \dots, \gamma_k \in \Gamma$, $k > 0$, set $\tilde{f}_D(\gamma_1\gamma_2\cdots\gamma_k) = f_D(u, (\gamma_1, t_1)(\gamma_2, t_2)\cdots(\gamma_k, t_k), t)$, for some arbitrary input $u \in \mathrm{PC}(\mathbb{R}_+, \mathbb{R}^m)$ and times $t_1, t_2, \dots, t_k, t \in \mathbb{R}_+$. Existence of a **HKR** ensures that the definition above does not depend on the* *choice of the time instances* $t_1, \ldots, t_k \in \mathbb{R}_+$ *and on the choice of continuous-valued input signal* $u \in \mathrm{PC}(\mathbb{R}_+, \mathbb{R}^m)$.

Note that the *existence of a* **HKR** *is necessary for existence of a LHS realization of the input–output map f.*

Proposition 3.57 *If* H *is a realization of* f, *and* H *is of the form* (3.1) *with* $h_0 = (q_0, x_0)$, $q_0 \in Q$, $x_0 \in \mathbb{R}^{n_{q_0}}$, *then* f *has a* **HKR** *of the following form;*

$$\begin{aligned} K^f_{\varepsilon}(t) &= C_{q_0} e^{A_{q_0} t} x_0, \quad G^f_{\varepsilon,1}(t) = C_{q_0} e^{A_{q_0} t} B_{q_0} \\ f_D(\varepsilon, 1) &= \lambda(q_0), \end{aligned}$$

and for each $v = \gamma_1\gamma_2 \cdots \gamma_k \in \Gamma^*$, $\gamma_1, \ldots, \gamma_k \in \Gamma$, $k > 0$, $l = 1, 2, \ldots, k+1$,

$$\begin{aligned} K^f_v(t_1, \ldots, t_{k+1}) &= C_{q_k} e^{A_{q_k} t_{k+1}} M_{q_k,\gamma_k,q_{k-1}} e^{A_{q_{k-1}} t_k} \cdots e^{A_{q_1} t_2} M_{q_1,\gamma_1,q_0} e^{A_{q_0} t_1} x_0 \\ G^f_{v,k+2-l}(t_l, \ldots, t_{k+1}) &= C_{q_k} e^{A_{q_k} t_{k+1}} M_{q_k,\gamma_k,q_{k-1}} e^{A_{q_{k-1}} t_k} \cdots e^{A_{q_l} t_{l+1}} \\ &\quad \times M_{q_l,\gamma_l,q_{l-1}} e^{A_{q_{l-1}} t_l} B_{q_{l-1}} \\ f_D(v) &= \lambda(q_k) \end{aligned}$$

where $q_i = \delta(q_{i-1}, \gamma_i)$, $i = 1, 2, \ldots, k$.

Next, we define the notion of *Markov-parameters* for those input–output maps of the form (3.4) which admit a **HKR**. Assume that f is an input–output map of the form (3.4) and assume that f admits a **HKR**. The Markov parameters of f are the high-order derivatives of f with the respect to the arrival times of the discrete events. This is analogous to the approach taken in realization theory of linear and nonlinear systems [6, 15, 19]. In order to define Markov parameters, we recall from Notation 3.44 the definition of the alphabet $\widetilde{\Gamma}$. The words over the alphabet $\widetilde{\Gamma}$ will be used to index the high-order derivatives of the continuous-valued component f_C. If $w \in \widetilde{\Gamma}^*$ is of the form $w = e^{\alpha_1}\gamma_1 e^{\alpha_2}\gamma_2 \cdots \gamma_k e^{\alpha_{k+1}}$, then we use w to index derivatives of the following type; derivative of order α_i at 0 is taken with respect to the arrival time of the event γ_i for all $i = 1, 2, \ldots, k$, and derivative of order α_{k+1} at 0 is taken with respect to the time after the arrival of the last event γ_k. We represent these derivatives as a map mapping a word $w \in \widetilde{\Gamma}^*$ to the corresponding derivative.

In order to present the definition of Markov parameters, recall that e_l is the jth standard unit basis vector in $\mathbb{R}^m$, i.e., the lth entry of e_l is 1 and all the other entries are 0, and that we identify $\mathbf{e}_l$ with the constant input function $\mathbb{R}_+ \ni t \mapsto e_l \in \mathbb{R}^m$. Similarly, $\mathbf{0}$ denotes the constant function $\mathbb{R}_+ \ni t \mapsto 0 \in \mathbb{R}^m$.

Definition 3.58 (*Markov-parameters*) Let f be an input–output map, and assume that f has a **HKR**. Define the maps $\mathbb{M}_{f,j} : \widetilde{\Gamma}^* \to \mathbb{R}^p$, $j = 0, 1, 2, \ldots, m$, as follows. Define the map $y^f_0 : \mathbf{U} \to \mathbb{R}^p$ as follows:

$$\forall (u, s, t) \in \mathbf{U} : y^f_0(u, s, t) = f_C(u, s, t) - f_C(\mathbf{0}, s, t)$$

With this notation, for every $\alpha \in \mathbb{N}$, define

$$\begin{aligned} \mathbb{M}_{f,0}(e^{\alpha}) &= \tfrac{d^{\alpha}}{dt^{\alpha}} f_C(\mathbf{0}, \varepsilon, t)|_{t=0}, \\ \mathbb{M}_{f,j}(e^{\alpha}) &= \tfrac{d^{\alpha}}{dt^{\alpha}} y_0^f(\mathbf{e}_j, \varepsilon, t)|_{t=0}, \quad j = 1, 2, \ldots, m \end{aligned} \tag{3.21}$$

For each word $w \in \widetilde{\Gamma}^*$ of the form $w = e^{\alpha_1}\gamma_1 e^{\alpha_2}\gamma_2 \cdots \gamma_k e^{\alpha_{k+1}} \in \widetilde{\Gamma}^*$, for some $\alpha_1, \alpha_2, \ldots, \alpha_{k+1} \in \mathbb{N}$ $\gamma_1, \gamma_2, \ldots, \gamma_k \in \Gamma$, $k > 0$, for all $j = 1, \ldots, m$,

$$\begin{aligned} \mathbb{M}_{f,0}(w) &= \frac{d^{\alpha_1}}{dt_1^{\alpha_1}} \frac{d^{\alpha_2}}{dt_2^{\alpha_2}} \cdots \frac{d^{\alpha_{k+1}}}{dt_{k+1}^{\alpha_{k+1}}} f_C(\mathbf{0}, (\gamma_1, t_1)(\gamma_2, t_2) \cdots (\gamma_k, t_k), t_{k+1})|_{t_1=t_2=\cdots=t_{k+1}=0}, \\ \mathbb{M}_{f,j}(w) &= \frac{d^{\alpha_1}}{dt_1^{\alpha_1}} \frac{d^{\alpha_2}}{dt_2^{\alpha_2}} \cdots \frac{d^{\alpha_{k+1}}}{dt_{k+1}^{\alpha_{k+1}}} y_0^f(\mathbf{e_j}, (\gamma_1, t_1)(\gamma_2, t_2) \cdots (\gamma_k, t_k), t_{k+1})|_{t_1=\cdots=t_{k+1}=0} \end{aligned} \tag{3.22}$$

Define the map $\mathbb{M}_f : \widetilde{\Gamma}^* \to \mathbb{R}^{p \times (m+1)}$ by

$$\forall w \in \widetilde{\Gamma}^* : \mathbb{M}_f(w) = \left[\mathbb{M}_{f,0}(w),\ \mathbb{M}_{f,1}(w),\ \ldots,\ \mathbb{M}_{f,m}(w)\right]$$

The value of the map $\mathbb{M}_f$ are called *Markov parameters* of f.

Since f admits a hybrid kernel representation, the derivatives on the right-hand side of (3.21)–(3.22) exist. Note that the Markov parameters can be defined as derivatives of the functions K_v^f, $G_{v,l}^f$, $v \in Q^*$, $l = 1, \ldots, |v| + 1$ as follows. For every $\alpha \in \mathbb{N}$,

$$\begin{aligned} \mathbb{M}_{f,0}(e^{\alpha}) &= \frac{d^{\alpha}}{dt^{\alpha}} K_{\varepsilon}^f(t)|_{t=0}, \\ \mathbb{M}_{f,j}(e^{\alpha}) &= \begin{cases} \dfrac{d^{\alpha-1}}{dt^{\alpha-1}} G_{\varepsilon}^f(t)|_{t=0}, & \text{if } \alpha > 0 \\ 0 & \text{if } \alpha = 0 \end{cases} . \quad j = 1, 2, \ldots, m \end{aligned} \tag{3.23}$$

For each word $w \in \widetilde{\Gamma}^*$ of the form $w = e^{\alpha_1}\gamma_1 e^{\alpha_2}\gamma_2 \cdots \gamma_k e^{\alpha_{k+1}} \in \widetilde{\Gamma}^*$, for some $\alpha_1, \alpha_2, \ldots, \alpha_{k+1} \in \mathbb{N}$ $\gamma_1, \gamma_2, \ldots, \gamma_k \in \Gamma$, $k > 0$,

$$\mathbb{M}_{f,0}(w) = \frac{d^{\alpha_1}}{dt_1^{\alpha_1}} \frac{d^{\alpha_2}}{dt_2^{\alpha_2}} \cdots \frac{d^{\alpha_{k+1}}}{dt_{k+1}^{\alpha_{k+1}}} K_{\gamma_1\gamma_2\cdots\gamma_k}^f(t_1, t_2, \ldots, t_k, t_{k+1})|_{t_1=t_2=\cdots=t_{k+1}=0}. \tag{3.24}$$

For each word $w \in \widetilde{\Gamma}^*$ of the form $w = e^{\alpha_1}\gamma_1 e^{\alpha_2}\gamma_2 \cdots \gamma_k e^{\alpha_{k+1}} \in \widetilde{\Gamma}^*$, for some $\alpha_1, \alpha_2, \ldots, \alpha_{k+1} \in \mathbb{N}$ $\gamma_1, \gamma_2, \ldots, \gamma_k \in \Gamma$, $k > 0$, $\mathbb{M}_{f,j}(w)$, $j = 1, \ldots, m$ can be expressed as follows. If the word w has no symbol e, i.e., $\alpha_1 = \alpha_2 = \cdots = \alpha_{k+1} = 0$, then

$$\mathbb{M}_{f,j}(w) = 0.$$

If w contains at least one occurrence of e, then let $l \in \{1, 2, \ldots, k+1\}$ be the smallest index such that $\alpha_l > 0$, i.e., $\alpha_1 = \cdots = \alpha_{l-1} = 0$ and $\alpha_l > 0$. In this case,

$$\mathbb{M}_{f,j}(w) = \frac{\mathrm{d}^{\alpha_l - 1}}{\mathrm{d}t_l^{\alpha_l - 1}} \frac{\mathrm{d}^{\alpha_{l+1}}}{\mathrm{d}t_{l+1}^{\alpha_{l+1}}} \cdots \frac{\mathrm{d}^{\alpha_{k+1}}}{\mathrm{d}t_{k+1}^{\alpha_{k+1}}} G^f_{\gamma_1, \gamma_2 \cdots \gamma_k, k+2-l}(t_l, t_{l+1}, \ldots, t_{k+1}) e_j |_{t_l = t_{l+1} = \cdots t_{k+1} = 0}. \tag{3.25}$$

Note that the functions appearing in the right-hand sides of (3.21)–(3.22) (alternatively, (3.23)–(3.25)) are entire analytic functions in $t_1, t_2, \ldots, t_{k+1}$, and *the maps* $\mathbb{M}_f$ *and* $\mathbb{M}_{f,j}$, $j = 1, 2, \ldots, m$ *completely determine the continuous part* f_C *of the input–output map* f. Note that Markov parameter of f are defined even when f cannot be realized by a LHS. All we need for the Markov parameters to exist is that f has a hybrid kernel representation. The Markov parameters and the matrices of an LHS realization can be related as follows.

Lemma 3.59 ([33, 43]) *The LHS H of the form* (3.1) *is a realization of* f *if and only if* f *has a* ***HKR*** *and for any word* $w \in \widetilde{\Gamma}^*$ *of the form* $w = ves$, $v \in \Gamma^*$, $s \in \widetilde{\Gamma}^*$,

$$\begin{aligned} \mathbb{M}_f(w) &= \left[\mathscr{O}(q_0, w)x_0,\ \mathscr{O}(\delta(q_0, v), s) B_{\delta(q_0, v)}\right], \\ f_D(w) &= \lambda(\delta(q_0, w)), \end{aligned} \tag{3.26}$$

and for any $w \in \Gamma^*$,

$$\begin{aligned} \mathbb{M}_f(w) &= \left[\mathscr{O}(q_0, w)x_0,\ 0\right], \\ f_D(w) &= \lambda(\delta(q_0, w)). \end{aligned} \tag{3.27}$$

The equations (3.26)–(3.27) can be rewritten as follows. For $w = \varepsilon$,

$$\mathbb{M}_f(\varepsilon) = \left[C_{q_0} x_0,\ 0\right].$$

Assume that $w = e^{\alpha}$, $\alpha > 0$. It then follows that

$$\mathbb{M}_f(e^{\alpha}) = \left[C_{q_0} A^{\alpha}_{q_0} x_0,\ C_{q_0} A^{\alpha-1}_{q_0} B_{q_0}\right]$$

Assume that $w = e^{\alpha_1}\gamma_1 e^{\alpha_2}\gamma_2 \cdots \gamma_k e^{\alpha_{k+1}}$ for some $k > 0$, discrete events $\gamma_1, \gamma_2, \ldots, \gamma_k \in \Gamma$, and indices $\alpha_1, \alpha_2, \ldots, \alpha_{k+1} \in \mathbb{N}$, and assume that e occurs in w. Let let $l \in \{1, 2, \ldots, k+1\}$ be such that $\alpha_1 = \ldots = \alpha_{l-1} = 0$ and $\alpha_l > 0$. It then follows that

$$\begin{aligned} \mathbb{M}_f(w) = & C_{q_k} A^{\alpha_{k+1}}_{q_k} M_{q_k, \gamma_k, q_{k-1}} \cdots M_{q_l, \gamma_l, q_{l-1}} A^{\alpha_l - 1}_{q_{l-1}} \\ & \times \left[A_{q_{l-1}} M_{q_{l-1}, \gamma_{l-2}, q_{k-2}} \cdots M_{q_1, \gamma_1, q_0} x_0,\ B_{q_{l-1}}\right] \end{aligned}$$

where $q_i = \delta(q_{i-1}, \gamma_i)$, $i = 1, \ldots, k$. Finally, if $w = \gamma_1 \cdots \gamma_k$, $\gamma_1, \ldots, \gamma_k \in \Gamma$, $k > 0$ then

$$\mathbb{M}_f(w) = \left[C_{q_k} M_{q_k,\gamma_k,q_{k-1}} \cdots M_{q_1,\gamma_1,q_0} x_0,\ 0\right],$$

where $q_i = \delta(q_{i-1}, \gamma_i)$, $i = 1, \ldots, k$.

Notice the analogy between Lemma 3.59 and Lemma 3.37 for LSSs. Similarly to the case of LSSs, Lemma 3.59 allows us to characterize the existence of an LHS realization as a finite rank condition of an infinite matrix constructed from Markov parameters. Similarly to the case of linear switched systems, we will call this matrix a Hankel matrix. In order to define this matrix, we first define a lexicography ordering on $\widetilde{\Gamma}^*$. This ordering is similar to the one from Definition 3.25 and it is defined as follows:

Definition 3.60 (*Lexicographic ordering*) Assume that $\Gamma = \{1, \ldots, \mathbf{G}\}$ and identify e with 0, i.e., assume that $\widetilde{\Gamma} = \{0, \ldots, \mathbf{G}\}$, where $e = 0$ and the letters $1, \ldots, \mathbf{G}$ correspond to the elements of Γ. We define a lexicographic ordering $\prec$ on $\widetilde{\Gamma}^*$ as follows. For any $v, s \in \widetilde{\Gamma}^*$, $v \prec s$ if either $|v| < |s|$ or $0 < |v| = |s|$, $v \neq s$ and for some $l \leq |s|$, $v_l < s_l$ with the usual ordering of integers and $v_i = s_i$ for $i = 1, \ldots, l-1$. Here v_i and s_i denote the ith letter of v and s, respectively.

Notation 3.61 $(\upsilon_1, \ldots, \upsilon_k, \ldots)$ *Note that* $\prec$ *is a complete ordering and* $\widetilde{\Gamma}^* = \{\upsilon_1, \upsilon_2, \ldots\}$ *with* $\upsilon_1 \prec \upsilon_2 \prec \ldots$. *Note that* $\upsilon_1 = \varepsilon$ *and for all* $0 < i \in \mathbb{N}$, $q \in \widetilde{\Gamma}$, $\upsilon_i \prec \upsilon_i q$.

Now we are ready to define the notion of Hankel matrix. The definition is analogous to that of for linear switched systems.

Definition 3.62 (*Hankel-matrix*) Consider the words $\upsilon_1, \ldots, \upsilon_k, \ldots$ defined in Notation 3.61. Define the *Hankel matrix* H_f of f as the following infinite matrix

$$H_f = \begin{bmatrix} \mathbb{M}_f(\upsilon_1\upsilon_1) & \mathbb{M}_f(\upsilon_2\upsilon_1) & \cdots & \mathbb{M}_f(\upsilon_k\upsilon_1) & \cdots \\ \mathbb{M}_f(\upsilon_1\upsilon_2) & \mathbb{M}_f(\upsilon_2\upsilon_2) & \cdots & \mathbb{M}_f(\upsilon_k\upsilon_2) & \cdots \\ \mathbb{M}_f(\upsilon_1\upsilon_3) & \mathbb{M}_f(\upsilon_2\upsilon_3) & \cdots & \mathbb{M}_f(\upsilon_k\upsilon_3) & \cdots \\ \vdots & \vdots & \cdots & \vdots & \cdots \end{bmatrix},$$

i.e., the $p \times (m+1)$ block of H_f in the block row i and block column j equals $\mathbb{M}_f(v_j v_i)$. The *rank of* H_f (denoted by rankH_f) is the dimension of the vector space spanned by the columns of H_f.

Notice that the classical Hankel matrix of linear systems or the Hankel matrix from Definition 3.39 is a special case of the Hankel matrix defined above. We will also be interested in a subset of columns of H_f which are indexed by elements of Γ^*. To this end, recall that $\Gamma^* \subseteq \widetilde{\Gamma}^*$ and hence the lexicographic ordering defined in Definition 3.60 can also be applied to Γ^*. This leads us to define the following notation.

Notation 3.63 $(\omega_1, \omega_2, \ldots,)$ *Let* $\omega_1 \prec \omega_2 \prec \cdots$ *be the complete ordering induced by* $\prec$ *on* Γ^*, *where* $\prec$ *is as in Definition 3.60. That is,* $\Gamma^* = \{\omega_1, \omega_2, \ldots\}$ *with* $\omega_1 \prec \omega_2 \prec \ldots$. *Note that* $\omega_1 = \varepsilon$ *and for all* $0 < i \in \mathbb{N}$, $\gamma \in \Gamma$, $\omega_i \prec \omega_i \gamma$.

Definition 3.64 Let $H_{f,O}$ be the following matrix

$$H_{f,0} = \begin{bmatrix} \mathbb{M}_{f,O}(\omega_1 v_1) & \mathbb{M}_{f,O}(\omega_2 v_1) & \cdots & \mathbb{M}_{f,O}(\omega_k v_1) & \cdots \\ \mathbb{M}_{f,O}(\omega_1 v_2) & \mathbb{M}_{f,O}(\omega_2 v_2) & \cdots & \mathbb{M}_{f,O}(\omega_k v_2) & \cdots \\ \mathbb{M}_{f,O}(\omega_1 v_3) & \mathbb{M}_{f,O}(\omega_2 v_3) & \cdots & \mathbb{M}_{f,O}(\omega_k v_3) & \cdots \\ \vdots & \vdots & \cdots & \vdots & \cdots \end{bmatrix},$$

where $\mathbb{M}_{f,O}:\widetilde{\Gamma}^* \to \mathbb{R}^{p \times m}$ is defined as

$$\forall w \in \widetilde{\Gamma}^* : \mathbb{M}_{f,O}(w) = \left[\mathbb{M}_{f,1}(w), \ \mathbb{M}_{f,2}(w), \ \cdots, \ \mathbb{M}_{f,m}(w)\right].$$

The number of distinct columns of $H_{f,O}$ is denoted by $card(H_{f,O})$, i.e., $card(H_{f,0})$ is the cardinality of the set of columns of $H_{f,O}$.[1]

The role of $H_{f,O}$ is the following. Consider the input–output map y_0^f defined by $y_0^f(u, w, t) = f_C(u, w, t) - f_C(\mathbf{0}, w, t)$, $\forall (u, w, t) \in \mathbf{U}$. If f is realized by a LHS, then y_0^f is completely determined by the discrete state component of the initial state which induces f. In fact, y_0^f can be considered as an *additional discrete output* of the LHS. Then $\mathbb{M}_{f,O}$ is an encoding of y_0^f. In fact, if f has a realization by a LHS H of the form (3.1), then the columns of $H_{f,0}$ encode the zero-state Markov parameters $\mathbb{ZM}_q$, $q \in Q$ of H. More precisely, if H is of the form (3.1) and H is a realization of f, then $\forall w \in \widetilde{\Gamma}^*$, $\mathbb{M}_{f,O}(\omega_i w) = \mathbb{ZM}_q(w)$, $q = \delta(q_0, \omega_i)$, $i = 1, 2, \ldots,$

So far, we have defined everything we needed for the continuous-valued components of f. Next, we define an counterpart of the Hankel matrix for the discrete-valued components of f.

Definition 3.65 (*Hankel table*) Without loss of generality, assume that $O = \{1, 2, \ldots, \mathbf{O}\}$ and recall Notation 3.63. Define the matrix $H_{f,D}$ as follows:

$$H_{f,D} = \begin{bmatrix} f_D(\omega_1 \omega_1) & f_D(\omega_2 \omega_1) & \cdots & f_D(\omega_k \omega_1) & \cdots \\ f_D(\omega_1 \omega_2) & f_D(\omega_2 \omega_2) & \cdots & f_D(\omega_k \omega_2) & \cdots \\ f_D(\omega_1 \omega_3) & f_D(\omega_2 \omega_3) & \cdots & f_D(\omega_k \omega_3) & \cdots \\ \vdots & \vdots & \cdots & \vdots & \cdots \end{bmatrix}.$$

Denote by $card(H_{f,D})$ the number of different columns of $H_{f,D}$, i.e., $card(H_{f,D})$ is the cardinality of the set of columns of $H_{f,D}$.

[1] Two columns are considered equal, if all their respective entries are equal.

The intuition behind the definitions above is the following. The Hankel matrix H_f contains all the information on the continuous-valued components of the input–output maps. The matrix $H_{f,D}$ contains all the information on the discrete-valued components of the input–output maps. Finally, $H_{f,O}$ contains information on the zero-state Markov parameters from Definition 3.47.

Now we are ready to state the main results on existence of a LHS realization.

Theorem 3.66 (Realization by LHSs, [33, 43, 47]) *There exists a LHS realization of the input–output map f, if and only if*

(1) f has a hybrid kernel representation, and
(2) $\text{rank} H_f < +\infty$, $card(H_{f,O}) < +\infty$, $card(H_{f,D}) < +\infty$.

Notice that if condition **(2)** of Theorem 3.66 is satisfied, then it is possible to construct a minimal LHS from the columns of H_f, $H_{f,0}$ and $H_{f,D}$. In fact, the proof of Theorem 3.66 is constructive. Furthermore, one can formulate a Kalman–Ho-like realization algorithm based on the proof of Theorem 3.66, see [33, 43].

The intuition behind the theorem is the following. The finite rank condition on the Hankel matrix makes sure that the elements of the Hankel matrix (i.e., the Markov-parameters) can be represented as products of matrices, as in Lemma 3.59. It is analogous to the finite rank condition for the Hankel matrix for linear switched systems. The condition that the number of columns of $H_{f,O}$ is finite is inspired by the following observation. If f has a realization by an LHS, then $w \in \widetilde{\Gamma}^*$, $\mathbb{M}_{f,O}(\omega_i w) = \mathbb{ZM}_q(w)$, $q = \delta(q_0, \omega_i)$, $i = 1, 2, \ldots,$ where $\mathbb{ZM}_q$ is the zero-state Markov parameter from Definition 3.47. That is, in this case there is a one-to-one correspondence between zero-state Markov parameters and blocks of columns of $H_{f,O}$. In particular, as there are at most as many zero-state Markov parameters as discrete states, the number of distinct columns of $H_{f,O}$ will be finite. Finally, if f has a realization by an LHS, then the ith column of $H_{f,D}$ corresponds to the map $\Gamma^* \ni v \mapsto \lambda(\delta(q_0, \omega_i v))$. Since there are at most finitely many such maps (in fact, there are at most as many such maps as discrete states), the number of distinct columns of $H_{f,D}$ will be finite. In fact, the discussion above implies that the rank of the Hankel matrix determines the number of continuous states and that the number of distinct columns of $H_{f,O}$ and $H_{f,D}$ determine the number of discrete states (in fact, these columns correspond to discrete states of minimal LHSs.) The precise relationship between these concepts is not straightforward though, see [47] for a detailed analysis. The interested reader might wish to compare the finiteness conditions for $H_{f,0}, H_{f,D}$ with the finiteness conditions which are required for realizability of a discrete-valued input–output map by a finite-state automaton [9, 12, 47].

3.6 Applications of Realization Theory

The goal of this section is to present existing and potential applications of hybrid realization theory to control theory of hybrid systems. Realization theory has a direct relevance for system identification and model reduction of hybrid systems, and has already been applied to these topics. We believe it could directly be relevant for control design, but it has not been applied for that purpose yet. Below describe these applications one by one.

3.6.1 System Identification

The goal of hybrid system identification is to estimate a hybrid state-space model of a dynamical system from its observed input–output behavior. System identification is crucial for improving applicability of hybrid systems: in practice, good models are difficult to get, and modeling takes up a large portion of practitioners' time.

First, realization theory, more precisely, algorithms based on computing realizations from Hankel matrices, can be used to formulate subspace identification for hybrid systems, see [37]. Moreover, realization theory can help analyzing the existing system identification algorithms, for example, by enabling identifiability analysis, finding identifiable parameterizations and by deriving conditions for experiment design.

Realization theory can be applied to identifiability analysis as follows. Let us consider a parameterized subset of hybrid systems $\{H(\theta)\}_{\theta \in \Theta}$, where Θ is a set of parameters and for each $H(\theta)$ is a hybrid system (state-space representation). To keep the discussion more concrete, we will assume that all $H(\theta)$, $\theta \in \Theta$ is an LSS. However, the discussion could potentially be extended to LHSs too. We call a parameterized subset $\{H(\theta)\}_{\theta \in \Theta}$ identifiable, if for any two distinct parameter values $\theta_1 \neq \theta_2 \in \Theta$, the input–output maps of the corresponding LSS $H(\theta_1)$ and $H(\theta_2)$ are different. It can be shown that it is sufficient to consider parameterizations which are minimal, i.e., for every parameter value $\theta \in \Theta$, the corresponding LSS $H(\theta)$ is minimal. Roughly speaking, we can apply a minimization algorithm to each $H(\theta)$, $\theta \in \Theta$ to obtain a minimal parameterization and the resulting parameterization is identifiable if and only if the original parameterization is identifiable. Intuitively, the reason for this that for the original parameterization to be identifiable, the parameter θ should completely be determined by the minimal subsystem, since the nonminimal parts of $H(\theta)$ do not influence the input–output behavior anyway. Since two minimal LSSs with the same input–output map are necessarily isomorphic and two isomorphic LSSs have the same input–output map, it then follows that for the parameterization to be identifiable, there should be no isomorphism between $H(\theta_1)$ and $H(\theta_2)$, $\theta_1, \theta_2 \in \Theta, \theta_1 \neq \theta_2$. If the matrices of $H(\theta)$ are polynomials in θ and Θ is an algebraic subset of $\mathbb{R}^d$ for some $d \in \mathbb{N}$, than the latter condition can be checked numerically. The ideas discussed above were worked out in [38, 40].

Experiment design is another important topic of system identification for which realization theory could be useful. We explain the main idea on the example of LSSs. Let us call a pair of input and switching signals persistently exciting, if the response of the system to this particular pair of input and switching signals allows us to identify the whole system, i.e., predict the system's response to any other input and switching signal. Realization theory can be used for characterizing persistence of excitation as follows. Recall that a basis of the Hankel matrix is sufficient for constructing a LSS realization of the input–output map. Hence, persistence of excitation can be reformulated as follows. *For which input and switching signals the corresponding response is sufficient to find the Markov parameters which form a basis of the Hankel matrix?* This line of research was elaborated in [37]. In fact, in [37] a identification algorithm was proposed which computes a LSS from experimental data in two steps: first, it computes Markov parameters from the experimental data, second, from the thus computed Markov parameters it constructs a finite submatrix of the Hankel matrix and applies the Kalman-Ho-like algorithm Algorithm 3.1 to it.

Finally, realization theory enables us to define a topology and distance on the space of hybrid systems. This allows us to compare various systems obtained from experimental measurements and it also enables us to find identifiable parameterizations. We explain the main idea on the example of LSSs. A good topology on systems is the one which compares not the matrices of the systems, but their input–output maps. For LSSs, this means that we have to define a distance and topology between equivalence classes of isomorphic minimal LSSs. This can be done, by noticing that each such equivalence class yields the same sequence of Markov parameters and the same Hankel matrix. A representative of such an equivalence class corresponds to a certain choice of a basis of the Hankel matrix. This can be used to prove that the space of equivalence classes forms an analytic (in fact Nash) manifold and it also allows to define a system of differentiable coordinate charts. In fact, the obtained topology is such that if the matrices of two systems are close, then the two systems will be close. However, the converse is not true. Furthermore, the local coordinates correspond to identifiable parameterizations. A particularly striking conclusion of the results is that there is no identifiable parameterization which could cover all the systems: note that the coordinate charts provide *local* parameterizations. This line of research is elaborated in [41, 48].

3.6.2 Model Reduction

The goal of model reduction is to replace a hybrid system with a lot of state variables by a hybrid system with a smaller number of state variables, such that the input–output map of the smaller hybrid system is close to the input–output map of the original one.

First of all, realization theory yields a minimization algorithm, which itself can be viewed as basic model reduction algorithms.

Second, realization theory allows to develop and analyze model reduction methods which extend balanced truncation and moment matching for linear systems [1]. As these extensions were developed for LSSs, in the discussion below we will concentrate on LSSs. However, we are hopeful that similar results could be achieved for LHSs. As for balanced truncation, realization theory can be used to define counterparts of Hankel singular values for LSSs and prove that this definition is independent of the choice of state-space representation. These Hankel singular values and the corresponding generalizations of grammians can then be used to obtain reduced order LSSs with analytic error bounds. This line of research was for pursued in [52]. As for moment matching, realization theory yields the notion of Markov parameters, which in turn enables model reduction approaches such as the moment matching approach for model reduction of linear systems. This approach has been worked out for LSSs [3]. The main idea is to replace the original LSS by another LSS of lower dimension, such that a number of Markov parameters of the original and lower dimensional LSS coincide. The intuition behind this method is that Markov parameters represent coefficients of Taylor series expansion of the output of a LSS. Hence, if the lower order LSS reproduces some of the Markov parameters of the original LSS, then the Taylor series coefficients of the corresponding outputs will be close and hence the outputs will be close. Unfortunately, there are no analytical error bounds yet for this method.

3.6.3 Control Design

The application of realization theory for hybrid systems to their control design is largely unexplored. Below we will point out some potential applications. In order to simplify the discussion, we will explain the ideas on the example of LSSs. However, the discussion can be extended to LHSs.

To begin with, recall that realization theory tells us that minimal LSSs are span-reachable and observable. Hence, the first step is to find out the solution of which control problems is ensured by span-reachability of LSSs. The second step is to find out if and how observability guarantees existence of an observer. Note that observability as it is used in realization theory means that any two stated can be distinguished by some input and switching signal. Hence, the task is to understand if there exist universal inputs and switching signals, i.e., inputs and switching signals which distinguish all states, and how to use such inputs for observer design. Note that this problem has been known for nonlinear systems and existence of universal inputs has been proven long time ago [11, 58]. What remains to show their existence for hybrid systems. Note that observer design usually arises together with control design, hence in this case it is natural to consider both of them at the same time. That is, the task is to make sure that the control law is rich enough to allow both meeting the control objectives and observing the hidden states.

If we succeed, then we will be able to say that the identified control problems will always have a solution for minimal LSSs, if they admit a solution at all. Moreover, due to the uniqueness of minimal LSSs which are input–output equivalent, the resulting controller will probably be independent of the choice of the state-space representation.

Another line of research is to relate stability and dissipativity of the input–output behavior with the corresponding properties of LSSs. Recall for example, that for linear systems, the stability of the transfer function implies the stability of its minimal state-space representations, or that dissipativity of the input–output behavior is equivalent to existence of a quadratic Lyapunov function for minimal state-space representations. It would be useful to extend this type of results to LSSs. We expect that these results will help to show that the cited properties are state-space representation independent. Moreover, they might yield algorithms for finding Lyapunov functions for LSSs, and necessary and sufficient conditions for existence of controllers and observers. Some preliminary steps to this direction were made in [52].

References

1. Anthoulas, C.A.: Approximation of Large-Scale Dynamical Systems. SIAM, Philadelphia (2005)
2. Bartosiewicz, Z.: Realizations of polynomial systems. Algebraic and Geometric Methods in Nonlinear Control Theory, vol. 29, pp. 45–54. Reidel, Dordrecht (1986)
3. Bastug, M., Petreczky, M., Wisniewski, R., Leth, J.: Model reduction by moment matching for linear switched systems. In: Proceedings of the American Control Conference, to appear
4. Berstel, J., Reutenauer, C.: Rational series and Their Languages. Springer, Berlin (1984)
5. Caines, P.E.: Linear Stochastic Systems. Wiley, New York (1998)
6. Callier, F.M., Desoer, C.A.: Linear System Theory. Springer, New York (1991)
7. Celle, F., Gauthier, J.P.: Realizations of nonlinear analytic input-output maps. Math. Syst. Theory **19**, 227–237 (1987)
8. Desai, U.B.: Realization of bilinear stochastic systems. IEEE Trans. Autom. Control **31**(2), 189–192 (1986)
9. Eilenberg, S.: Automata Languages and Machines. Academic Press, New York (1974)
10. Frazho, A.E.: On stochastic bilinear systems. Modelling and Application of Stochastic Processes, pp. 215–241. Kluwer Academic, Boston (1986)
11. Gauthier, J.-P.A.: Observability (deterministic systems) and realization theory. In: Meyers, R.A. (ed.) Mathematics of Complexity and Dynamical Systems, pp. 1195–1204. Springer, New York (2011)
12. Gécseg, F., Peák, I.: Algebraic Theory of Automata. Akadémiai Kiadó, Budapest (1972)
13. Grossman, R.L., Larson, R.G.: An algebraic approach to hybrid systems. Theoret. Comput. Sci. **138**, 101–112 (1995)
14. Isidori, A.: IEEE Trans. Autom. Control. Direct construction of minimal bilinear realizations from nonlinear input-output maps **18**, 626–631 (1973)
15. Isidori, A.: Nonlinear Control Systems. Springer, New York (1989)
16. Isidori, A., D'Alessandro, P., Ruberti, A.: Realization and structure theory of bilinear dynamical systems. SIAM J. Control **12**(3), 517–535 (1974)
17. Jakubczyk, B.: Existence and uniqueness of realizations of nonlinear systems. SIAM J. Control Optim. **18**(4), 455–471 (1980)
18. Jakubczyk, B.: Construction of formal and analytic realizations of nonlinear systems. In: Isodori, A., Hinrichsen, D. (eds.) Feedback Control of Linear and Nonlinear Systems. Lecture Notes in Control and Information Sciences, vol. 39, pp. 147–156 (1982)
19. Jakubczyk, B.: Realizations of nonlinear systems: three approaches. In: Fliess, M., Hazewinkel, M. (eds.) Proceedings of the Conference on the Algebraic and Geometric Methods in Non-linear Control Theory, Paris, 1985, Reidel, Dordrecht (1986)
20. Kalman, R.E.: Advanced theory of linear systems. Topics in Mathematical System Theory, pp. 237–339. McGraw-Hill, New York (1969)
21. Kalman, R.E.: Introduction to the algebraic theory of linear dynamical systems. In: Mathematical Systems Theory and Economics, I, II (Proceedings of an International Summer School). Lecture Notes in Operations Research and Mathematical Economics, Vols. 11, 12, pp. 41–65, Springer, Berlin (1969)

22. Kamen, E.W., Hafez, K.M.: Algebraic theory of linear time-varying systems. SIAM J. Control Optim. **17**(4), 500–510 (1979)
23. Katayama, T.: Subspace Methods for System Identification. Springer, London (2005)
24. Kuich, W., Salomaa, A.: Semirings, Automata Languages. Springer, Berlin (1986)
25. Liberzon, D.: Switching in Systems and Control. Birkhäuser, Boston (2003)
26. Lindquist, A., Picci, G.: On the stochastic realization problem. SIAM J. Control Optim. **17**(3), 365–389 (1979)
27. Lindquist, A., Picci, G.: Realization theory for multivariate stationary gaussian processes. SIAM J. Control Optim. **23**(6), 809–857 (1985)
28. Olsder, G.: On the characteristic equation and minimal realizations for discrete-event dynamic systems. In: Bensoussan, A., Lions, J. (eds.) Analysis and Optimization of Systems. Lecture Notes in Control and Information Sciences, vol. 83, pp. 187–201. Springer, Berlin (1986). doi:10.1007/BFb0007557
29. Paoletti, S., Roll, J., Garulli, A., Vicino, A.: Input/ouput realization of piecewise affine state space models. In: 46th IEEE Conference on Decision and Control (2007)
30. Petreczky, M.: Realization theory for bilinear hybrid systems. In: 11th IEEE Conference on Methods and Models in Automation and Robotics (2005)
31. Petreczky, M.: Hybrid formal power series and their application to realization theory of hybrid systems. In: Proceedings of International Symposium on Mathematical Theory of Networks and Systems (2006)
32. Petreczky, M.: Realization theory for discrete-time piecewise-affine hybrid systems. In: Proceedings of International Symposium on Mathematical Theory of Networks and Systems (2006)
33. Petreczky, M.: Realization Theory of Hybrid Systems. Ph.D. thesis, Vrije Universiteit, Amsterdam (2006)
34. Petreczky, M.: Realization theory for linear switched systems: formal power series approach. Syst. Control Lett **56**, 588–595 (2007)
35. Petreczky, M.: Realization theory of linear and bilinear switched systems: A formal power series approach: part i. ESAIM Control Optim. Calc. Var. **17**, 410–445 (2011)
36. Petreczky, M.: Realization theory of linear and bilinear switched systems: a formal power series approach: part ii. ESAIM Control Optim. Calc. Var. **17**, 446–471 (2011)
37. Petreczky, M., Bako, L.: On the notion of persistence of excitation for linear switched systems. In: 50th IEEE Conference on Decision and Control (CDC) (2011)
38. Petreczky, M., Bako, L., Lecouche, S.: Minimality and identifiability of sarx systems. In: 16th IFAC Symposium on System Identification (SYSID), Brussel (2012)
39. Petreczky, M., Bako, L., van Schuppen, J.H.: Realization theory for discrete-time linear switched systems. Automatica **49**(11), 3337–3344 (2013)
40. Petreczky, M., Bako, L., van Schuppen, J.H.: Identifiability of discrete-time linear switched systems. In: Hybrid Systems: Computation and Control, pp. 141–150, ACM (2010)
41. Petreczky, M., Peeters, R.: Spaces of nonlinear and hybrid systems representable by recognizable formal power series. In Proceedings of MTNS2010, pp. 1051–1058 (2010)
42. Petreczky, M., Pommet, J.-B.: Realization theory of nonlinear hybrid systems. In: Taming Heterogenity and Complexity of Embedded Control (CTS-HYCON Workshop on Embedded and Hybrid Control). International Scientific and Technical Encyclopedia (ISTE), London (2006)
43. Petreczky, M., van Schuppen, J.H.: Realization theory for linear hybrid systems, Part I: Existence of a realization. Technical report, Centrum voor Wiskunde en Informatica (CWI), Amsterdam, 2008. Available at https://sites.google.com/site/mihalypetreczky/
44. Petreczky, M., van Schuppen, J.H.: Realization theory for linear hybrid systems, Part II: Spanreachability, observability and minimality. Technical report, Centrum voor Wiskunde en Informatica (CWI), Amsterdam, 2008. Available at https://sites.google.com/site/mihalypetreczky/
45. Petreczky, M., van Schuppen, J.H.: Realization theory of discrete-time linear hybrid system. In: 15th IFAC Symposium on System Identification (2009)
46. Petreczky, M., Van Schuppen, J.H.: Partial-realization of linear switched systems: a formal power series approach. Automatica **47**(10), 2177–2184 (2011)

47. Petreczky, M., van Schuppen, J.H.: Realization theory for linear hybrid systems. IEEE Trans. Autom. Control **55**, 2282–2297 (2010)
48. Petreczky, M., Vidal, R.: Metrics and topology for nonlinear and hybrid systems. In: Hybrid Systems: Computation and Control, vol. 441, LNCS,Springer (2007)
49. Petreczky, M., Vidal, R.: Realization theory of stochastic jump-markov linear systems. In: Proceedings 46th IEEE Conference on Decision and Control (2007)
50. Petreczky, M., Vidal, R.: Realization theory of discrete-time semi-algebraic hybrid systems. In: Hybrid Systems: Computation and Control (HSCC), vol. 4981, pp. 386–400, LNCS (2008)
51. Petreczky, M., Wisniewski, R., Leth, J.: Balanced truncation for linear switched systems. Nonlin. Anal. Hybrid Syst. **10**, 4–20 (2013)
52. Petreczky, M., Wisniewski, R., Leth, R.: Balanced truncation for linear switched systems. Nonlin. Anal. Hybrid Syst. **10**, 4–20 (2013)
53. De Schutter, Bart, Blondel, Vincent, de Vries, Remco, De Moor, Bart: On the boolean minimal realization problem in the max-plus algebra. Syst. Control Lett. **35**(2), 69–78 (1998)
54. Sontag, E.D.: Polynomial Response Maps. Lecture Notes in Control and Information Sciences, vol. 13. Springer, Berlin (1979)
55. Sontag, E.D.: Realization theory of discrete-time nonlinear systems: part I - the bounded case. IEEE Trans Circuits Syst. **26**(4), 342–356 (1979)
56. Sontag, E.D.: Realization and input/output relations: the analytic case. In: Proceedings of the 28th Conference on Decision and Control (1989)
57. Sun, Z., Ge, S.S.: Switched Linear systems : Control and Design. Springer, London (2005)
58. Sussmann, H.J.: Single-input observability of continuous-time systems. Math. Syst. Theory **12**(1), 371–393 (1978)
59. Tóth, R.: Identification and Modeling of Linear Parameter-Varying Systems. Lecture Notes in Control and Information Sciences, vol. 403. Springer, Germany (2010)
60. van der Schaft, A.J.: On realization of nonlinear systems described by higher-order differential equations. Math. Syst. Theory **19**(3), 239–275 (1987)
61. van der Schaft, A.J., Schumacher, H.: An Introduction to Hybrid Dynamical Systems. Springer, London (2000)
62. Wang, Y., Sontag, E.: Algebraic differential equations and rational control systems. SIAM J. Control Optim. **30**, 1126–1149 (1992)
63. Weiland, S., Juloski, A.L., Vet, B.: On the equivalence of switched affine models and switched ARX models. In: 45th IEEE Conference on Decision and Control (2006)

Chapter 4
Nonlinear Observer for Autonomous Switching Systems with Jumps

Mohamed Djemai, Noureddine Manamanni and Jean Pierre Barbot

Abstract This work deals with nonlinear observer synthesis for a particular class of hybrid dynamic systems (HDS): autonomous switching systems with jumps. The jumps can result from the system's dynamics or from the diffeomorphism, which makes it possible to lead the system to an observability canonical form. In this paper, our contribution relates to the design of a second order sliding mode based observer ("Super Twisting Algorithm"). It allows for estimating both continuous and discrete state related to the active dynamic. On the other hand, these observers ensure a finite time convergence of the estimation error.

4.1 Introduction

Most of the complex systems can be modeled as HDS for which each discrete mode has one or more continuous dynamic [1, 2]. In this work, we are interested in the hybrid observer synthesis for a particular class of HDS: the autonomous switching systems with state jumps.

In the literature, the problem of the observers design was largely discussed and several methods were proposed, but there is always a lack of general results in hybrid systems theory. In the case of switched systems, it seems reasonable to design a switching observer by synthesizing an observer for each mode of the system [3]. If the active mode of the system is known, it must activate only the corresponding

M. Djemai (✉)
LAMIH, CNRS UMR 8201, University of Valenciennes and Hainaut Cambresis, 59313 Valenciennes, France
e-mail: mohamed.djemai@univ-valenciennes.fr

N. Manamanni
CReSTIC EA3804, Université de Reims Champagne Ardenne, UFR SEN, Moulin de la Housse, 51687 Reims, France
e-mail: noureddine.manamanni@univ-reims.fr

J.P. Barbot
ENSEA, 6 Avenue du Ponceau, 95014 Cergy-Pontoise, France
e-mail: barbot@ensea.fr

M. Djemai and M. Defoort (eds.), *Hybrid Dynamical Systems*, Lecture Notes in Control and Information Sciences 457, DOI 10.1007/978-3-319-10795-0_4

mode of the observer. However, this approach does not guarantee that the estimation error converges, even if it converges for each mode separately. This problem is dual with that well-known that switching between stable dynamics can lead to unstability [4, 5].

A fundamental problem for this type of observer relates to the knowledge of the active mode of the switching system. In [4, 6], the authors supposed that the active mode is known, on the other hand in [7, 8], this assumption is not considered, however they consider that the common quadratic Lyapunov function decreases when the active mode of the observer differs from the active mode of the switching system. The disadvantage of this result is that the linear fields of the vector of the switching system must be the same one at the boundaries of the switching instant. This assumption is rarely met in the physical systems [9].

In [10], linear observers design for a class of linear hybrid systems was approached. Two prototypes of observers based on the prediction errors were proposed. The first one is based on the observation of the time discrete system. The second one estimates the continuous state at each time instant from the initial conditions. A methodology for the hybrid observer synthesis, without re-initialisation of the continuous state, for a class of HDS was proposed in [11]. The structure of the hybrid observer is composed of a discrete observer called "location observer" and of a continuous observer. Thus, in spite of the abundant literature on the synthesis of linear observers for hybrid systems, few works concern with the synthesis of non-linear hybrid observers for hybrid systems (see for example [12]), nevetheless the state jumps occur. Another problem involved in the convergence of the estimation error in the case of switching systems, relates to the re-initialisation of the estimated states of the switched observer when the mode changes. This allows to avoid jumps in the estimated states which lead to the consequence of a loss of information or an instability if the observer is not robust. Thus, it will be shown that by updating correctly the estimated states, the convergence of the estimation error can be guaranteed. The idea of the proposed method consists in using the step-by-step observer developed for a class of triangular form systems in [13] as follows; the $(n-1)$ first steps allows to reconstruct the state vector and after that it is possible to know in which discrete state q the system is. Finally a reinitialization procedure can be proposed. A complete scheme of the observer is depicted in Fig. 4.1. The main contribution of this work concerns the design of high order sliding mode based observer for nonlinear autonomous switched systems with jumps.

Remark 4.1 The discrete location block is composed with the decision functions where are calculated all $\sigma_q(\hat{x})$ and $f_q(\hat{x})$ functions and transferred to the block of location identification logic where conditions of Theorem 4.5 will be used to estimate the discret location $\hat{q}$.

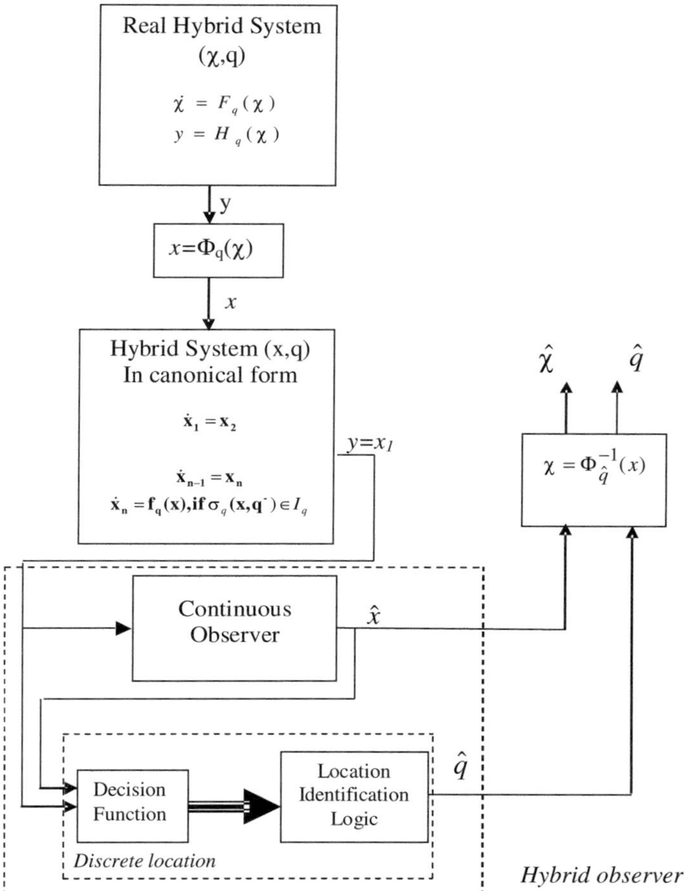

Fig. 4.1 The hybrid observer structure with diffeomorphism

4.2 Problem Statement

Let us consider the switched autonomous nonlinear systems

$$\dot{\chi} = F_q(\chi) \quad \text{for} \quad \chi \in \Omega_q(\chi) \tag{4.1}$$
$$y = H_q(\chi)$$
$$\chi(t_c^+) = r_{pq}^o \chi(t_c^-) + \varepsilon_{pq}^o \tag{4.2}$$

with $\chi \in \Re^n$, the state vector, $y \in \Re$, is the output, Ω_q is the domain evolution of the dynamic F_q, and r_{pq}^o and ε_{pq}^o are jump parameters of the continuous state at the switching instant t_c, where : t_c^+ is the time just after the switch, while t_c^- represents the time just before the switch (reset condition [14]).

In the case when a jump occur, and to guarantee the system's observability, we consider the following assumptions:

Assumption 4.2 : We assume throughout this chapter that

(a) All the evolution duration *dwell time* of each subsystem τ_q of (4.6) are measurable.
(b) Each subsystem is locally regularly weakly observable.[1] That is, for all $q = 1...p$ the codistribution:

$$\left\{dH_q, dL_{F_q}H_q, \ldots, dL_{F_q}^{(n-1)}H_q\right\}$$

is of $rank\ n$

The Assumption 4.2(a) means that the time τ_q between two switches is supposed sufficiently large so that the observer converges towards the real system before the next jump of the state. This assumption means that the system with Zeno phenomena will not be considered. The duration τ_q is a so-called *dwell time* [16] verifying the following conditions:

$$\min_q\{\tau_q\} > \tau_{\min} > 0$$

Now, we put the system in the well-known observability canonical form. To do this, we need the following set of diffeomorphisms which can be defined for $q = 1, \ldots, p$, as:

$$\begin{aligned} x = \Phi_q(\chi) &= \left(y, \dot{y}, \ldots, y^{(n-1)}\right)^T \\ &= \left[H_q(\chi), L_{F_q}H_q(\chi), \ldots, L_{F_q}^{n-1}H_q(\chi)\right]^T \end{aligned} \tag{4.3}$$

Remark 4.3 In this work, are considered only the cases where the dimension of the vector state x, hold

$$\dim(x) = n, \forall\ q$$

the case where $\dim(x)_{q=q_i} \neq \dim(x)_{q=q_j}$, for $i \neq j$ will be studied in a future work.

The diffeomorphism (4.3) leads to the following canonical form:

$$\begin{cases} \dot{x}_1 = x_2 \\ \vdots \\ \dot{x}_{n-1} = x_n \\ \dot{x}_n = f_q(x) \quad \text{if} \quad x \in I_q(x) \end{cases} \tag{4.4}$$

$$y = h_q(x) = x_1$$

$$x(t_c^+) = r_{pq}x(t_c^-) + \varepsilon_{pq} \tag{4.5}$$

[1] The regularity condition means that the $n-1$ derivatives are enough to recover all the state vector x. The localy weakly observabilty was introduced in [15].

where the vector field $f_q(x)$ is sufficiently smooth, $h_q(x) = x_1$ is the output, I_q is the domain evolution of the dynamic f_q. The jump parameters r_{pq} and ε_{pq} characterize the jump from the discrete location p to q and they are justified by those given in (4.2) or by the diffeomorphism (4.3). In fact, each diffeomorphism $\Phi_q(\chi)$ is obtained by using the output and its successive derivatives, with the assumption that q is well known.

Remark 4.4 The canonical form is used in order to obtain the estimation $\hat{x}$ without using any information about the last row of the dynamic. Consequently, we do not need the discrete location information q.

As the system (4.1–4.2) or (4.4–4.5), the jump appears on the continuous state when the system switches between two subsystems q and p, then just after the switching instant, at $t = t_c^+$ some observation errors appear on the continuous time observer.

In order to overcome this problem, we use a *re-initialisation procedure* which should change at the same time, the value of the observed continuous state $\hat{x}$ and the state x of the real system. Obviously, this re-initialisation procedure implies to know exactly the discrete location $\hat{q}$ (thus we must estimate q without any delay).

Here, it is important to note that for system (4.1–4.2) it is necessary to know which subsystem is active (the discrete location q) in order to reconstruct $\hat{\chi}$. For the system (4.4–4.5) it is not necessary to know q in order to observe the continuous state $\hat{x}$. This important property of (4.4–4.5) is the basis of our hybrid observer scheme.

Roughly speaking, extending our previous works [13, 17], to the switching systems with jumps, we have to take care of the reset procedure.

Hereafter, we design the observer after the transformation of the original system into a canonical form. In fact, when a change of location occurs, the continuous dynamics suddenly change and may be followed by a jump of the initial conditions of the observer i.e., *re-initialisation procedure*. The identification of this jump is one of the hybrid observer tasks. One can represent that by two cars running in a same direction. If the first driver changes direction abruptly, the other (which play the role of the observer) follows it with a delay. Re-initialisation procedure makes it possible to solve this problem, or at least allows to decrease this delay to the maximum. Roughly speaking, our re-initialisation procedure predicts and anticipates the abrupt change of direction. It is similar to the case that the first driver annonces his intention to change direction.

4.3 Recalls on Observability Study

Observability of nonlinear systems was largely studied in the literature (see for example [18]). More recently, several researchs concerned the study of hybrid systems observability in general and more particularly, the observability of switched systems. In [19] algebraic and geometrical observability conditions were established as well as in the case of linear and nonlinear switching systems. We recall that the class of the considered systems in this paper is the HDS (4.1–4.2). Thus, the observability of the systems will be established by using the observability conditions developed in [19] for the particular class of piecewise continuous dynamic systems.

As the considered system is rewritten in the observability canonical form (4.4–4.5) (see for example [20] for non autonomous systems), each sub system is observable and if we know which sub system is active, we can then conclude the whole observability of the system. Let us recall the main result given in [19] by one of authors and consider the dynamic systems formed with two dynamics inter-connected by a switch function:

$$\begin{cases} \dot{\chi} = F_1(\chi) \text{ and } y = H_1(\chi) \text{ if } \chi \in \Omega_1 = \{x/\sigma(x) \leq 0\} \\ \dot{\chi} = F_2(\chi) \text{ and } y = H_2(\chi) \text{ if } \chi \in \Omega_2 = \{x/\sigma(x) > 0\} \end{cases} \tag{4.6}$$

where $F_i(x)$ are smooth vector fields, $H_i(x)$ are smooth outputs and $\sigma(x)$ is a smooth switching function.

Under conditions of Assumption 4.2, if we know which the subsystem evolves, we can conclude the observability of the global system (4.6). Using the observability canonical form, each subsystem of (4.6) can be written as:

$$\begin{cases} \dot{z}_i^1 = z_{i+1}^1 & \text{for } i = 1:n-1 \\ \dot{z}_n^1 = f_1(z_1^1, z_2^1, \ldots, z_n^1) \end{cases} \tag{4.7}$$

if $\sigma_1 := \sigma(z_1^1, z_2^1, \ldots, z_n^1) \leq 0$, and

$$\begin{cases} \dot{z}_i^2 = z_{i+1}^2 & \text{for } i = 1:n-1 \\ \dot{z}_n^2 = f_2(z_1^2, z_2^2, \ldots, z_n^2) \end{cases} \tag{4.8}$$

if $\sigma_2 := \sigma(z_1^2, z_2^2, \ldots, z_n^2) > 0$.
where $(z^j,\ j = 1:2)$ are the observability coordinates given by:

$$z_{i+1}^j = L_{F_j}^{(i)} H_j \qquad \text{for } 0 \leq i \leq n-1$$

where $L_{F_j}^{(i)} H_j$ is the ith Lie derivative of H_j in the direction of F_j.

One approach to analyze the observability of (4.6), presented in [19], is based on the comparison of f_1 and f_2 on the one hand and σ_1 and σ_2 on the other hand. For this, we need to evaluate such functions in terms of the same variables. These variables are given naturally by the output y and its successive time derivatives $y^{(i)} = \frac{d^i y}{dt^i}$ for $i = 1:n-1$.

Let us consider the two submanifolds:

$$\mathscr{M} = \left\{ v \in \Re^n / f_1(v) = f_2(v) \right\}$$
$$\mathscr{S} = \left\{ v \in \Re^n / \sigma_1(v) = \sigma_2(v) \right\}$$

and finally, the submanifold of common singularities of subsystems of system (4.6):

$$\mathscr{L} = \{\chi \in \Re^n / F_1(\chi) = F_2(\chi) = 0\}$$

The main result that we recall here is given in this theorem.

Theorem 4.5 *(i) If* $\mathscr{M}$ *is a discrete set then system* (4.6) *is observable for any switch* σ *for which we have* $\sigma(\mathscr{L}) \leq 0$ *or else* $\sigma(\mathscr{L}) > 0$.

(ii) If dynamics (4.7) *and* (4.8) *are transverse to* $\mathscr{M}$ *except on a discrete subset then the system is observable for any switch* σ *for which we have* $\sigma(\mathscr{L}) \leq 0$ *or else* $\sigma(\mathscr{L}) > 0$.

(iii) If $\mathscr{S} = \Re^n$ *then system* (4.4) *is observable.*

The reader can see [19] for proof and more details. He can find also some algebraic sufficient conditions to analyze the observability of piecewise linear systems.

4.4 Hybrid Observer

In [17], a step-by-step sliding mode observer was mainly used for the following reasons: finite time convergence, ability to take into account the natural switching of structure and its robustness to external disturbances. Nevertheless, some difficulties occur due to the chattering phenomena. They induce a bad decision-making during the switch between two subsystems. This problem was circumvented by using a low-pass filter in the computation of the equivalent vector; unfortunately, this solution generates a delay [17]. To solve this problem, we propose a suitable solution for switching systems, which consists in using a high-order sliding mode observer (Robust Differentiator) (see [3, 21]).

4.4.1 Robust Differentiator "Super Twisting Algorithm"

The "Super Twisting Algorithm" (Fig. 4.2) is given by the following structure [22]:

$$\sum_{\text{obs}} = \begin{cases} u(e_1) = u_1 + \lambda_1 |e_1|^{\frac{1}{2}} \operatorname{sign}(e_1) \\ \dot{u}_1 = \alpha_1 \operatorname{sign}(e_1) \\ \lambda_1, \alpha_1 > 0 \end{cases} \tag{4.9}$$

where $e_1 = x_1 - \hat{x}_1$ and λ_1, α_1 are positive parameters, and u_1 is the differentiator output where:

$$\operatorname{sign}(e_1) = \begin{cases} +1 & \text{if } e_1 > 0 \\ -1 & \text{if } e_1 < 0 \\ \in [-1, 1] & \text{if } e_1 = 0 \end{cases}$$

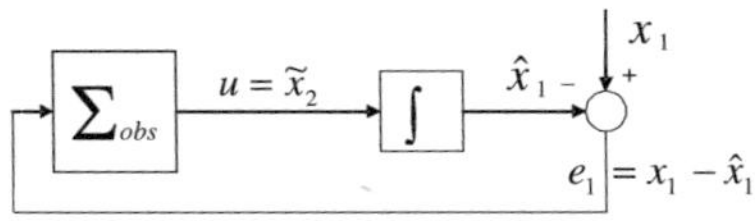

Fig. 4.2 Super Twisting Algorithm Structure

The step-by-step exact differentiator applied to (4.4), leads to the following form:

$$\begin{cases} \dot{\hat{x}}_1 = \tilde{x}_2 + \lambda_1 \left|\tilde{e}_1\right|^{1/2} \text{sign}(\tilde{e}_1) \\ \dot{\tilde{x}}_2 = \alpha_1 \text{sign}(e_1) \\ \dot{\hat{x}}_2 = E_1 \left[\tilde{x}_3 + \lambda_2 \left|\tilde{e}_2\right|^{1/2} \text{sign}(\tilde{e}_2)\right] \\ \dot{\tilde{x}}_3 = E_1 \alpha_2 \text{sign}(\tilde{e}_2) \\ \dot{\hat{x}}_3 = E_2 \left[\tilde{x}_4 + \lambda_3 \left|\tilde{e}_3\right|^{1/2} \text{sign}(\tilde{e}_3)\right] \\ \vdots \\ \dot{\tilde{x}}_{n-1} = E_{n-3} \alpha_{n-2} \text{sign}(\tilde{e}_{n-2}) \\ \dot{\hat{x}}_{n-1} = E_{n-2} \left[\tilde{x}_n + \lambda_{n-1} \left|\tilde{e}_{n-1}\right|^{1/2} \text{sign}(\tilde{e}_{n-1})\right] \\ \dot{\tilde{x}}_n = E_{n-2} \alpha_{n-1} \text{sign}(\tilde{e}_{n-1}) \\ \dot{\hat{x}}_n = E_{n-1} \left[\tilde{\theta} + \lambda_n \left|\tilde{e}_n\right|^{1/2} \text{sign}(\tilde{e}_n)\right] \\ \dot{\tilde{\theta}} = E_{n-1} \alpha_n \text{sign}(\tilde{e}_n) \end{cases} \tag{4.10}$$

where $\tilde{e}_i = \tilde{x}_i - \hat{x}_i$, with $\tilde{x}_1 = x_1$ for $i = 1, \ldots, n$, and the E_i for $i = 1, \ldots, n-1$ are defined as

$$E_i = 0 \quad \text{if } \tilde{e}_i = \tilde{x}_i - \hat{x}_i \neq 0, \text{else } E_i = 1 \tag{4.11}$$

The structure of the step-by-step differentiator for a system of order n in canonical form is given in Fig. (4.3), where each bloc B_i for $i = 1, \ldots, n-2$, is only valid when $E_i = 1$ (E_i are defined in (4.11)).

4.4.2 Convergence Analysis

The convergence of the observation error is obtained in $(n-1)$ step in finite time. The idea consists in using the step-by-step observer such as described hereafter: The $(n-1)$ first steps consist in reconstructing the state vector and after that, under conditions of Theorem 4.5, recovering in which state (location) q, the system evolves. Another feature of the differentiator (4.9) is the fact that the output $\mathbf{u}_1$ does not depend directly on discontinuous functions but on an integrator output. So high frequency chattering, which can be very harmful for the system (see for example chaotic system known for its extreme sensitivity to noise [17]), can be avoided. These properties are importants since the switching function can be obtained in a continuous way and without delays and chattering.

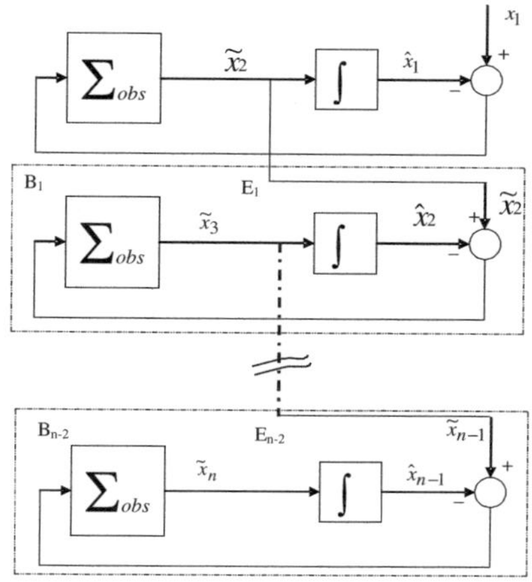

Fig. 4.3 Structure of step by step differentiator

Theorem 4.6 *Consider the system* (4.4), *assumed to be bounded state in finite time* $t < \infty$, *and the observer* (4.10) *based on the Super Twisting Algorithm* (4.9). *For any bounded initial conditions* $x(0)$, $\hat{x}(0)$, *there exists a choice of* λ_i *and* α_i *such that the state observer* $\hat{x}$ *converges in finite time* $T_{fs} \ll \tau_q$ *to* x *and* $\tilde{\theta}$ *converge also in finit time to* $f_q(x)$.

Proof The proof is given in [3] for the case of $n = 2$. Figure 4.4 illustrates the finite time convergence behavior of the proposed observer. The demonstration is based on the error trajectory for each quadrant in the worst cases.

In the case of $n > 2$, the convergence is ensured step by step following this order: $(\dot{e}_1 = e_2, e_1) \to (0, 0)$ in finite time T_1 in the first step (Fig. 4.5). $(\dot{e}_2 = e_3, e_2) \to (0, 0)$ in finite time T_2 in the second step. And $(\dot{e}_i = e_i + 1, e_i) \to (0, 0)$ in finite time T_i in the step i. Finally, $(\dot{e}_{n-1} = e_n, e_{n-1}) \to (0, 0)$ in finite time T_{n-1} in the step $(n - 1)$. The finite time of convergence of the full state x is:

$$T_{fs} = \sum_{j=1}^{n-1} T_j \tag{4.12}$$

The choice of α_i and λ_i are such that

$$T_{fs} \ll \tau_q$$

and then we have the time to converge before the next jump.

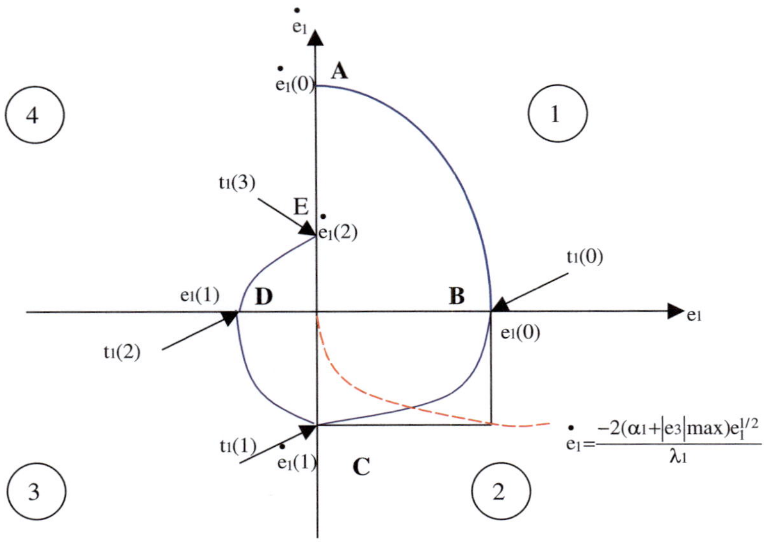

Fig. 4.4 Majoring curve for finite time convergence observer

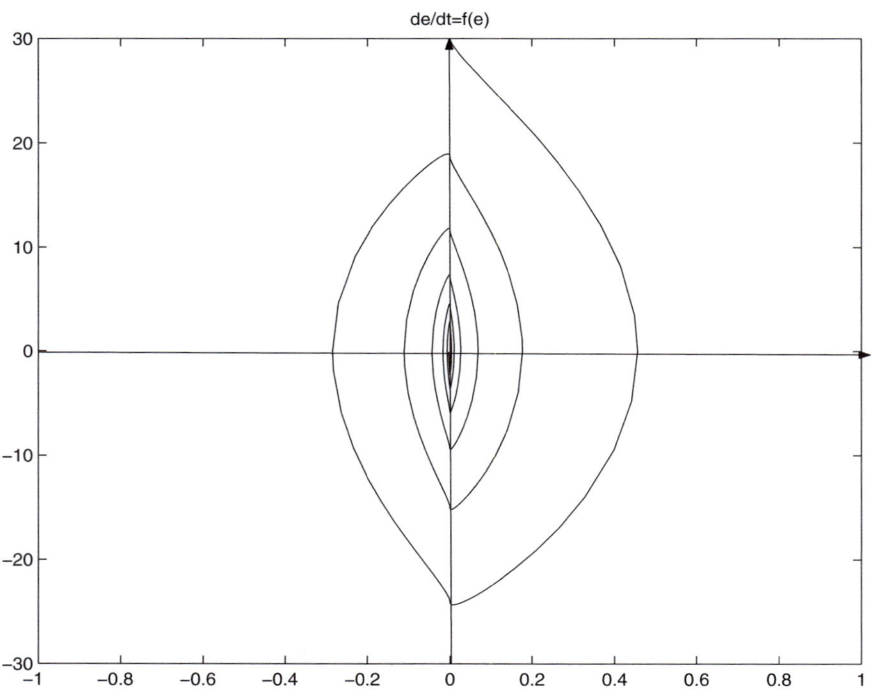

Fig. 4.5 Finite time convergence observer

4.4.3 Discrete Time Observer

In this section, we design a discrete part of the proposed observer. Let us consider the system (4.4), the task of the discrete time observer is to locate which dynamic of the system is in evolution? In some cases, the knowledge of the system's output is sufficient to estimate the current location. If this is not the case, some additional information obtained by using the continuous part of our observer, may be useful or are necessary to estimate the current location.

In our case, the discrete-observer receives as input: the observed state $\hat{x}$, the output y and the information E_{n-1}. Its task is to provide an estimation $\hat{q}$ of the discrete location q of the hybrid plant at the current time. Contrarily to the general case; here, the continuous observer does not need to know the discrete location q. This is the main property of the canonical form (4.4). Indeed, the second order sliding mode observer (4.10) has to know only the output $y = x_1$ and also the $f_q(x)$ upper bound, noted g^+.

Thus, one can announce the following result:

Proposition 4.7 *1. If the observer is sufficiently fast (i.e., $T_{fs} \ll \tau_q$), then for $t \geq T_{fs}$, one has $\sigma_q(\hat{x})$ converges towards $\sigma_q(x)$, (i.e., $\sigma_q(\hat{x}) = \sigma_q(x)$) and $f_q(x) = f_q(\hat{x})$ for a given q. Then under conditions of Theorem 4.5, the knowledge of the state estimation and the use of a location identification Logic, leads to conclude the estimation of $\hat{q}$.*

2. *Any wrong re-initialisation may occur under* conditions of Theorem 4.5 and the *particular design of the hybrid observer* (4.10) *with re-initialisation procedure only authorized when $E_{n-1} = 1$.*

Proof 1. As the system is given in the observability canonical form, the use of a sliding mode observer (4.10) allows us to reconstruct in finite time the state vector by using only $y = x_1$ and it is not necessary to know the discrete location q. Now to estimate q, we use the estimation of x and then evaluate all $\sigma_q(x)$ and $f_q(x)$. Thus, by using the logic identification (with respect to i, ii, iii of Theorem 4.5), we can give an exact estimation $\hat{q}$ of q.

2. As $E_{n-1} = 1$ ensures that $\hat{x} = x$ and $\tilde{\theta} = f_q(x)$ then under conditions of Theorem 4.5, we have exactly q and detect the switch.

Remark 4.8 Note that, at the switching time, the observer error convergence can be affected particularly if some jumps occur. Indeed, the terms E_i which are defined such as $E_i = 0$ if $\hat{x}_i \neq x_i$, $E_i = 1$ otherwise can be equal to zero in the reinitialization phase. This means that there is a jump phenomenon in the state trajectory (x_i). Hence, to avoid the divergence of the observation error, one can define un switching zone (Fig. 4.6) such as $|\sigma_i(\tilde{x})| < \varepsilon$ with ε as small as possible; and as soon as this zone is reached one preserves the value of the state $\hat{x}$ at this moment and then use it just after applying the reinitialization procedure $\hat{x}_i(t_c^+) = R_e(\hat{x}_i(t_c^-))$ Where t_c is the switching time, t_c^- is the time just before switching and t_c^+ is the time just after switching.

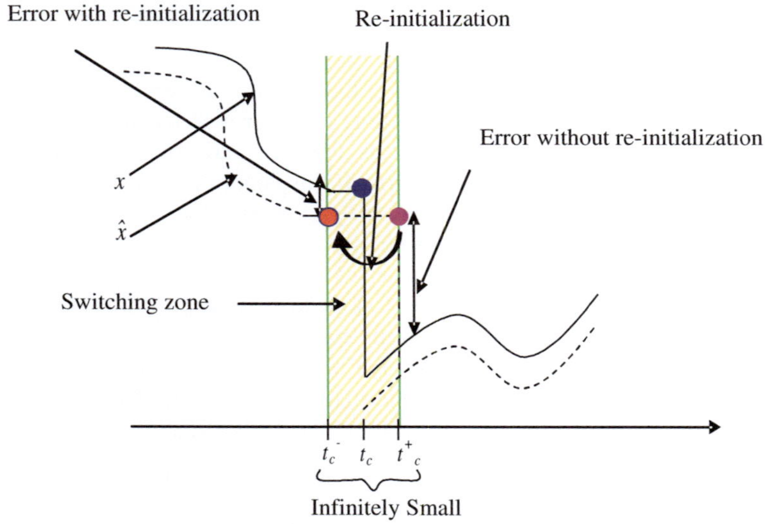

Fig. 4.6 Procedure of re-initialization

4.5 Illustrative Examples

4.5.1 Example 1

In this section, we present an example to highlight the performances of the designed sliding mode observer for hybrid systems. Let us consider the following switching system [23], in which we introduce state jumps

$$\begin{cases} \dot{x} = A_i x \qquad i = 1, 2 \\ y = Cx \\ x_2(t_c^+) = x_2(t_c^-) + \varepsilon \end{cases} \tag{4.13}$$

where $A_1 = \begin{bmatrix} 1 & -5 \\ 0 & 1 \end{bmatrix}$, $A_2 = \begin{bmatrix} 1 & 0 \\ 5 & 1 \end{bmatrix}$, and $C = [1, -2]$, $\varepsilon = 0.1$, t_c is the switching instant. The switching functions are given by:

$$s_{1,2}(x) = 2x_1 + x_2 = \begin{pmatrix} 2 & 1 \end{pmatrix} x = 0$$
$$s_{2,1}(x) = x_1 - 2x_2 = \begin{pmatrix} 1 & -2 \end{pmatrix} x = 0$$

where $x \in \Omega_1$ such that

$$\Omega_1 = \left\{ \begin{array}{c} x \in R^2/(x_1 - 2x_2 \geqslant 0 \wedge 2x_1 + x_2 > 0) \\ \vee (x_1 - 2x_2 \leq 0 \wedge 2x_1 + x_2 < 0) \end{array} \right\}$$

and $x \in \Omega_2$ such that

$$\Omega_2 = \left\{ \begin{array}{c} x \in R^2/(x_1 - 2x_2 > 0 \wedge 2x_1 + x_2 \leq 0) \\ \vee (x_1 - 2x_2 < 0 \wedge 2x_1 + x_2 \geqslant 0) \end{array} \right\}$$

Remark 4.9 The original system without jumps presented in [23] is a zeno system. The introduction of jumps on the state removes this phenomena by introducing a limit cycle.

In order to use the designed observer (4.10), the system (4.13) should be rewritten in the observability canonical form. To perform this let us apply the following diffeomorphisms (Fig. 4.7): $z = \Phi_i(x) := (z_1, z_2)^T := (Cx, CA_i x)^T = T_i x$ for $i = 1, 2$, with

$$T_1 = \begin{bmatrix} 1 & -2 \\ 1 & -7 \end{bmatrix} \quad \text{and} \quad T_2 = \begin{bmatrix} 1 & -2 \\ -9 & -2 \end{bmatrix}$$

And finally with $\bar{A} = T_i A_i T_i^{-1}$ for $i = 1, 2$, we find for both the subsystems the same observability canonical form:

$$\begin{cases} \dot{z} = \bar{A} z \\ y = z_1 \end{cases} \quad \text{with} \quad \bar{A} = \begin{bmatrix} 0 & 1 \\ -1 & 2 \end{bmatrix}$$

$$s_{1,2}(z) = 3z_1 - z_2 = 0 \quad \text{and} \quad s_{2,1}(z) = z_1 = 0$$

with $z \in I_1$ such that

$$I_1 = \left\{ \begin{array}{c} z \in R^2/(z_1 \geqslant 0 \wedge 3z_1 - z_2 > 0) \\ \vee (z_1 \leq 0 \wedge 3z_1 - z_2 < 0) \end{array} \right\}$$

and $z \in I_2$ such that

$$I_2 = \left\{ \begin{array}{c} z \in R^2/(z_1 > 0 \wedge 3z_1 - z_2 \leq 0) \\ \vee (z_1 < 0 \wedge 3z_1 - z_2 \geqslant 0)) \end{array} \right\}$$

The system expressed in canonical form is given by:

$$\begin{cases} \dot{z}_1 = z_2 \\ \dot{z}_2 = f_i(z) = -z_1 + 2z_2 \quad i = 1, 2. \\ y = z_1, \text{with } s_{1,2} = 3z_1 - z_2, s_{2,1} = z_1. \\ z_2^+ = 0.25 z_2^- + 5\varepsilon \\ z_2^+ = -z_1 + 20\varepsilon \end{cases}$$

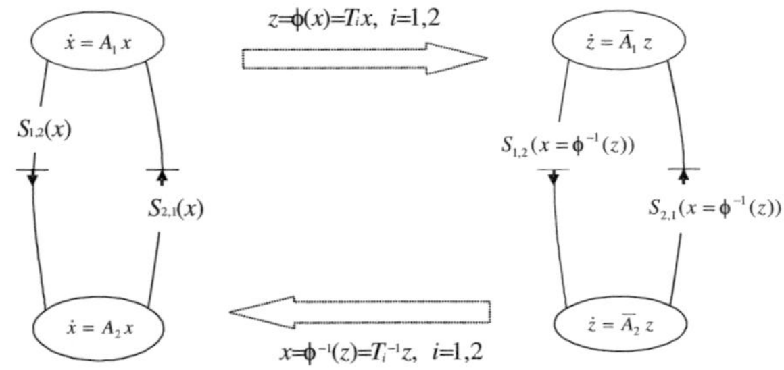

Fig. 4.7 Diffeomorphisms and switches

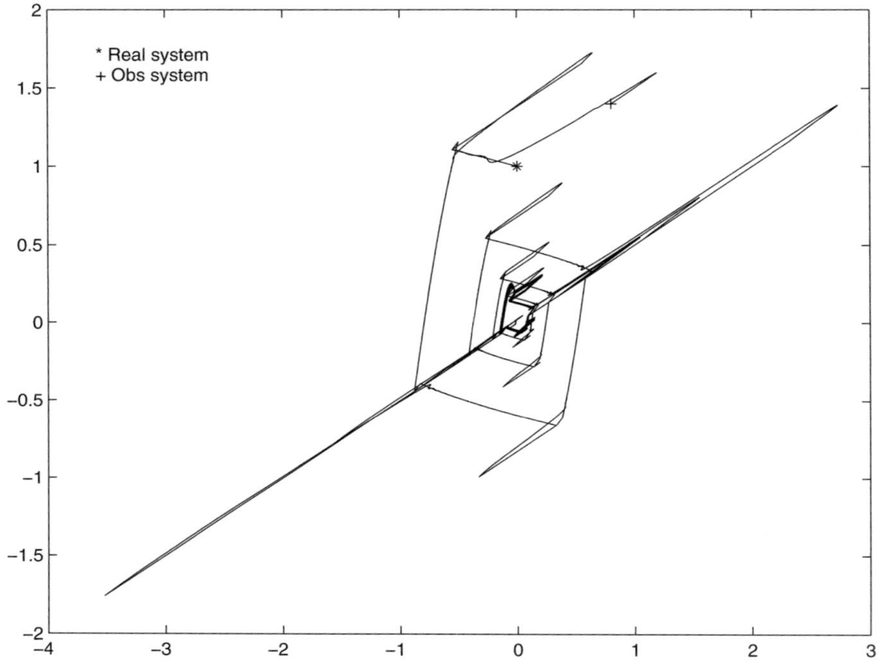

Fig. 4.8 Phase portrait without re-initialisation

note that $z_1^- = z_1^+ = z_1$,no jump on z_1 it only occurs on z_2.

The associated observer (4.10) is expressed in the following form

$$\dot{\hat{z}}_1 = \hat{z}_2 + \lambda_1 \left|z_1 - \hat{z}_1\right|^{1/2} \operatorname{sign}(z_1 - \hat{z}_1)$$
$$\dot{\hat{z}}_2 = \alpha_1 \operatorname{sign}(z_1 - \hat{z}_1).$$

and the re-initialisation is obtained as follows : $\hat{z}_2^+ = 0.25\hat{z}_2^- + 5\varepsilon$ and $\hat{z}_2^+ = -\hat{z}_1 + 20\varepsilon$. Note that $z_1 = z_1^- = z_1^+$, i.e., z_1 does not contain jump.

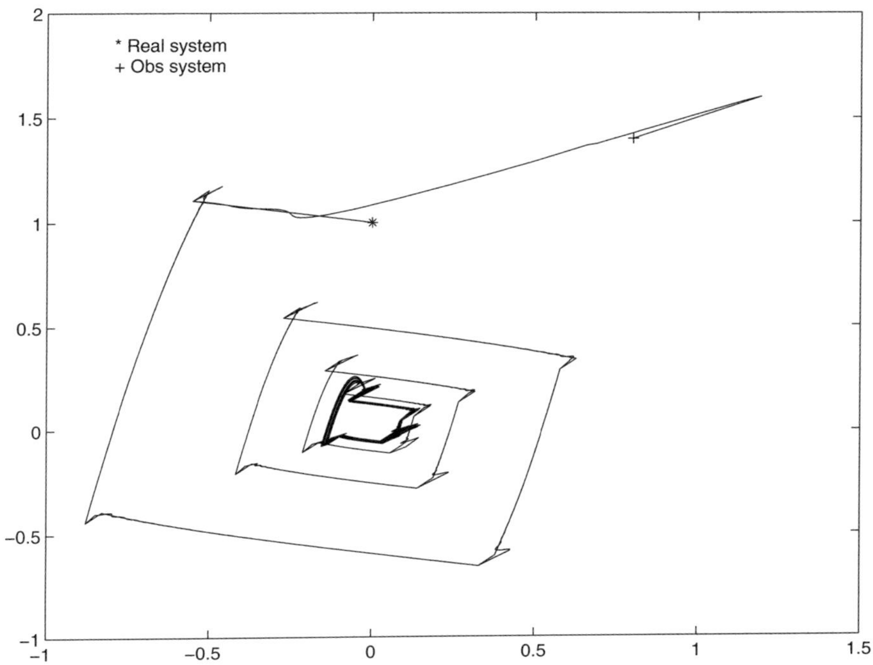

Fig. 4.9 Phase portrait with re-initialisation

Figures 4.8 and 4.9 illustrate respectively the phase portrait of the real and estimated stated with and without re-initialisation. The estimation is better preserved in the case of re-initialisation along the trajectory.

Figure 4.10 highlights the efficiency of our re-initialized observer. We see clearly that, the error goes to zero quickly when the re-initialisation is used.

4.5.2 Example 2

Working on hopping [24] juggling [25] etc. illustrates ways in which discontinuous dynamical systems can arise in control and are suggestive of the type of system that can be modelled as hybrid switching systems. However, these applications sometimes involve other complications and for that reason it seems to be worthwhile to consider some simple problems from mechanics that nicely illustrates the depth and breadth of effects what one can expect from this particular class of hybrid systems. To do this, the considered application in this paper is a spring-mass-friction system. The stiction phenomenon, due to the presence of friction force is common in many mechanical systems. The friction force between two surfaces is tied to their instantaneous relative velocity. The continuous trajectory of velocity and position is subject

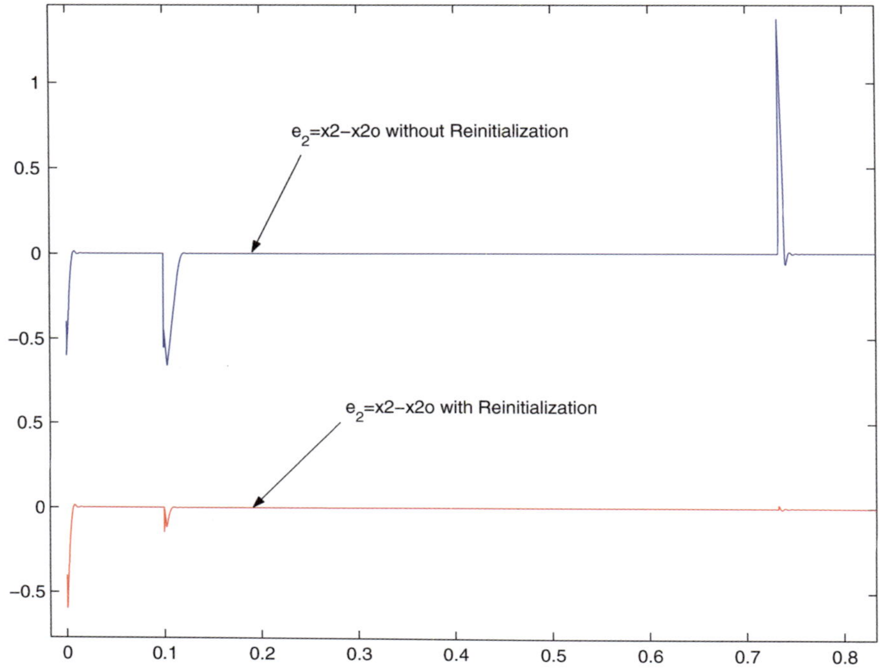

Fig. 4.10 Error $e_2 = z_2 - \hat{z}_2$ in both cases with and without re-initialisation

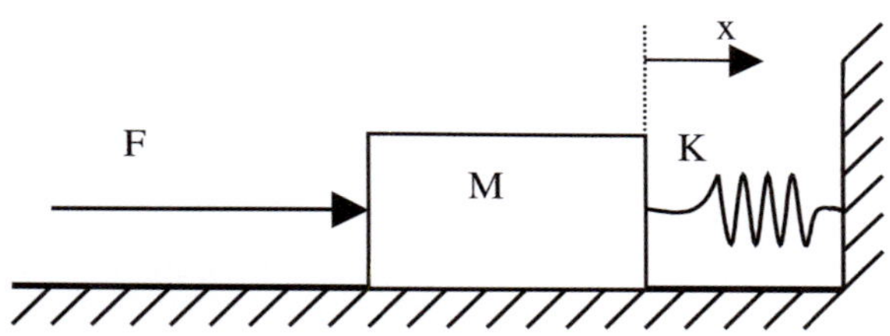

Fig. 4.11 Spring-mass-friction mechanical system

to abrupt changes in acceleration, however, corresponding to transition between the discrete states "stuck" and "sliding".

Let us consider the following mechanical spring-mass-friction system [26] which consists in a block sliding surface and compressing a spring under the influence of a user designated input force (Fig. 4.11). In the absence of friction, this behaves like a classical spring-mass system with the study state position proportional to the applied force. When friction between block and surface is taken into account, it tends to resist to the motion. The friction force changes with velocity and tends to be greatest when stationary. This results in a motion with alternately "*sticks*" and "slips" as the overall force balance requires.

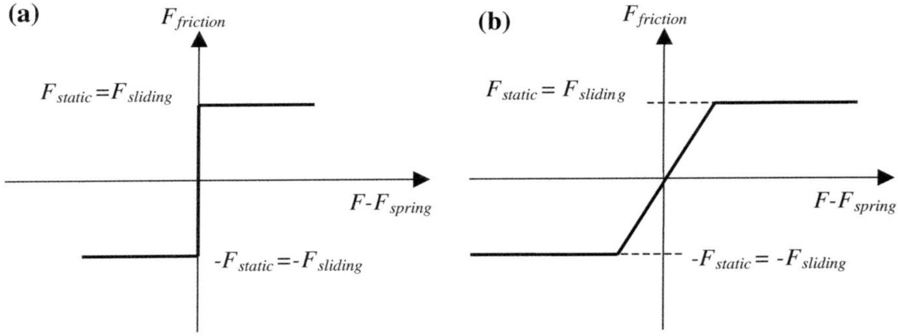

Fig. 4.12 Friction without sliding, **a** $x_2 \neq 0$, **b** $x_2 = 0$

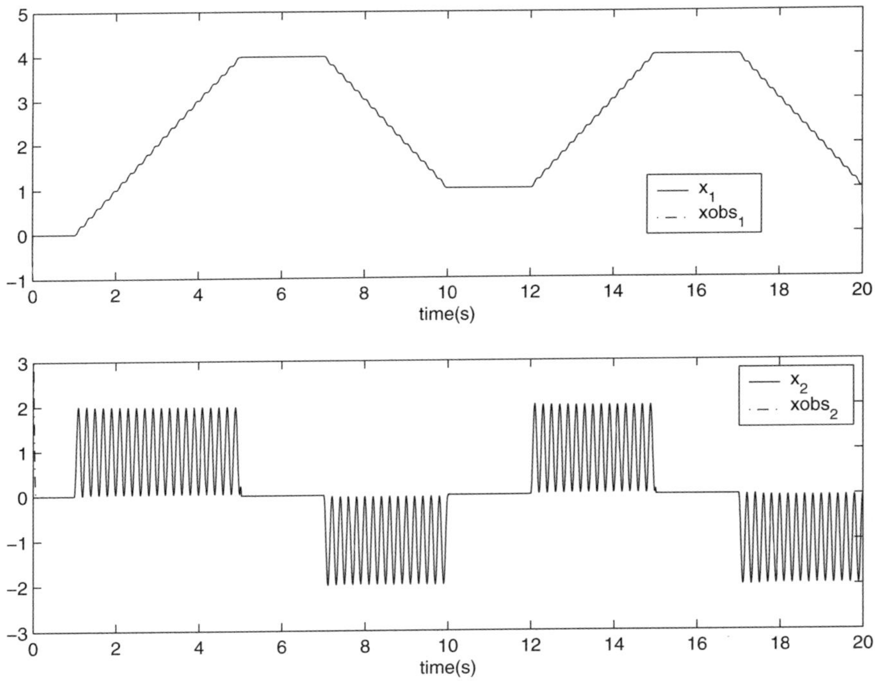

Fig. 4.13 The real states x and observed states $\hat{x}$

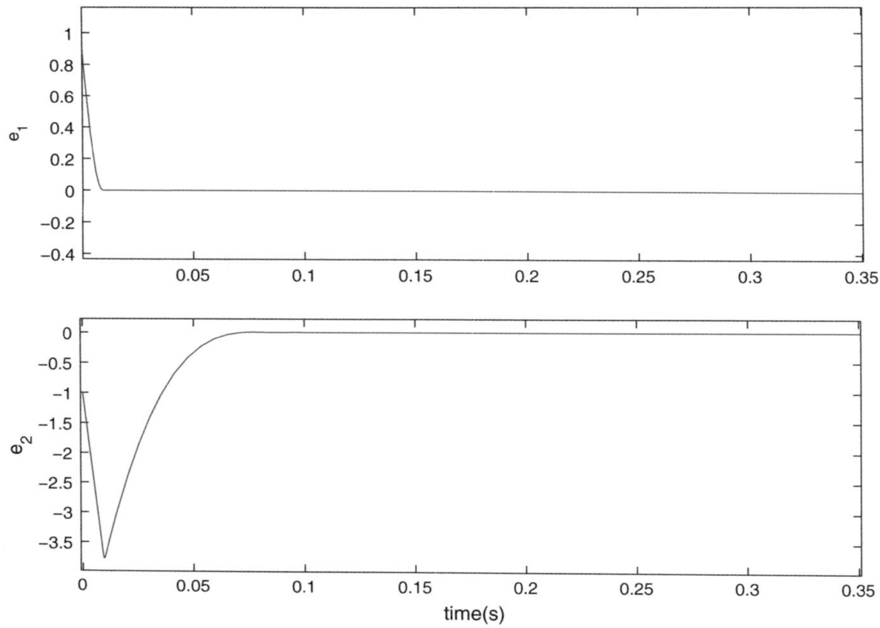

Fig. 4.14 Zoom : Observation error : $e_1 = x_1 - \hat{x}_1$ and $e_2 = x_2 - \hat{x}_2$

The equation of motion is

$$M\ddot{x} = F - F_{\text{spring}} - F_{\text{friction}}$$

choosing $x = x_1$ as the position, $\dot{x} = x_2$ is the velocity and $\ddot{x} = \dot{x}_2$ is the acceleration, the system can be written as

$$\begin{cases} \dot{x}_1 = x_2 \\ \dot{x}_2 = \frac{1}{M}\left(F - F_{\text{spring}} - F_{\text{friction}}\right) \\ \quad\;\; = f_q(x_1, x_2) \quad q = \{1, 2, 3\} \end{cases}$$

where M is the block mass, F is the input force. The force operated by the spring is given by $F_{\text{spring}} = Kx$ with K, the spring rate. The following logic determines $F_{\text{stationary}} = F - F_{\text{spring}} = F_{\text{sum}}T.$, the instantaneous force such that $\dot{x} = 0$. Whenever the velocity is nonzero, an impulsive force would be needed to make it zero instantaneously. This always exceeds the capacity, F_{sliding}, so the latter magnitude is used. When the velocity is already zero, however, $F_{\text{stationary}}$ is the force which maintains this condition by making the acceleration zero. The friction force F_{friction} is given as follows:

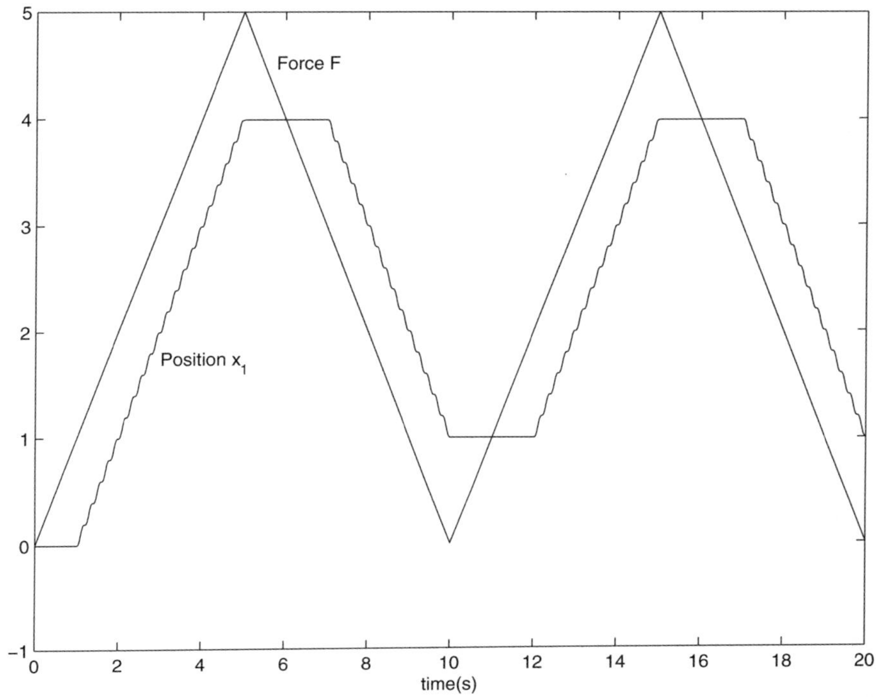

Fig. 4.15 The input force $u = F$ and the output $y = x_1$

$$F_{\text{friction}} = \begin{cases} \text{sign}(x_2) F_{\text{sliding}}, & x_2 \neq 0 \\ F_{\text{sum}}, & |F_{\text{sum}}| < F_{\text{static}}, \quad x_2 = 0 \\ \text{sign}(F_{\text{sum}}) F_{\text{static}}, & |F_{\text{sum}}| \geq F_{\text{static}}, \quad x_2 = 0 \end{cases}$$

with $F_{\text{sum}} = F - Kx_1$, and $F_{\text{static}} = F_{\text{sliding}} = 1\,\text{N}$,
The observer is designed as follows:

$$\begin{cases} \dot{\hat{x}}_1 = \hat{x}_2 + \lambda_1 \left|x_1 - \hat{x}_1\right|^{1/2} \text{sign}(x_1 - \hat{x}_1) \\ \dot{\hat{x}}_2 = f_q(x_1, \tilde{x}_2) + \alpha_1 \ \text{sign}(x_1 - \hat{x}_1) \qquad \text{if } \ \sigma_q(x) \text{ is verified } q = 1, 2, 3. \end{cases}$$

$$\tilde{x}_2 = \hat{x}_2 + \lambda_1 \left|x_1 - \hat{x}_1\right|^{1/2} \text{sign}(x_1 - \hat{x}_1)$$

$$\text{sign}(e) = \begin{cases} +1 & \text{if } \ e > 0 \\ -1 & \text{if } \ e < 0 \\ \in [-1, 1] & \text{if } e = 0 \end{cases}$$

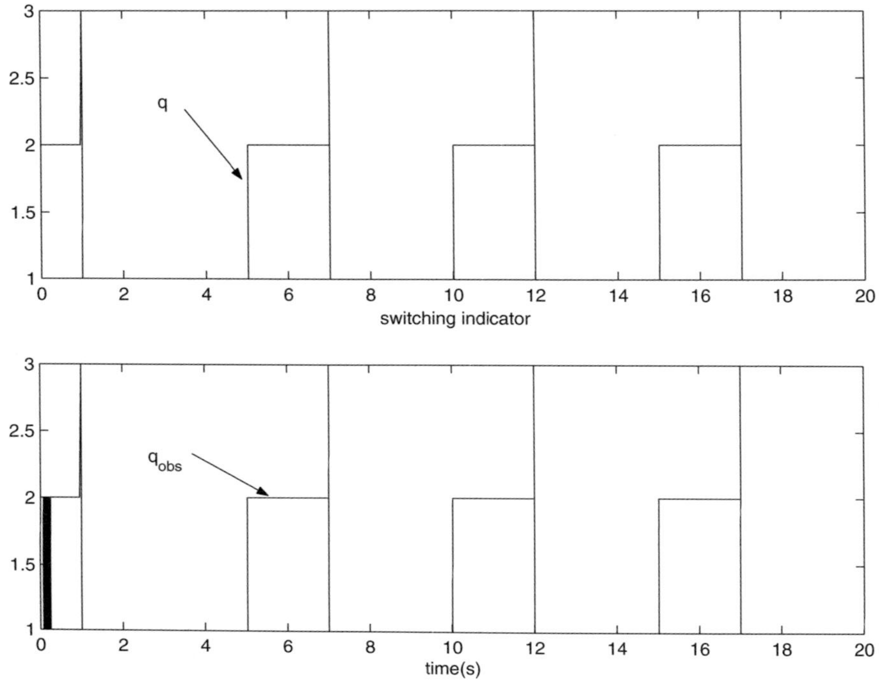

Fig. 4.16 The discrete location q and its estimation $\tilde{q}$

Case 1 Simulations were done using the following default parameters: $M = 0.001$ Kg, $K = 1$ N/m, $F_{\text{static}} = 1$ N, $F_{\text{sliding}} = 1$ N. The input force ramps linearly from zero to 5 N and back to zero, with a period of 5 s. The initial conditions are $(x_{10}, x_{20}) = (0, 0)$, while those of the observer are equal to $(\hat{x}_{10}, \hat{x}_{20}) = (-1, 1)$.

In this case, the natural frequency of the system is given by $w_n = \sqrt{\frac{K}{M}}$ =31.6 rad/s, and both the static F_{static} and sliding F_{sliding} forces are equal as illustrated in Fig. 4.12.

Figures 4.13, 4.14, 4.15, 4.16 and 4.17 illustrate the simulation results.

- In Fig. 4.13, we see that $\hat{x}_1$ and $\hat{x}_2$ reaches x_1 and x_2 respectively in finite time. In fact, a zoom of the error dynamic given in Fig. 4.14 confirms the above result and shows the convergence of e towards zero.
- Fig. 4.15, depicts the time histories of the input force and the resulting position $x_1$of the mass. The input force F must exceed that of static friction in order to

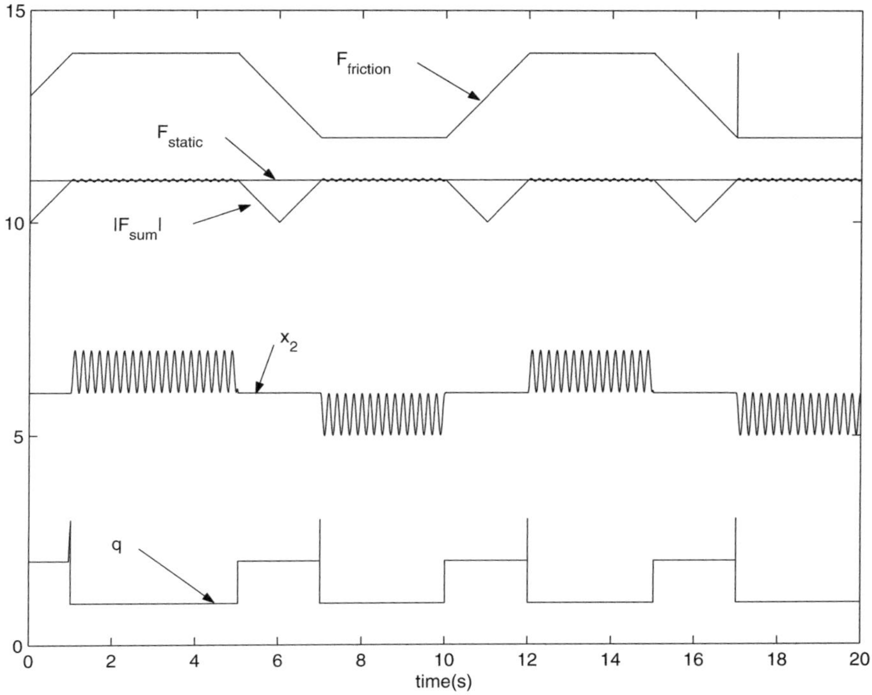

Fig. 4.17 The discrete location q and switching conditions F_{friction}, x_2, and F_{sum} and F_{static}

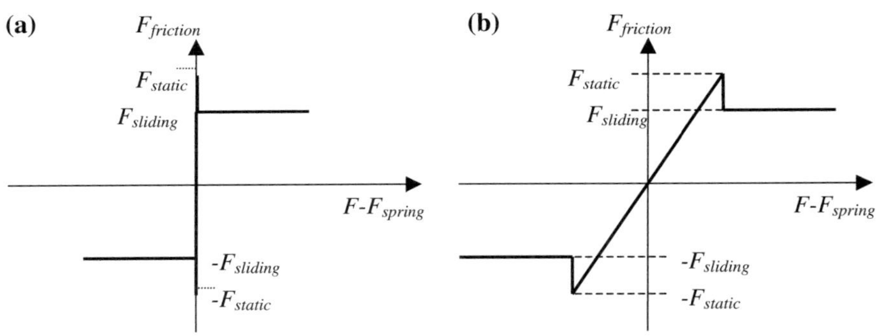

Fig. 4.18 Friction when $F_{\text{static}} \neq F_{\text{sliding}}$ **a** $x_2 \neq 0$, **b** $x_2 = 0$

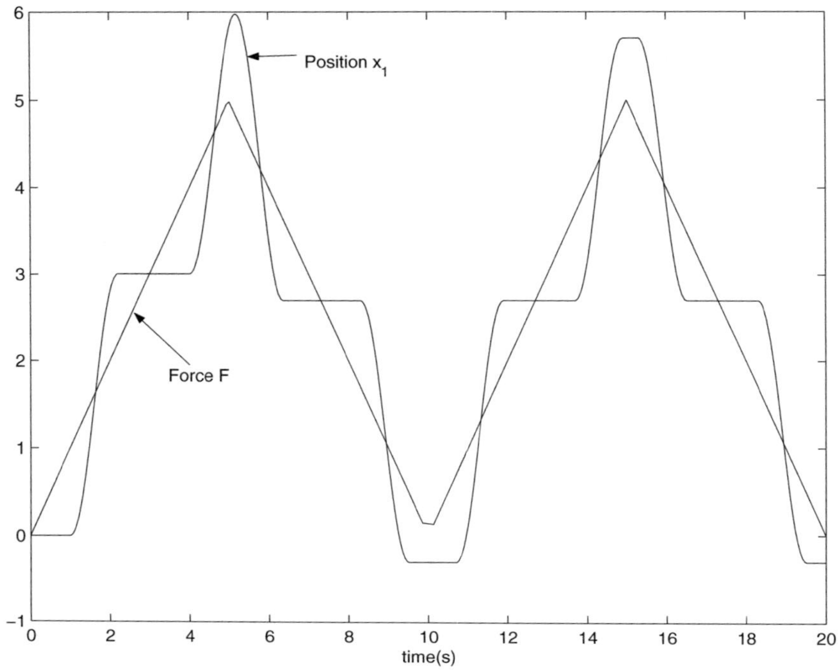

Fig. 4.19 The input force $u = F$ and the output $y = x_1$

begin motion at $t = 1$. For $t \in]1s, 5s[$, the position tracks the spring force less the kinematic friction force, with small oscillations showing changes in velocity at the natural frequency w_n. The input force begins to decrease. The mass immediately comes to a halt and sticks until $t = 7$. When the input force again exceeds the static friction force, the same movement occur in the inverse direction.

- In Fig. 4.16, one can remark the good results of the identification of the discrete location q and its estimation $\tilde{q}$ note q_{obs} in the figure. The discrete location takes values 1, 2, or 3 whether the system is in movement or not and depending on the friction force value. We notice that the system gives a good estimation, except in the transient part where the observer did not converge yet. It is thus important to ensure that the observer must be as fastest as possible.
- In Fig. 4.17, we represent the discrete location q with the switching functions conditions with the friction force F_{friction}. One can easily verify the correspondence with the physical behavior of the considered mechanical system with the switching conditions defined by the velocity and the considered friction force values.

Case 2 Simulation were done using the following parameters: $M = 0.1\,\mathrm{Kg}$, $K = 1\,\mathrm{N/m}$, $F_{\mathrm{static}} = 1\,\mathrm{N}$, $F_{\mathrm{sliding}} = 0.14\,\mathrm{N}$. In this case, the static F_{static} and sliding F_{sliding} forces are not equal and the friction force F_{friction} is represented in Fig. 4.18.

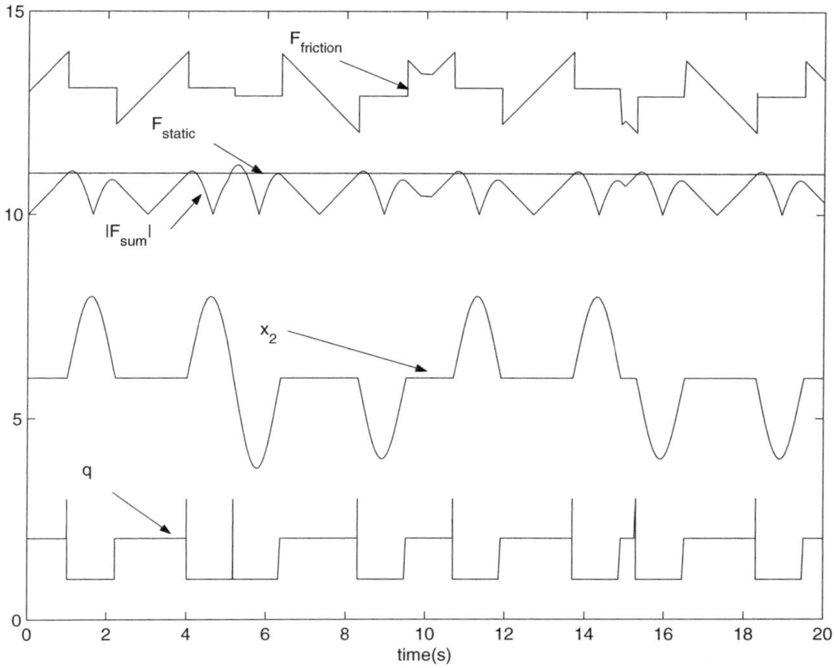

Fig. 4.20 The discrete location and its function decisions

As in the previous case, Figs. 4.19, 4.20, 4.21, 4.22, and 4.23 illustrate the simulation results. With kinetic friction lower in magnitude than static friction, abrupt discontinuities in acceleration occur at the stuck-to sliding and sliding to stuck state transition. As the velocity reaches zero, the acceleration is often nonzero. If the stuck state in entered, however, the acceleration becomes zero immediately.This highly nonlinear behavior is typical in many systems, making it difficult to precisely control and observer position and velocity design. In this case, also both continuous and discrete observer give good performances and despite the new condition on the static F_{static} and sliding F_{sliding} forces, the convergence remains ensured and the discrete location well estimated.

Thus, one can conclude that in both cases, the hybrid observer works correctly and gives a satisfactory results.

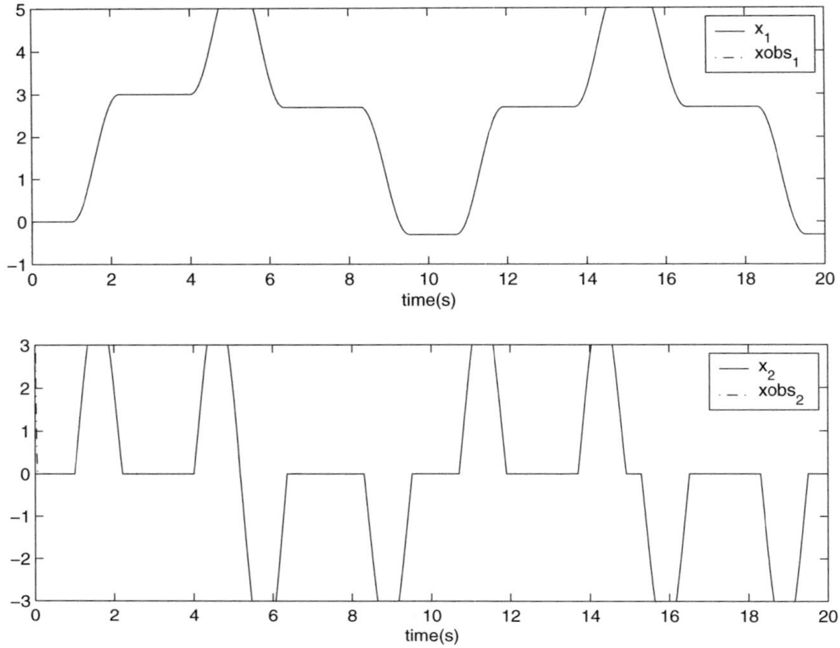

Fig. 4.21 The real states x and observed states $\hat{x}$

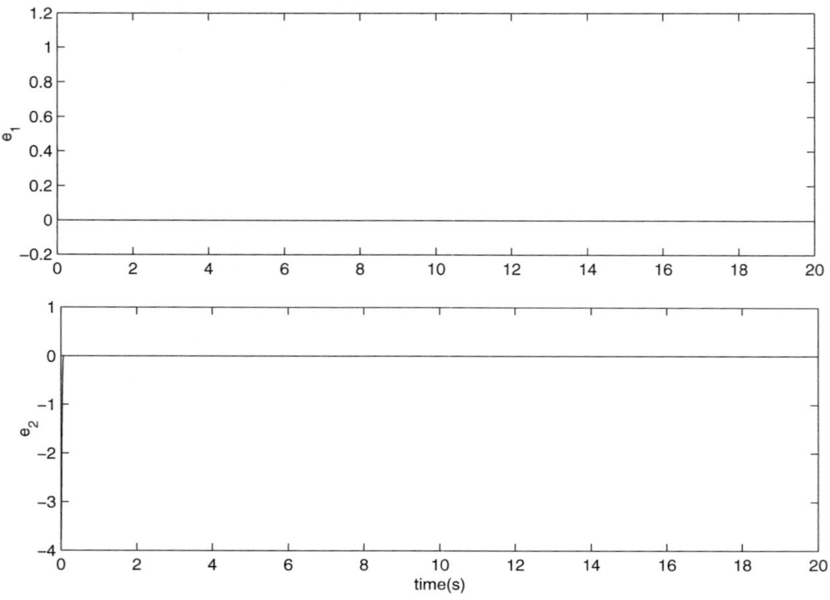

Fig. 4.22 Zoom : Observation error: $e_1 = x_1 - \hat{x}_1$ and $e_2 = x_2 - \hat{x}_2$

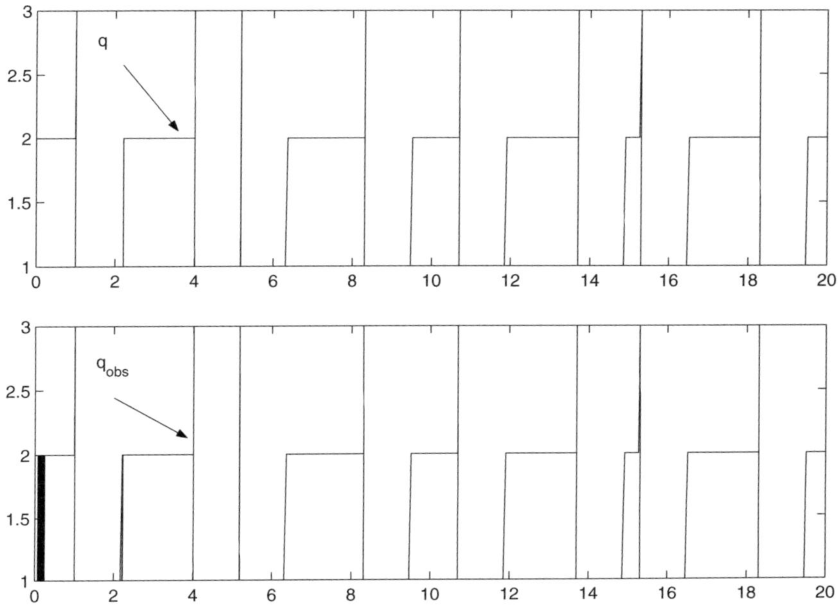

Fig. 4.23 The discrete location q and its estimation $\hat{q}$

References

1. Antsaklis, P.J. (ed.): Special issue on hybrid control systems. Proc. IEEE **43**(4), pp. 452–587 (1998)
2. Branicky, M.S.: Multiple Lyapunov functions and other analysis tools for switched and hybrid systems. IEEE Trans. Autom. Control **43**(2), 461–474 (1998)
3. Saadaoui, H., Manamani, N., Djemai, M., Barbot, J.P., Floquet, T.: Exact differentiation and sliding mode observer for switched Lagrangian systems. In: Nonlinear Analysis: Theory, Methods & Applications, special issue: Hybrid Systems and Applications, pp. 1–20 (2005)
4. DeCarlo, R., Branicky, M., Pettersson, S., Lennartson, B.: Perspectives and results on the stability and stabilizability of hybrid systems. Proc. IEEE **88**(7), 1069–1082 (2000)
5. Pettersson, S., Lennartson, B.: Hybrid system stability and robustness verification using linear matrix inequalities. Int. J. Control **75**(16/17), 1335–1355 (2002)
6. Alessandri, A., Coletta, P.: Switching observers for continuous time and discrete-time linear systems. In: Proceedings of the American Control Conference, Arlington, Virginia, pp. 2516–2521 (2001)
7. Juloski, A. L., Heemels, W. P. M. H., Boers, Y., Verschure, F.: Two approaches to state estimation for a class of piecewise affine systems. In: Proceedings of the 42nd IEEE Conference on Decision and Control. Maui, Hawaii, USA. pp. 143–148 (2003)
8. Juloski, A.L., Heemels, W.P.M.H., Weiland, S.: Observer design for a class of piecewise affine systems. In: Proceedings of the 41st IEEE Conference on Decision and Control, Las Vegas, Nevada, USA, pp. 2606–2611 (2002)
9. Pettersson, S.: Observer design for switched systems using multiple quadratic Lyapunov functions. In: Proceedings of the 2005 IEEE International Symposium on, Mediterrean Conference on Control and Automation 27–29 June 2005 pp. 262–267 (2005)

10. De la Sen, M., Luo, N.: Design of linear observers for a class of linear hybrid systems. Int. J. Syst. Sci. **31**(9), 1077–1090 (2000)
11. Balluchi, A., Benvenutiz, L., Di Benedetto, M.D., Sangiovanni Vincentelliy, A.L.: A Hybrid Observer for the Driveline Dynamics. ECC-2001 (2001)
12. Lin, L., Linawati, Lie Josa, Ambikarajah, E.: A hybrid state estimation scheme for power systems. Asian Pacific Conf. **1**, 555–558 (2002)
13. Boukhobza, T., Djemai, M., Barbot, J. P.: Nonlinear sliding observer for systems in output and output derivative injection form. In: Proceedings of the IFAC World Congres San Francisco (1996)
14. Lygeros, J., Johansson, H.K., Simć, S.N., Zhang, J., Sastry, S.S.: Dynamical properies of hybrid automata. EEE Trans. Autom. Control **48**(1), 2–17 (2003)
15. Hermann, R., Krener, A.J.: Nonlinear controllability and observability. IEEE Trans. Autom. Control, 22 **9**, 728–740 (1977)
16. De Santis, E., Di Benedetto, M.D., Girasole, G.: Digital Idle Speed Control of Automative Engines using Hybrid Models. IFAC World Congress, Prague (2005)
17. Djemai, M., Manamani, N., Barbot, J.P.: Sliding mode observer for triangular input hybrid system. In: Proceedings of the IFAC World Congres Prague (2005)
18. Bemporad, A., Ferrari-Trecate, G., Morari, M.: Observability and controllability of piecewise affine and hybrid systems. IEEE Trans. Autom. Control **45**(10), 1864–1876 (2000)
19. Boutat, D., Benali, A., Barbot, J.P.: About the observability of piecewise dynamical systems. NOLCOS-2004. (CD-ROM) (2004)
20. Fliess, M.: Generalized controller canonical forms for linear and nonlinear dynamics. IEEE Trans. Autom. Control, 35 **9**, 994–1000 (1990)
21. Levant, A.: Robust exact differentiation via sliding mode terchnique. In Autom. **34**(3), 379–384 (1998)
22. Fridman L., Levant, A.: Sliding modes of higher order as a natural phenomenon in control theory. In: Garofalo, F., Glielmo, L. (eds.) Robust control via variable structure & Lyapunov techniques. Lecture Notes in control and Information Science 217. Springer, London, p. 107 (1996)
23. Petterson, S.: Sythesis of switched linear systems. In: Proceedings of the 42nd IEEE-CDC, Hawaii, USA (2003)
24. Raibert, M.: Legged Robot that Balance. MIT Press, Cambridge (1986)
25. Buhler, M., Koditschek, D.: From stable to chaotic juggling: theory, simulation and experiments. In: Proceeding of the IEEE Conference on Robotics and Automation (1990)
26. Scientific Software, Using simulink and stateflow in automative applications, Simulink-Stateflow Technical Examples. The MathWorks Inc. (1998)
27. Levant, A.: Higher order sliding modes and arbitrary-order exact robust differentiation. In: Proceedings of the European Control Conference (2001)

Chapter 5
On the Observation Analysis and Observer Design for a Class of Hybrid Continuous-Discrete Dynamic System

Noureddine Manamanni, Mohamed Djemai and Jean Pierre Barbot

Abstract This chapter deals with observability conditions and state observer design for a class of hybrid systems whose the continuous part combines continuous and discrete dynamics. The main contribution of the work lies in the performed observability conditions for this class of systems and the design of a hybrid observer to reconstruct both continuous and discrete states starting only from the knowledge of a continuous output. Firstly, a high-order sliding mode based observer is used to estimate the continuous state and to generate a discrete output. Secondly, starting from this discrete output, a discrete state reconstructor is designed. An illustrative example is provided to show the efficiency of the proposed observer.

5.1 Introduction

Hybrid dynamical systems (HDS) was widely studied in literature these last decades in science and engineering (see, for example, [1]). A common definition is that they are modeled or defined as a combination of both time continuous systems and discrete event ones. Unlike the stability and stabilization problems [1, 11, 13, 24, 25], the observability and observer design problem for hybrid systems with nonlinear dynamics, received less attention and only some recent studies are available in literature concerning the observability of HDS and nonlinear observer design (see, for example, [3, 4, 10–12] and references therein). This chapter addresses

N. Manamanni (✉)
CReSTIC EA3804, Université de Reims Champagne Ardenne,
UFR SEN, Moulin de la Housse, 51687 Reims, France
e-mail: noureddine.manamanni@univ-reims.fr

M. Djemai
LAMIH, CNRS UMR 8201, University of Valenciennes and Hainaut-Cambresis,
59313 Valenciennes, France
e-mail: mohamed.djemai@univ-valenciennes.fr

J.P. Barbot
ENSEA, 6 Avenue du Ponceau, 95014 Cergy-Pontoise, France
e-mail: barbot@ensea.fr

M. Djemai and M. Defoort (eds.), *Hybrid Dynamical Systems*, Lecture Notes in Control and Information Sciences 457, DOI 10.1007/978-3-319-10795-0_5

observability conditions and state observer design for a particular class of hybrid systems whose the continuous part combines continuous and discrete state dynamics and called hybrid systems in state. Such models can represent a very broad range of systems for which the control and/or the observer design problems remain still open. Indeed, some systems may be controlled by discrete system when computers are used in the control procedure for example. Also, in some applications such as electrical drive, this kind of representation can be met after discretization of a part of the dynamics which will be controlled by a discrete system (system with commutation). In the case of electric motors, the use of the pulse width modulation (PWM) is generally used. The PWM being discrete by definition, since the system works by commutation and makes it possible to control the electrical part of the engine (see, for example, [15, 28]). In this case, the control strategy consists in discretizing the electrical part of the dynamics (which will be controlled by the orders of commutations) and to keep the mechanical part in continuous time.

Moreover, observer design problem was largely discussed and several methods were proposed for linear and nonlinear systems in both discrete and continuous cases. In the case of switched systems, some results were given by designing an observer for each mode of the system [26] and relate to the knowledge of the active mode of the switching system ([2, 5–7, 9–12, 14, 26, 27]). Nevertheless, there is always a lack of general results in the class of hybrid systems where the continuous part combines both continuous and discrete dynamics and few of them concern the case of hybrid dynamic systems HDS as defined in (5.1)–(5.3).

This chapter concerns hybrid observer design for the following particular class of HDS:

$$\dot{x} = F(x, Z(k)) \tag{5.1}$$

$$Z(k+1) = A_d Z(k) + \beta_d(v(k), y(k)) \tag{5.2}$$

$$y = H(x) = \left[h_1(x), \ldots, h_p(x)\right] \tag{5.3}$$

where $x \in \mathscr{N}$, with $\mathscr{N}$ an open set of $\Re^n$, is the continuous state vector, $Z = (z_1, \ldots, z_m)^T \in \mathscr{M}$ with $\mathscr{M}$ an open set of $\Re^m$, is the discrete state vector with $Z(t) = Z(t_k) = Z(k)$ for $t \in [t_k,\ t_{k+1}[\ \forall\, k > 0$. $A_d \in \Re^{m \times m}$, is constant matrix with appropriate dimension and $F(x, Z(k))$, $H(x)$ and β_d are vectors functions of appropriate dimensions, $y(k) = y(t = t_k)$ is the output and $v(k) \in \Re^{\nu}$ a known input.

For such systems, one can enumerates several open problems in designing observer for all continuous and discrete states. For example, the observer may be designed starting from the knowledge either of a part of the discrete states, or of a part of the continuous states, or of a combination of both.

In this work, the hybrid observer is designed in the case where the continuous output subsystem is available and all the switching instants t_k, i.e., the moments when the discrete system switches are also available and supposed to be known. Nevertheless, it is possible to overcome the previous assumption by using some technics (see, for example, [18]).

Note that, in the following, the considered class of HDS is supposed without jumps for the continuous subsystem, i.e., $x(t_k^+) = x(t_k^-)$ where t_k is the time when the discrete state switches from $Z(k-1)$ to $Z(k)$ and t_k^+ and t_k^- are the instants just before and just after the switching time, respectively.

The chapter is organized as follows: In Sect. 5.2, some definitions and observability conditions are proposed for both continuous and discrete subsystems, followed by a general condition for the whole hybrid system. Section 5.3 is devoted to the hybrid observer design. The chapter ends with an illustrative example in Sect. 5.1.

5.2 Hybrid Observability Conditions

Throughout the chapter, one considers a class of HDS of form (5.1)–(5.3) whose the continuous part combines continuous and discrete dynamics and supposed to be bounded for both continuous and discrete states and verifing the following assumption:

Assumption 5.1 The period $T_k = t_{k+1} - t_k$ for the dynamics (5.2) is supposed to check the following condition:

$$T_k > \tau_m > 0 \qquad \forall\, k \geq 0 \tag{5.4}$$

By consequence, the particular problem of systems with Zeno phenomena is not considered.

Let us now introduce some observability definitions for the considered subsystems of (5.1)–(5.3).

Definition 5.2 A pair of points x^0 and x^1 in $\mathcal{N}$ are uniformly distinguishable if for all input sequences $Z(k)\ \forall\, k \geq 0$, generated by (5.2), the system (5.1) generates solutions $x^0(t)$ and $x^1(t)$ satisfying $x^i(0) = x^i$ such that $x^i(t) \in \mathcal{N}\ \forall t \in [t_k,\ t_{k+1}[$, $k \geq 0$ and $h(x^0(t)) \neq h(x^1(t))$ for some $t \in [t_k,\ t_{k+1}[$, $k \geq 0$.

We denote by $\mathcal{I}(x^0; U \subseteq \mathcal{N})$ the set of all points $x^1 \in U$, an open neighborhood of x^0, that are not uniformly distinguishable from x^0.

Definition 5.3 The system (5.1) is uniformly observable at $x^0 \in N$ if $\mathcal{I}(x^0; \mathcal{N}) = x^0$ and is (Locally Uniformly) LU-observable at $x^0 \in \mathcal{N}$ if, there exist an open neighborhood U of x^0, such that $\mathcal{I}(x^0; U) = x^0$.

Definition 5.4 The dynamics (5.1) is (Locally Regularly Weakly Uniformly)[1] LRWU-observable[2] at x^0, for all $t \in [t_k,\ t_{k+1}[\ \forall\, k > 0$, and $Z(k)$ is constant during the same interval if there exist $U(x^0)$, a neighborhood of x^0 and p integers

[1] The LRWU-observability leads to the LU-observability.

[2] This is a modified version of observability conditions given in [20].

$(k_1, \ldots, k_p)$, independent from $Z(k)$ that form, after a possible reordering of the outputs, the smallest p-tuple with respect to the lexicographic ordering, such that for all $x \in U(x^0)$ and $Z(k)$ constant for $t \in [t_k,\ t_{k+1}[\ \forall\, k \geq 0$.

(i) $k_1 \geq k_2 \geq \ldots \geq k_p \geq 0$,
(ii) $\sum_i^p k_i = n$,(k_i are observability indices of (5.1))
(iii) The rank observability condition is verified $\forall\ Z(k)$ and $x(t)$ with $t \in [t_k,\ t_{k+1}[$, $k \geq 0$

$$\text{rank}\begin{pmatrix} dh_1 \\ dL_F h_1 \\ \vdots \\ dL_F^{(k_1-1)} h_1 \\ \vdots \\ dh_p \\ \vdots \\ dL_F^{(k_p-1)} h_p \end{pmatrix}_{|x(t)} = n \tag{5.5}$$

Remark 5.5 Note that, the choice of k_i is not unique and the assumption that the observability indices k_i are not affected by $Z(k)$ is a strong one.

It is not obvious that the condition (5.5) ensures that $n-1$ derivatives are sufficient to reconstruct all the components of the continuous state vector x starting from the output y_c and using the observability canonical form (OCF) and a diffeomorphism $\zeta = \Phi_Z(x)$ which allows obtaining this form (it will be defined here after).
Under condition (5.5), (5.1)–(5.3) may be written in OCF [17], $\forall t \in [t_k,\ t_{k+1}[$ and $\forall\, k$ thanks to the diffeomorphism $\zeta = \Phi_Z(x)$ defined as:

$$\begin{aligned} \zeta &= (\zeta_1^T, \ldots, \zeta_i^T, \ldots, \zeta_p^T)^T \\ &= \Big[h_1(x), L_F h_1(x), \ldots, L_F^{k_1-1} h_1(x), \\ &\quad \ldots, h_p(x), L_F h_p(x), \ldots, L_F^{k_p-1} h_p(x)\Big]^T \end{aligned} \tag{5.6}$$

with $\zeta_i = (\zeta_{1,i}, \zeta_{2,i}, \ldots, \zeta_{k_i,i})^T$, for $i = 1, \ldots, p$.

Note that, the diffeomorphism $\Phi_{z(k)}(x)$ is parameterized by $Z(k)$ and is discontinuous at each switching instant t_k. Hence, to guarantee the linearity of the OCF, with respect to $Z(k)$, the following assumption is needed

Assumption 5.6 The continuous dynamics (5.1) should verify the following conditions:

(i) $\exists$ at least $j \in \{1, \ldots, p\}$ such that

$$\frac{\partial \left(L^{k_j}_{F(x=\Phi_Z^{-1}(\zeta),Z)} h_j \right)\Big|_{\zeta}}{\partial Z} \neq 0, \tag{5.7}$$

(ii) For all $i = 1, \ldots, p$, one has

$$\frac{\partial^2 \left(L^{k_i}_{F(x=\Phi_Z^{-1}(\zeta),Z)} h_i \right)\Big|_{\zeta}}{\partial Z^2} = 0, \tag{5.8}$$

Obviously, Assumption 5.6 is sufficient for the considered problem and can be more relaxed with respect to the left invertibility problem [8, 21]. Hence, subsystem (5.1)–(5.3), under conditions (5.5)–(5.7)–(5.8) and diffeomorphism (5.6) can be rewritten in OCF linear with respect to $Z(k)$ as:

$$\begin{cases} \dot{\zeta}_{1,i} = \zeta_{2,i} \\ \dot{\zeta}_{2,i} = \zeta_{3,i} \\ \quad \vdots \\ \dot{\zeta}_{k_i,i} = \left(L^{k_i}_{F(x,Z)} h_i(x) \right)\Big|_{x=\Phi_Z^{-1}(\zeta)} \\ \quad = \Gamma_i(\zeta) + \Lambda_i(\zeta) Z(k) = \Gamma_i(\zeta) + \sum_{j=1}^{m} \Lambda_{ij}(\zeta) z_j(k) \\ y_i = \zeta_{1,i} \quad i = 1, \ldots, p \end{cases} \tag{5.9}$$

$$\text{where}: \Gamma_i(\zeta) = \left(L^{k_i}_{F(x,0)} h_i \right)\Big|_{x=\Phi_0^{-1}(\zeta)} \quad i = 1, \ldots, p$$

$$\Lambda_{i,j}(\zeta) = \frac{\partial \left(L^{k_i}_{F(x=\Phi_Z^{-1}(\zeta),Z)} h_i \right)\Big|_{\zeta}}{\partial z_j}, \quad j = 1, \ldots, m$$

Remark 5.7 The subsystem (5.9) given in OCF is U-observable with respect to $Z(k)$, but the original subsystem (5.1) needs the knowledge of $Z(k)$ and $\Phi_Z^{-1}(\zeta)$ to be able to estimate x.

We note that the system (5.9) can present jumps even if it is the case for the original one. In fact,

$$x(t_k^+) = x(t_k^-) = x(t_k) \tag{5.10}$$

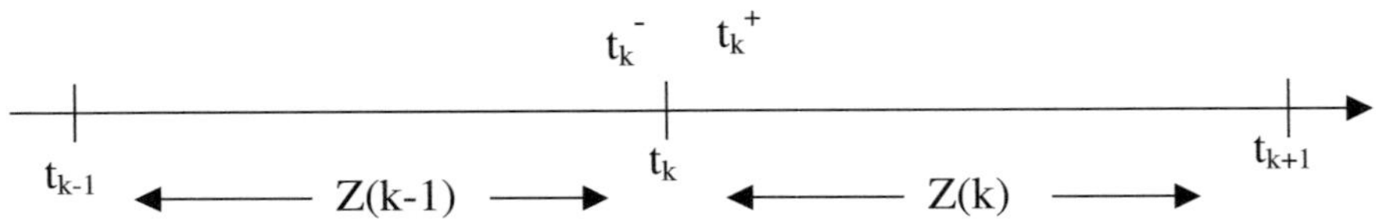

Fig. 5.1 t_k is the switching time and t_k^+ and t_k^- are instants just after and just before t_k

using (5.6), for $t \in [t_{k-1}, t_k[$ and $t \in [t_k, t_{k+1}[$, one has

$$\zeta(t) = \Phi_{z(k-1)}(x(t)) \quad \forall\, t \in [t_{k-1}, t_k[\tag{5.11}$$

$$\zeta(t) = \Phi_{z(k)}(x(t)) \quad \forall\, t \in [t_k, t_{k+1}[\tag{5.12}$$

Now, starting from these Eqs. (5.11), (5.12) and (5.10), at the sampling instants t_k (see Fig. 5.1), one obtains:

$$\zeta(t_k^+) = \Phi_{z(k)}\left[\Phi^{-1}_{z(k-1)}(\zeta(t_k^-))\right] = \mathscr{R}_{[z(k-1),z(k)]}(\zeta(t_k^-)) \tag{5.13}$$

Note that, this relation gives the reset equation of $\zeta(t)$ at time t_k^+ just after the switching time.
The following proposition summarizes the previous result:

Proposition 5.8 *Under conditions* (5.5)–(5.7)–(5.8), *the subsystem* (5.1) *is transformable by diffeomorphism parametrized by* $Z(k)$ *into the OCF* (5.9).

As the subsystem (5.9) is given in OCF, it is U-observable with respect to $Z(k)$ [19]. Nevertheless, in order to reconstruct the original state x of (5.1), the knowledge of $\Phi_Z^{-1}(\zeta)$ is necessary and consequently $Z(k)\ \forall\ k$. To overcome this difficulty, a software sensor from the continuous dynamics (written in OCF), in order to obtain information with respect to the discrete states $Z(k)$ is used.
Thus, from (5.9), all the last row of the subdynamics of ζ_i ($\zeta_{k_i,i}$ for $i = 1, \ldots, p$), may be written as:

$$\dot{\theta} = \Gamma(\zeta) + \Lambda(\zeta)Z(k) \tag{5.14}$$

with $\theta = \left[\zeta_{k_1,1}, \ldots, \zeta_{k_p,p,}\right]^T$, and $\Gamma(\zeta) = \left[\Gamma_1(\zeta), \ldots, \Gamma_p(\zeta)\right]^T$ is a column vector of dimension p, and

$$\Lambda(\zeta)|_{x=\Phi_Z^{-1}(\zeta)} = \left[\Lambda_1^T(\zeta), \ldots, \Lambda_m^T(\zeta)\right]^T$$

is a $p \times m$ matrix and $\Lambda_i(\zeta)$ is a line vector of dimension m defined as:

$$\Lambda_i(\zeta) = \left[\Lambda_{i,1}(\zeta), \ldots, \Lambda_{i,m}(\zeta)\right], \quad i = 1, \ldots, p$$

For the sake of simplicity, and because the continuous states in ζ coordinates are also known after the convergence of the continuous observer, we choose to consider each $\Lambda_{i,j}(\zeta)$ at time t_k^-, i.e., $\Lambda_{i,j}(\zeta(t_k)) = \Lambda_{i,j}(\zeta(t_k^-))$, to guarantee the finite time convergence of the continuous observer. The discrete dynamics (5.2) remains linear in $Z(k)$, and will be completed by an output $Y_d(k) = \Lambda(\zeta(k))Z(k)$:

$$Z(k+1) = A_d Z(k) + \beta_d(v(k), y(k)) \tag{5.15}$$

$$Y_d(k) = \Lambda(\zeta(k))Z(k) \tag{5.16}$$

where $\Lambda(\zeta(k)) = \Lambda(\zeta(t = t_k^-))$.

The discrete subsystem (5.15)–(5.16) will now be used to design a discrete observer. As $\Lambda(\zeta(k))$ is a $(p \times m)$ constant matrix only at time $t \in [t_k, t_{k+1}[$ but may change at each t_k $\forall\, k > 0$, we need to define the distinguishability and observability of (5.15)–(5.16).

Definition 5.9 A pair of points Z^0 and Z^1 in $\mathscr{M}$ is uniformly distinguishable with respect to $v(k)$ $\forall\, k > 0$, and $\forall\, \zeta(0)$ if the system (5.15)–(5.16) generates, $\forall\, \zeta(0)$ and $v(k)$, solutions $Z^0(k)$ and $Z^1(k)$ satisfying $Z^i(0) = Z^i$ with $Z^i(k) \in \mathscr{M}$, and there exists k such that $\Lambda(\zeta(k))Z^0(k) \neq \Lambda(\zeta(k))Z^1(k)$.

We denote by $\mathscr{I}(Z^0; \mathscr{M})$ all points $Z^1 \in \mathscr{M}$ that are not uniformly distinguishable from Z^0.

Definition 5.10 The system (5.15)–(5.16) is uniformly observable at $Z^0 \in \mathscr{M}$ if $\mathscr{I}(Z^0; \mathscr{M}) = \{Z^0\}$.

Definition 5.11 The system (5.15)–(5.16) is, in a delayed way, U-observable at $Z^0 \in \mathscr{M}$, if there exist a bounded $\mu \geq 0$ such that, for all $\zeta(0)$ and $v(k)$, $\forall\, k < \mu$ the system is uniformly observable at $Z(\mu)$.

The above definition means that, even if the reconstruction of $Z(0)$ is not guaranteed, then the reconstruction of $Z(\mu)$ may be possible. In fact, to reconstruct a state vector Z at time μ starting from Eqs. (5.15)–(5.16) with the following notations $Y_d^{\mu} := Y_d(\mu)$, $\zeta_{\mu} := \zeta(\mu)$, $Z_{\mu} := Z(\mu)$, and $\beta_{d,\mu} := \beta_d(v(\mu), y(\mu))$, one obtains:

$$\begin{aligned}
Y_d^{\mu} &= \Lambda(\zeta_{\mu})Z_{\mu} \\
Y_d^{\mu+1} &= \Lambda(\zeta_{\mu+1})\left[A_d Z_{\mu} + \beta_{d,\mu}\right] \\
&\vdots \\
Y_d^{\mu+N} &= \Lambda(\zeta_{\mu+N})A_d^N Z_{\mu} + \Lambda(\zeta_{\mu+N})\sum_{i=1}^{N} A_d^{N-i}\beta_{d,\mu+i-1}
\end{aligned}$$

which can be rewritten as:

$$\begin{bmatrix} Y_d^{\mu} \\ Y_d^{\mu+1} \\ \vdots \\ Y_d^{\mu+N} \end{bmatrix} = \begin{bmatrix} \Lambda(\zeta_{\mu}) \\ \Lambda(\zeta_{\mu+1})A_d \\ \vdots \\ \Lambda(\zeta_{\mu+N})A_d^N \end{bmatrix} Z_{\mu} + \begin{bmatrix} D_{\mu} \\ D_{\mu+1} \\ \vdots \\ D_{\mu+N} \end{bmatrix} \tag{5.17}$$

where $D_{\mu+j} = \Lambda(\zeta_{\mu+j}) \sum_{i=1}^{j} A_d^{j-i} \beta_{d,\mu+i-1}$ for $j = 1, \ldots, N$, is a known expression depending only on $(v(.), y(.))$.

Proposition 5.12 (Observability of Z(μ)) *The discrete system* (5.15)–(5.16) *is observable at* $Z(\mu) \in \mathscr{M}$ *for a fixed* $\zeta(\mu) \in \mathscr{N}$ *and for a fixed set of input* $v(k)$ *,*$k > \mu$ *if and only if* $(\Lambda(.), A_d)$*, verifies the following observability conditions:*

There exists a bounded integer $\tilde{N}$ *such that:*

$$\text{rank}\left\{ \bar{\Lambda}_{\mu} = \left(\begin{bmatrix} \Lambda(\zeta_{\mu}) \\ \Lambda(\zeta_{\mu+1})A_d \\ \vdots \\ \Lambda(\zeta_{\mu+\tilde{N}})A_d^{\tilde{N}} \end{bmatrix} \right) \right\} = m \tag{5.18}$$

Proof

- Sufficiency: From (5.17), verifying (5.18) and using the Left Pseudo Inverse, one obtains $Z(k)$.
- Necessary: If for all integer $\tilde{N}$, rank $\{\bar{\Lambda}_{\mu}\} < m$, then Ker $\{\bar{\Lambda}_{\mu}\}$ is at least of dimension 1, and consequently the unobservable space is at least of dimension one.

The construction of the matrix $\bar{\Lambda}_{\mu}$ is based on FIFO (First Input First Output) principle.

Corollary 5.13 (Uniform Observability of Z(μ)) *The discrete system* (5.15)–(5.16) *is U-observable at* $Z(\mu) \in \mathscr{M}$*,* $\forall\ \zeta(0) \in U$ *and* $\forall\ v(k)$ *for* $k > \mu$*, if and only if* $(\Lambda(.), A_d)$*, verifies the following observability conditions:*

There exists a bounded integer $N(\zeta(0), v(k))$*, noted* N *such that:*

$$\text{rank}\left\{ \bar{\Lambda}_{\mu} = \left(\begin{bmatrix} \Lambda(\zeta_{\mu}) \\ \Lambda(\zeta_{\mu+1})A_d \\ \vdots \\ \Lambda(\zeta_{\mu+N})A_d^{N} \end{bmatrix} \right) \right\} = m \tag{5.19}$$

Remark 5.14 If $\mu = 0$, the previous corollary leads to the U-Observability

Obviously, for causality reasons, $\mu + N$ is the present instant usually noted k and one doesn't know Z at time k but at time $k - N$ and then Eq. (5.17) will be rewritten as

$$\begin{bmatrix} Y_d^{k-N} \\ Y_d^{k-N+1} \\ \vdots \\ Y_d^k \end{bmatrix} = \begin{bmatrix} \Lambda(\zeta_{k-N}) \\ \Lambda(\zeta_{k-N+1})A_d \\ \vdots \\ \Lambda(\zeta_k)A_d^N \end{bmatrix} Z_{k-N} + \begin{bmatrix} D_{k-N} \\ D_{k-N+1} \\ \vdots \\ D_k \end{bmatrix} \tag{5.20}$$

Now, from Definitions (5.3), (5.4), (5.10) and (5.11), we are able to propose an observability definition for the whole HDS system defined in (5.1)–(5.3).

Definition 5.15 A pair of continuous and discrete state $\left(x^0(0), z^0(0)\right)$ and $\left(x^1(0), z^1(0)\right)$ in $\mathscr{N} \times \mathscr{M}$ are uniformly distinguishable if for all input $v(k)$, the system (5.1)–(5.2) generates solutions $\left(x^0(t), z^0(t)\right)$ and $\left(x^1(1), z^1(1)\right)$ such that $h(x^0(t)) \neq h(x^1(t))$ for some $t \neq +\infty$.

We denote by $\pounds(x^0, z^0; U \times \mathscr{M})$ all sets of couple $\left(x^1, z^1\right) \in U \times \mathscr{M}$ that are not distinguishable from$\left(x^0, z^0\right)$.

Definition 5.16 The class of HDS (5.1)–(5.3) is (Locally Uniformly) LU-observable at

$$\left(x^0, z^0\right) \in \mathscr{N} \times \mathscr{M} \quad \text{if} \quad \pounds(x^0, z^0; U \times \mathscr{M}) = \left(x^0, z^0\right)$$

Definition 5.17 The hybrid system (5.1)–(5.3) is in delayed way (Locally Uniformly) LU-Observable at (x^0, Z^0) if there exists a bounded integer $k > 0$, such that, the system (5.1)–(5.3) is LU-Observable at $(x(k), Z(k))$

Proposition 5.18 *If the continuous dynamics* (5.1) *verify conditions* (5.5)–(5.7)–(5.8) *and it is (Locally Regularly Weakly) LRW-Observable at* x^0*, for all* $t \in [t_k, t_{k+1}[$ $\forall$ *k and the subsystems* (5.15)–(5.16) *is uniformly observable at* $z^0 \in M$ *, then the whole HDS defined in* (5.1)–(5.3) *is (Locally Uniformly) LU-observable at* $\left(x^0, z^0\right)$ *and it is LU-observable, in delayed way, at* $\left(x^{t_\mu}, z^\mu\right)$ *for* $\mu > 0$*, if the subsystems* (5.15)–(5.16) *is uniformly observable in delayed way at* $z^\mu \in \mathscr{M}$*.*

In the next, a high-order sliding modes observer combined with a discrete state reconstructor will be designed for this class of HDS which is LU-Observable (or in a delayed way LU-Observable). The advantages of such observer, with respect to other ones which may be used, are its finite time convergence, and its ability to consider the variable structure systems.

5.3 Hybrid Observer Design

The hybrid observer is designed in the following way (Fig. 5.2).

Once the diffeomorphism carried out, a sliding mode-based observer (Super twisting algorithm), is firstly applied to the continuous system transformed in the OCF (5.9). Secondly, after the estimation of all the continuous states of the system in the

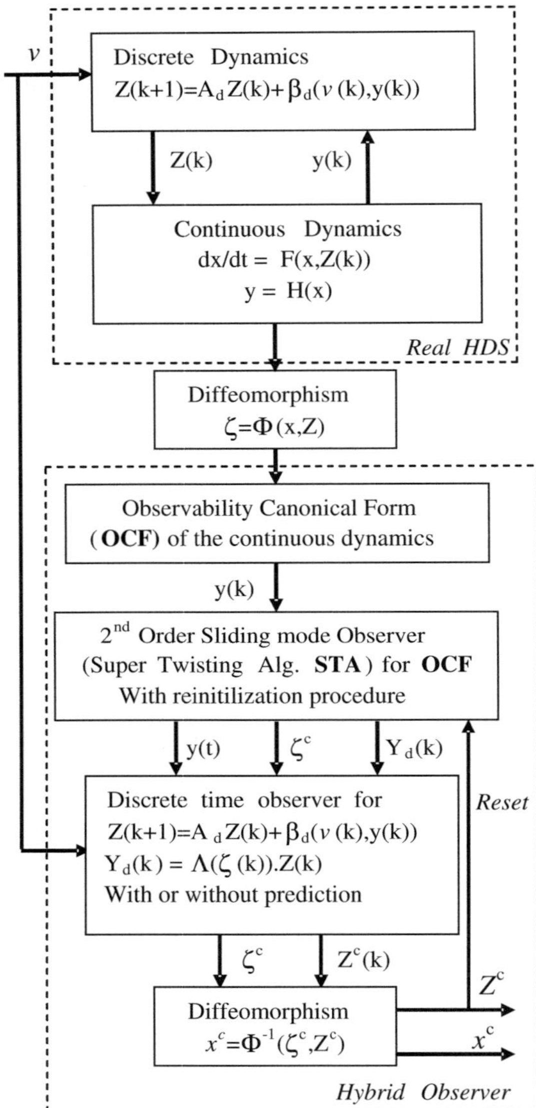

Fig. 5.2 Structure of the hybrid observer continuous and discrete states with : $Z^c = \hat{Z}$, $\zeta^c = \hat{\zeta}$ et $x^c = \hat{x}$

ζ coordinates in $n-1$ first steps, the last nth step allows reconstructing the whole dynamics and by consequence obtaining $Y_d(k) = \Lambda(\tilde{\zeta}(t_k^-))Z(k) = \Lambda(\tilde{\zeta}(k^-))Z(k)$. The discrete observer consists in using $Y_d(k)$ as an output for the discrete dynamics in order to design a discrete observer to estimate $\hat{Z}(k)$.

5.3.1 Super Twisting Algorithm

In [16], a step by step sliding mode observer was mainly employed, for a class of switching systems, for its finite time convergence, ability to take into account the natural switching of structure and its robustness to external disturbances. Nevertheless, some difficulties may occur due to the chattering phenomena. They induce a bad decision-making during the switch between two subsystems. This problem was circumvented by using a low-pass filter in the computation of the equivalent vector; unfortunately, this solution generates a delay [16]. To solve this problem, we have proposed a solution which consists in using a high-order sliding mode observer (robust differentiator) [22].

In [26], a super twisting algorithm (STA) was used for switched system. At first, the same observer is exploited in this chapter. In our case, a STA is used for the continuous part of the system. In fact, the continuous dynamics will be supposed as a forced commutation, since it is the discrete dynamics which makes the difference while $Z(k)$ change its value.

The STA is given by the following structure [22]:

$$\sum_{\text{obs}} = \begin{cases} u(e_1) = u_1 + \lambda_1 |e_1|^{\frac{1}{2}} \operatorname{sign}(e_1) \\ \dot{u}_1 = \alpha_1 \operatorname{sign}(e_1) \\ \lambda_1, \alpha_1 > 0 \end{cases} \tag{5.21}$$

where $e_1 = \zeta_{1,i} - \hat{\zeta}_{1,i}$, λ_1, α_1 are positive parameters, and u_1 is the differentiator output where:

$$\operatorname{sign}(.) = \begin{cases} +1 & \text{if } (.) > 0; \\ -1 & \text{if } (.) < 0; \\ \in [-1, 1] & \text{if } (.) = 0 \end{cases}$$

The step-by-step exact differentiator applied to each subsystem (5.9), leads to the following form. (See for more details the synthesis of such observers [3, 26] and its application for autonomous switched systems)

$$\begin{aligned}
\dot{\hat{\zeta}}_{1,i} &= \tilde{\zeta}_{2,i} + \lambda_{1,i} \left|\tilde{e}_{1,i}\right|^{1/2} \operatorname{sign}(\tilde{e}_{1,i}) \\
\dot{\tilde{\zeta}}_{2,i} &= \alpha_{1,i} \operatorname{sign}(e_{1,i}) \\
\dot{\hat{\zeta}}_{2,i} &= E_{1,i} \left[\tilde{\zeta}_{3,i} + \lambda_{2,i} \left|\tilde{e}_{2,i}\right|^{1/2} \operatorname{sign}(\tilde{e}_{2,i})\right] \\
&\vdots \\
\dot{\hat{\zeta}}_{k_p-1,i} &= E_{k_p-2,i} \Big[\tilde{\zeta}_{k_p,i} + \ldots \\
&\quad + \lambda_{k_p-1,i} \left|\tilde{e}_{k_p-1,i}\right|^{1/2} \operatorname{sign}(\tilde{e}_{k_p-1,i})\Big] \\
\dot{\tilde{\zeta}}_{k_p,i} &= E_{k_p-2,i} \alpha_{k_p-1,i} \operatorname{sign}(\tilde{e}_{k_p-1,i}) \\
\dot{\hat{\zeta}}_{k_p,i} &= E_{k_p-1,i} \left[\tilde{\theta}_i + \lambda_{k_p,i} \left|\tilde{e}_{k_p,i}\right|^{1/2} \operatorname{sign}(\tilde{e}_{k_p,i})\right] \\
\dot{\tilde{\theta}}_i &= E_{k_p-1,i} \alpha_{k_p,i} \operatorname{sign}(\tilde{e}_{k_p,i})
\end{aligned} \tag{5.22}$$

for $i = 1, \ldots, p$, where $\tilde{e}_{j,i} = \tilde{\zeta}_{j,i} - \hat{\zeta}_{j,i}$, with $\tilde{\zeta}_{1,i} = \zeta_{1,i}$ for $j = 1, \ldots, k_p$, and the vector $\tilde{\theta}_i$ represent the estimation of the dynamic $\Gamma_i(\zeta) + \Lambda_i(\zeta)Z(k)$ and $E_{j,i}$ for $j = 1, \ldots, k_i - 1$ are defined as

$$E_{j,i} = 0 \text{ if } \left|\tilde{e}_{j,i}\right| \geq \varepsilon > 0, \text{ else } E_{j,i} = 1 \tag{5.23}$$

See [26] for more details concerning the convergence proof and the definition of $E_{j,i}$. We also have to take into account the jumps (5.13) due to the diffeomorphism and use the reinitialisation procedure $\forall\, t = t_k, k > 0$.

$$\hat{\zeta}^{+}(t_k) = \Phi_{\hat{z}(k)}\left[\Phi^{-1}_{\hat{z}(k-1)}(\hat{\zeta}(t_k^{-}))\right] \tag{5.24}$$

The structure of the step-by-step differentiator for a system of order n in a canonical form is given in Fig. 5.3, where each block $B_{j,i}$ for $j = 1, \ldots, k_i - 1$, and $i = 1, \ldots, p$ is only valid when $E_{j,i} = 1$ ($E_{j,i}$ are defined in (5.23)) and the last block $B_{k_i,i}$ gives the vector $\tilde{\theta}_i$

The convergence of the observation error for all the continuous states is obtained in $(n-1)$ steps in finite time and allows to reconstruct the state vector. To obtain the discrete output $Y_d = \Lambda(\tilde{\zeta})Z(k)$, let us consider:

$$\tilde{\theta}(k) = \Gamma(\tilde{\zeta}(k^{-})) + \Lambda(\tilde{\zeta}(k^{-}))Z(k)$$

where k^{-} is the time $t = t_k^{-}$ (just before switching instant) then at time $t = t_k$ one obtains:

$$Y_d(k) = \Lambda(\tilde{\zeta}(k^{-}))Z(k) = [\tilde{\theta}(k) - \Gamma(\tilde{\zeta}(k^{-}))] \tag{5.25}$$

Thanks to the canonical form, the observation of ζ does not need the knowledge of $Z(k)$. In fact, the sliding mode observer based on the STA needs only to know $y_i = \zeta_{i,1}$. This is why the hybrid observer design is carried out in two sccessives stages, starting with a continuous observer. The hybrid observer design requires the following condition:

Condition 5.19 The sliding mode observer is designed such that the convergence time τ_c^k during the interval $[t_k,\ t_{k+1}[\ \forall\, k \geq 0$ verifies the following condition:

$$\tau_m > \max\left\{\tau_c^k\right\} > 0 \qquad \forall\, k \geq 0 \tag{5.26}$$

with τ_m is defined in (5.4).

Condition (5.4) is justified by the fact that the convergence time τ_c^k of the designed observer for each constant value of $Z(k)$ is not the same and also ensures that the designed observer has the time to estimate the continuous state during $[t_k,\ t_{k+1}[$ $\forall\, k \geq 0$. The following proposition summarizes the properties of the observer (5.22) (see [26] for more details).

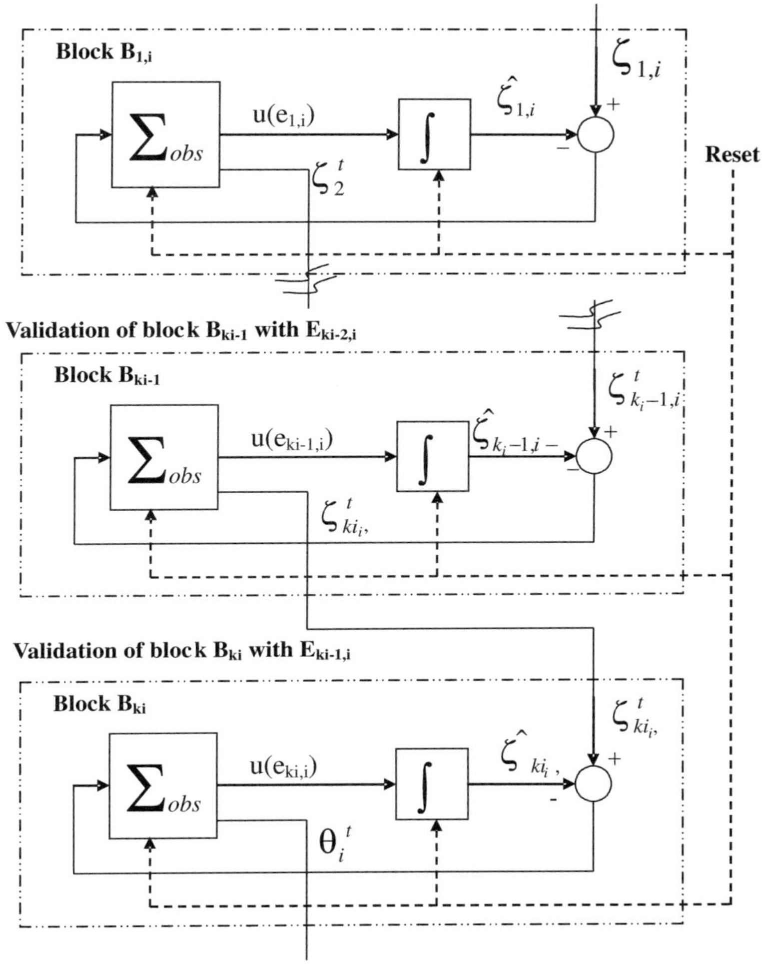

Fig. 5.3 General structure of the STA observer for system in OCF with $\zeta^t = \tilde{\zeta}$

Proposition 5.20 *Let us consider the system* (5.9), *supposed to be bounded for all* $t \in [t_k, t_{k+1}[\ \forall\, k > 0$, *and a sliding mode observer* (5.22). *For any initial condition* $\zeta(k)$, $\hat{\zeta}(k)$, *and for all* $Z(k)$, *there exist bounded* $\alpha_{i,j}$ *and* $\lambda_{i,j}$ *such that* $\hat{\zeta}$ *converges in finite time (smaller than* τ_m*) towards* ζ, *i.e.,* $\exists\, t = \tau_c^k > 0$ *such that* $\tilde{\zeta} = \zeta$, *and* $\tilde{\theta}$ *converges towards* $\Gamma(\tilde{\zeta}) + \Lambda(\tilde{\zeta})Z(k)$, *and by consequence reconstruct a discrete output* $Y_d(k) = \Lambda(\tilde{\zeta})Z(k)$ *as in* (5.25).

All switching instants are known by both observers and the reset information (5.24) are transmitted from the discrete observer to the continuous one.

5.3.2 Discrete State Observer

In this section, we will design the observer of the discrete subsystem to estimate the discrete state. Let us consider the system (5.1)–(5.3) and the observation structure given in the previous section, the discrete observer receives as input : the observed state $\tilde{\zeta}$, the output $y(k)$, the information $E_{k_i-1,i}$ for $i = 1, \ldots, p$ and $v(k)$. Its task is to provide an estimation $\hat{Z}$ of the discrete state Z of the hybrid plant starting from both the reconstructed discrete output $Y_d(k)$ given in (5.25) and the sampled continuous state $\tilde{\zeta}(k)$. To design the discrete observer, (5.1) should verify the observability condition (5.18). After the estimation of the continuous state $\tilde{\zeta}$ and a discrete output $Y_d = \Lambda(\tilde{\zeta}(k))Z(k)$, the following system will be used to estimate $\hat{Z}$:

$$
\begin{aligned}
Z(k+1) &= AZ(k) + \beta_d(v(k), y(k)) \\
Y_d &= \Lambda(\tilde{\zeta}(k))Z
\end{aligned}
\tag{5.27}
$$

If condition (5.18) is verified, it is possible to design an observer for the time-varying system (5.27) to obtain $\hat{Z}(k)$ which converges to $Z(k)$. Note that the designed method performed in [2, 23], for finite time convergent observers for linear time-varying systems, can be used in our case.

The estimation $\hat{Z}(k - N)$ can also be obtained from left invertibility on the Eq. (5.20). Thus, one obtains:

$$
\hat{Z}_{k-N} = \begin{bmatrix} \Lambda(\zeta_{k-N}) \\ \vdots \\ \Lambda(\zeta_k) A_d^N \end{bmatrix}^{-1*} \begin{pmatrix} Y_d^{k-N} - D_{k-N} \\ \vdots \\ Y_d^k - D_k \end{pmatrix}
\tag{5.28}
$$

where $(.)^{-1*}$ represents a pseudo inverse of (.).

Obviously, as it is mentioned before, for causality reasons and in order to estimate the discrete state at time k, a discrete time predictor of N -steps will be used as follows:

Starting from (5.27) and denoting $\beta_{d,k-N} = \beta_d(v(k-N), y(k-N))$, and $Z_k = Z(k)$, one obtains:

$$
\begin{aligned}
\hat{Z}_{k-N+1} &= A_d \hat{Z}_{k-N} + \beta_{d,k-N} \\
\hat{Z}_{k-N+2} &= A_d^2 \hat{Z}_{k-N} + A_d \beta_{d,k-N+1} + \beta_{d,k-N} \\
&\vdots \\
\hat{Z}_k &= A_d^N \hat{Z}_{k-N} + \sum_{i=1}^{N} A_d^{N-i} \beta_{d,k-N+i-1}
\end{aligned}
\tag{5.29}
$$

which gives the prediction of the discrete state Z at time k starting from the knowledge of $Z(k - N)$.

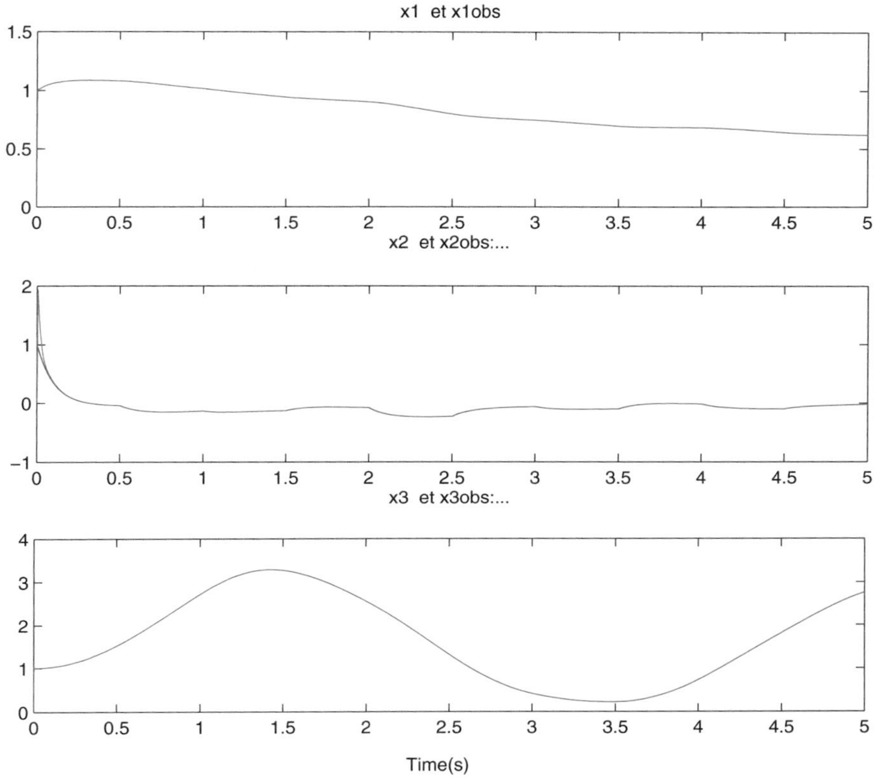

Fig. 5.4 Real and observed continuous states

5.4 Illustrative Example

In this section, an illustrative example will be used to show the efficiency of the proposed method and the designed observer (Fig. 5.4).

Example 5.21 Let us consider the system as defined in (5.1)–(5.3):

$$\begin{aligned} \dot{x} &= F(x, Z(k)) \\ Z(k+1) &= A_d Z(k) + \beta_d(y(k)) \\ y &= y_c = H(x) = \left[h_1(x), \ldots, h_p(x)\right] \end{aligned}$$

where $x = (x_1, x_2, x_3, x_4)^T$, $Z = (z_1, z_2, z_3)^T$, $y_1 = x_1$, $y_2 = x_3$

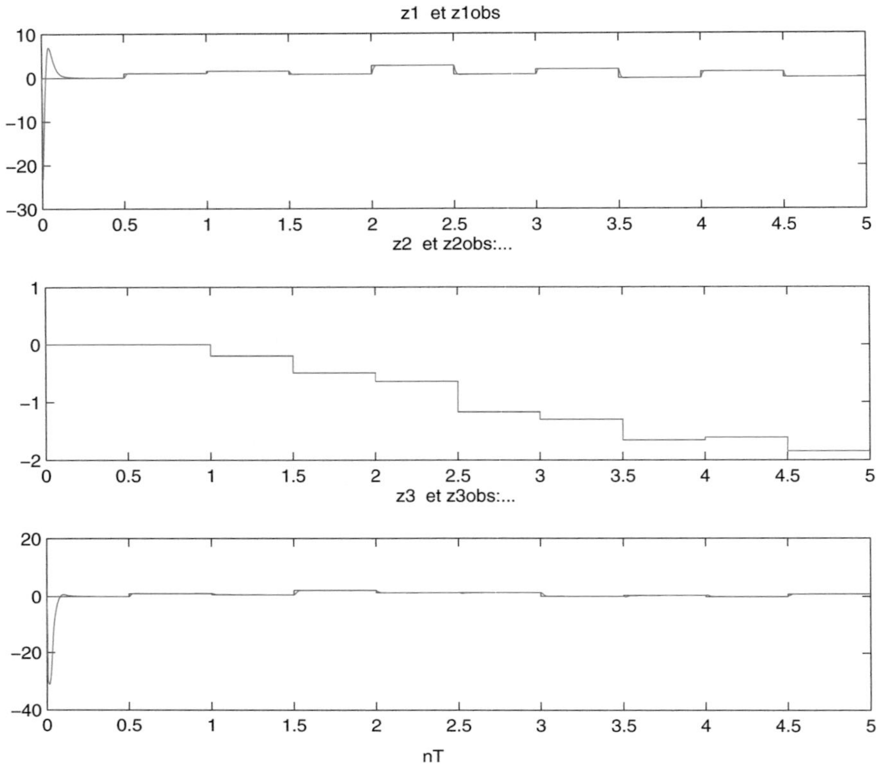

Fig. 5.5 Real and observed discrete states

$$F(x, Z(k)) = \begin{pmatrix} x_2 + z_1(k) \\ -10x_2 - 5\sin(0.5x_3) - x_1^2 z_1(k) \\ x_4 + z_1(k) \\ -x_3^2 + 5x_1 + x_1 z_2(k) \end{pmatrix}$$

and

$$A_d = \begin{pmatrix} 0.50 & 0 & 1 \\ -0.2 & 0.1 & 0 \\ -0.20 & 0 & -0.1 \end{pmatrix}, \quad \beta_d = \begin{pmatrix} y_1(k) \\ 0 \\ y_2(k) \end{pmatrix}$$

Using the following diffeomorphism $\zeta = \Phi_Z(x)$ as $\zeta_1 = x_1$; $\zeta_2 = x_2 + z_1(k)$; $\zeta_3 = x_3$ and $\zeta_4 = x_4 + z_1(k)$, one obtains (Fig. 5.5)

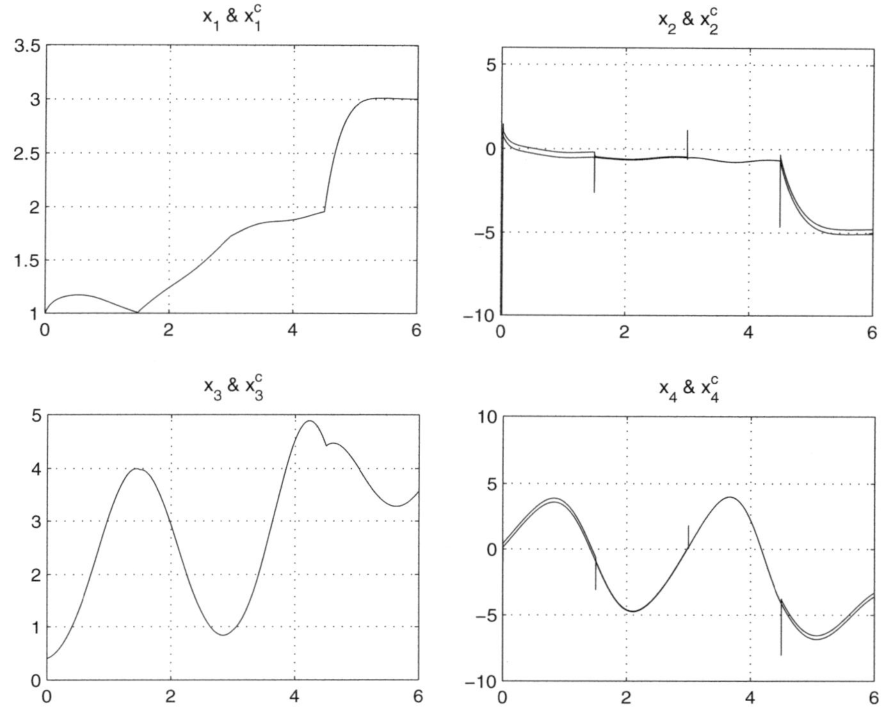

Fig. 5.6 Real x and observed $\hat{x}$ continuous states

$$
\begin{aligned}
\dot{\zeta}_1 &= \zeta_2 \\
\dot{\zeta}_2 &= -10\zeta_2 - 5\sin(0.5\zeta_3) - \zeta_1^2 z_1(k) + 10z_1(k) \\
\dot{\zeta}_3 &= \zeta_4; \\
\dot{\zeta}_4 &= -\zeta_3^2 + 5\zeta_1 + \zeta_1 z_2(k) - z_1(k) \\
Z(k+1) &= A_d Z(k) + \beta_d(y(k)) \\
y_1 &= \zeta_1; \quad \text{and} \quad y_2 = \zeta_3
\end{aligned}
$$

with $Z(t) = Z(k)$ for all $t \in [Tk, T(k+1)[$. To simplify the study, we consider a constant sampling period $T = t_{k+1} - t_k = 0.5s$. The associated continuous observer is then given as in (5.22):

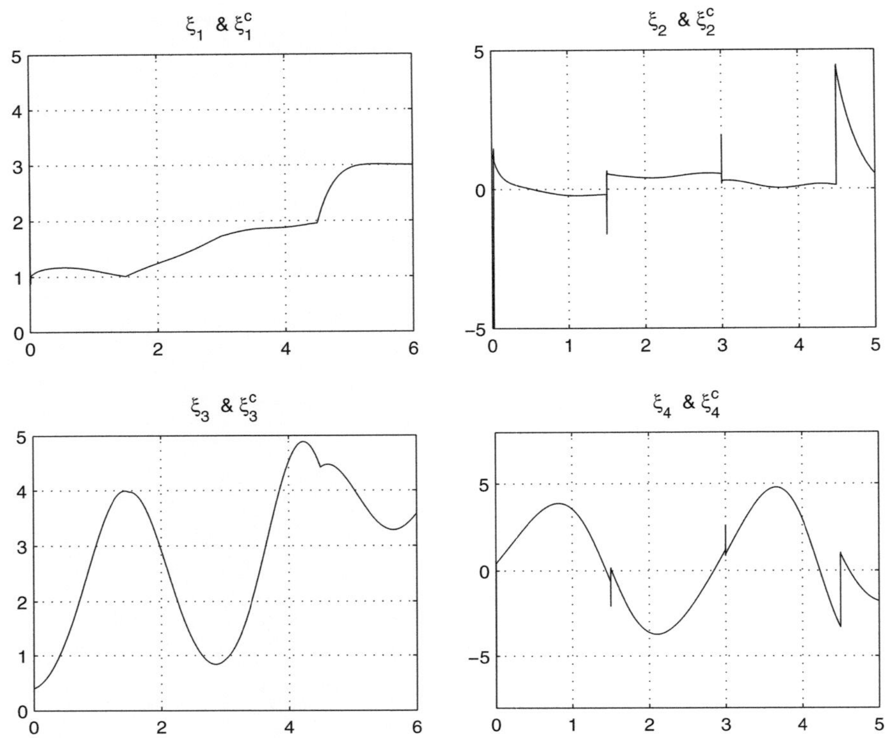

Fig. 5.7 Real ξ and observed $\hat{\xi}$ continuous states

$$
\begin{aligned}
\dot{\hat{\zeta}}_1 &= \tilde{\zeta}_2 + \lambda_1 \left|\tilde{e}_1\right|^{1/2} \operatorname{sign}(\tilde{e}_1) \\
\dot{\tilde{\zeta}}_2 &= \alpha_1 \operatorname{sign}(e_1) \\
\dot{\hat{\zeta}}_2 &= E_1 \left[\tilde{\theta}_1 + \lambda_2 \left|\tilde{e}_2\right|^{1/2} \operatorname{sign}(\tilde{e}_2)\right] \\
\dot{\tilde{\theta}}_1 &= E_1 \alpha_2 \operatorname{sign}(\tilde{e}_2) \\
\dot{\hat{\zeta}}_3 &= \tilde{\zeta}_4 + \lambda_3 \left|\tilde{e}_3\right|^{1/2} \operatorname{sign}(\tilde{e}_3) \\
\dot{\tilde{\zeta}}_4 &= \alpha_3 \operatorname{sign}(e_3) \\
\dot{\hat{\zeta}}_4 &= E_2 \left[\tilde{\theta}_2 + \lambda_4 \left|\tilde{e}_3\right|^{1/2} \operatorname{sign}(\tilde{e}_3)\right] \\
\dot{\tilde{\theta}}_2 &= E_2 \alpha_4 \operatorname{sign}(\tilde{e}_3)
\end{aligned}
$$

where $E_1 = 0$ if $\tilde{e}_1 = \zeta_1 - \hat{\zeta}_1 \neq 0$, else $E_1 = 1$ and $E_2 = 0$ if $\tilde{e}_3 = \zeta_3 - \hat{\zeta}_3 \neq 0$, else $E_2 = 1$.

After a finite time, one obtains $\hat{\zeta}$ and then $Y_d = \Lambda(\zeta)Z(k)$ in the following way:

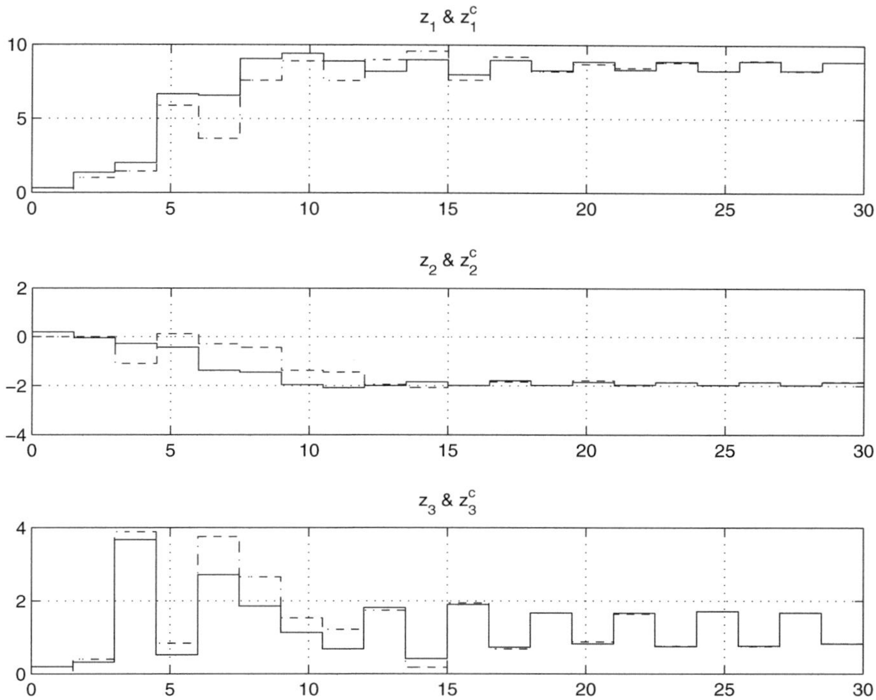

Fig. 5.8 Real and observed discrete states

$$
\begin{aligned}
-\zeta_1^2(k)\tilde{z}_1(k) &= (\tilde{\theta}_1(k) - 10\zeta_2(k) - 0.5\sin(\zeta_3(k)))\\
\zeta_1(k)\tilde{z}_2(k) &= (\tilde{\theta}_2(k) + \zeta_3^2(k) - 5\zeta_1(k))
\end{aligned}
$$

and thus, if the output $Y_d(k)$ is considered, one ca use a pseudo inverse or left inversion to obtain $\hat{Z}(k)$.

The reinitialization is realized as follow:

$$
\begin{aligned}
\zeta(t) &= \Phi_{z(k-1)}(x(t)) \quad \forall\, t \in [t_{k-1}, t_k[\\
&= I_d.x(t) + R.z_1(k-1)\\
\zeta(t) &= \Phi_{z(k)}(x(t)) \quad \forall\, t \in [t_k, t_{k+1}[\\
&= I_d.x(t) + R.z_1(k-1)
\end{aligned}
$$

where I_d is the (4×4) identity matrix and $R = (0\ 1\ 0\ 1)^T$.

One obtains then using (5.24):

$$
\hat{\zeta}^+(t_k) = \Phi_{\hat{z}(k)}\left[\Phi^{-1}_{\hat{z}(k-1)}(\hat{\zeta}(t_k^-))\right]
$$

which gives

$$\hat{\zeta}_1^+(t_k) = \hat{\zeta}_1(t_k^-); \quad \hat{\zeta}_2^+(t_k) = \hat{\zeta}_2(t_k^-) - z_1(k-1) + z_1(k)$$
$$\hat{\zeta}_3^+(t_k) = \hat{\zeta}_3(t_k^-); \quad \hat{\zeta}_4^+(t_k) = \hat{\zeta}_4(t_k^-) - z_1(k-1) + z_1(k)$$

Now, $\hat{Z}(k-1)$ is given by Eq. (5.28) and to obtain $\hat{Z}(k)$, one uses (5.29) to realize the predictor with $N = 1$:

$$\hat{Z}_k = A_d \hat{Z}_{k-1} + \beta_{d,k-1}$$

Finally, using Φ_Z^{-1}, one obtains the estimation of the continuous state in the original coordinates. The simulation results are depicted in Figs. 5.6, 5.7 and 5.8.

5.5 Conclusion

In this chapter, an original result was proposed concerning observability conditions for a particular class of hybrid dynamical systems. The considered class called hybrid in the state, considers hybrid models combining both continuous state variables and discrete time ones. Then, a high-order sliding mode observer was proposed combined with a discrete state reconstructor for the considered LU-observable class of hybrid systems.

References

1. Antsaklis, P.J. (ed.): In: Proceedings of the IEEE Special Issue on Hybrid Systems: Theory and Applications, vol. 43, no. (2) (2000)
2. Babaali, M., Egerstedt, M., Kamen, E.W.: An observer for linear systems with randomly-switching measurement equations. In: Proceedings of the American Control Conference, Denver, Colorado, 4–6 June 2003
3. Barbot, J.-P., Saadaoui, H., Djemaï, M., Manamanni, N.: Nonlinear observer for autonomous switching systems with jumps. In: Proceedings of the Nonlinear Analysis: Hybrid Systems, vol. 1, pp. 537–547 (2007)
4. Barbot, J., Djemai, M., Manamanni, N.: State observer and observability conditions for a class of hybrid continuous-discrete dynamic system. In: Proceedings of the 46th IEEE Conference on Decision and Control (CDC 07), pp. 708–713. IEEE New Orleans, December 2007
5. Baglietto, M., Battistelli, G., Scardovi, L.: Active mode observability of switched linear systems. In: Proceedings of the 45th IEEE-CDC, San Diego, USA, pp. 145–150 (2006)
6. Balluchi, A., Benvenuti, L., Di Benedetto, M.D., Sangiovanni Vincentelliy, A.L.: A Hybrid Observer for the Driveline Dynamics. ECC, Porto (2001)
7. Balluchi, A., Benvenuti, L., Di Benedetto, M.D., Sangiovanni-Vincentelli, A.L.: Observability for hybrid systems, In: Proceedings of the 42nd IEEE-CDC, Maui, Hawaii USA, Dec 2003
8. Barbot, J.P., Floquet, T.: State and unknown input estimation for linear discrete-time systems. In: Proceedings of the IFAC World Congress, Prague (2005)

9. Bemporad, A., Ferrari-Trecate, G., Morari, M.: Observability and controllability of piecewise affine and hybrid systems. IEEE Trans. Autom. Control **45**(10), 1864–1876 (2000)
10. Baglietto, M., Battistelli, G., Scardovi, L.: Active mode observability of switching linear systems. Automatica **43**, 1442–1449 (2007)
11. Di Benedetto, M.D., Di Gennaro, S., D'Innocenzo, A.: Discrete state observability of hybrid systems. Int. J. Robust Nonlinear Control **19**(14), 1564–1580 (2009)
12. De Santis, E., Di Benedetto, M.D., Pola, G.: Observability of internal variables in interconnected switching systems. In: Proceedings of the 45th IEEE-CDC, pp. 4121–4126 (2006)
13. DeCarlo, R., Branicky, M., Pettersson, S., Lennartson, B.: Perspectives and results on the stability and stabilizability of hybrid systems. Proc. IEEE **88**(7), 1069–1082 (2000)
14. De Santis, E., Di Benedetto, M.D.: Observability and observer-based control of hybrid systems. Int. J. Robust Nonlinear Control **19**(14), 1519–1520 (2009)
15. Djemai, M., Barbot, J.P.: singularly perturbed method for the control design of a synchronous motor with its PWM inverter. In: Proceedings of the IEEE-Conference on Control Application, CCA'95, New York, USA (1995)
16. Djemai, M., Manamani, N., Barbot, J.P.: Sliding mode observer for triangular input hybrid system, IFAC World Congres Prague (2005)
17. Fliess, M.: Generalized controller canonical forms for linear and nonlinear dynamics. IEEE Trans. Autom. Control **35**(9), 994–1000 (1990)
18. Fliess, M., Join, C., Mboup, M., Sira-ramirez, H.: Analyse et représentation de signaux transitoires : application a la compression, au débruitage et á la détection de ruptures. In: Proceedings of the IEEE 20éme colloque sur le traitement du signal et de l'image GRETSI, Louvain-la-Neuve, Belgique (2005)
19. Gauthier, J.P., Bornanrd, G.: Observability for any u(t) of a class of nonlinear systems. IEEE Trans. Autom. Control **26**(4), 922–926 (1981)
20. Hermann, R., Krener, A.J.: Nonlinear controllability and observability. IEEE Trans. Autom. Control **22**(9), 728–740 (1977)
21. Hirschorn, R.M.: Invertibilty of multivariable nonlinear control systems. IEEE Trans. Autom. Control **24**, 855–865 (1979)
22. Levant, A.: Robust Exact Differentiation via sliding mode technique. Automatica **34**(3), 379–384 (1998)
23. Menold, P.H., Findeisen, R., Allgöwer, F.: The peaking Finite time convergent observers for linear time-varying systems. In: Proceedings of the 42nd IEEE-CDC Conference. Maui, Hawaii USA, December 2003
24. Mansouri, B., Manamanni, N., Guelton, K., Djemai, M.: Robust pole placement controller design in LMI region for uncertain and disturbed switched systems. Nonlinear Anal.: Hybrid Syst. (Elsevier) **2**(4), 1136–1143 (2008)
25. Pettersson, S., Lennartson, B.: Hybrid system stability and robustness verification using linear matrix inequalities. Int. J. Control **75**(16/17), 1335–1355 (2002)
26. Saadaoui, H., Manamanni, N., Djemaï, M., Barbot, J.P., Floquet, T.: Exact differentiation and sliding mode observer for switched Lagrangian systems. In: Proceedings of the Nonlinear Analysis: Theory, Methods and Appl., Elsevier, vol. 6, pp. 1050–1069 (2005)
27. Vidal, R., Chiuso, A., Soatto, S., Sastry, S.S.: Observability of linear hybrid systems. In: Proceedings of the Hybrid Systems: Comput. and Cont. Lecture Notes in Computer Science, Springer Verlag, vol. 2623, pp. 526–539 (2003)
28. Vas, P.: Sensorless Vector and Direct Torque Control (Monographs in Electrical and Electronic Engineering). Oxford Science Publications, Oxford (1998)

Chapter 6
State Estimation on Switching Systems via High-Order Sliding Modes

Hector Ríos, Jorge Davila and Leonid Fridman

Abstract In this chapter, the problem of continuous and discrete state estimation for switched nonlinear systems is solved using High-Order Sliding-Mode techniques. In the first part of this chapter, the systems with exogenous switchings are studied. The solvability of the observation problem, for the continuous and discrete states, is proposed using structural properties of the system. This structural properties are exploited to design the observers. The high-order sliding-mode techniques are introduced to guarantee finite time convergence to zero of the estimation error for the continuous state. The discrete state is reconstructed using the information of the equivalent output injection. The continuous state observation, discrete state and unknown input reconstruction for switched systems with autonomous switchings are realized in the second part of the chapter. In this part, the structural properties of the systems are exploited to ensure the observability of the continuous and discrete states. Finite-time convergence to zero of the estimation error, for continuous and discrete states, are achieved. The information of the equivalent output injection is used for unknown input reconstruction. Simulation results are provided to illustrate the methods exposed in the chapter.

6.1 Introduction

Conventional dynamic systems are usually divided into two main classes: continuous-time systems, for which the time variable is a subset of the real numbers and the evolution of the state is described by an ordinary differential equation, and discrete-

H. Ríos (✉) · L. Fridman
Division of Electrical Engineering, Department of Control and Robotics,
National Autonomous University of Mexico, 04510 Mexico, D.F., Mexico
e-mail: hectorrios@comunidad.unam.mx

L. Fridman
e-mail: lfridman@unam.mx

J. Davila
Section of Graduate Studies and Research, ESIME-UPT National Polytechnic Institute,
07340 Mexico, D.F., Mexico
e-mail: jadavila@ipn.mx

M. Djemai and M. Defoort (eds.), *Hybrid Dynamical Systems*, Lecture Notes in Control and Information Sciences 457, DOI 10.1007/978-3-319-10795-0_6

time systems whose states are indexed by integer numbers and whose state evolution is typically described by a difference equation.

Hybrid systems comprise a special family of systems where the continuous dynamic is triggered by discrete events. In this sense the hybrid behavior demonstrates an interconnected nature with properties of continuous and discrete-time systems. Therefore, a challenging problem of analysis and design arises.

An important class of hybrid systems is comprised of switching systems. Such systems are dynamical systems consisting of a finite number of subsystems and a logical rule that orchestrates switching or leaps between these subsystems. Mathematically, these subsystems are usually described by a collection of indexed differential or difference equations. One convenient way to classify switched systems is based on the nature of the switching signal. In this sense, the systems under study are divided into systems with exogenous switching signal and systems with autonomous switchings.

The switched systems have been widely studied during the last decades since they can be used to describe a wide range of physical and engineering systems. Most of the attention have been focused on the problems of stability and stabilization with extensive and satisfactory results (see, for example, [5, 11, 23, 24, 33]). The techniques developed for switched system have been applied to solve complex problems, for example, in [9], a hybrid observer is applied to estimate the state of mechanical oscillators.

In the context of the observation problem for switched systems, i.e., the estimation of the continuous and discrete states, the observer design is of great interest for many control areas. This problem has been studied by many authors using different kind of approaches. The main difference is related to the knowledge of the active discrete state or operating mode: some approaches consider only continuous state uncertainty with known operating mode, while others assume that both the operating mode and the continuous state are unknown. In [1] and [7], a Luenberger observer approach and a high-order sliding-mode observer for linear systems are proposed for the known operating mode case. Another work considering the operating mode is known and that the output and input are available is the one presented by [28]. In this paper, the problem of the simultaneous state and input estimation for hybrid systems when subject to input disturbances is addressed by an algorithm based on the moving horizon estimation method. Considering that the continuous state is known, an algorithm for reconstructing the discrete state in nonlinear uncertain switched systems is presented in [27] based on sliding-mode control theory. On the other hand, for the unknown operating mode case, in [6], based on a property of strong detectability and using a LMI approach, are designed two state observers for some classes of switched linear systems with unknown inputs. Considering that the output and an initial state are available, in [32] necessary and sufficient conditions for a switched system to be left invertible are proposed, i.e., condition for recovering the switching signal and the input uniquely. In the same context, a nonlinear finite-time observer to estimate the capacitor voltage for multicellular converters, which have a switched behavior, is proposed by [16]. On graph theoretic approach, assuming only the knowledge of the system structure, in [10], the authors deal with the observability of the discrete state, the internal state and the input of switching structured linear systems with unknown input. In [14], based on the nonhomogeneous high-order sliding-mode approach, a robust observer for the unknown and exogenous switching signal is proposed to solve the problem of continuous and discrete state estimation for a class of nonlinear switched systems.

The problem of observability definitions is also intensively studied in the literature. For instance, [4, 8, 12] analyze the observability of hybrid systems, where the discrete state depends on the state trajectories. For the detectability case, in [15], the detectability of linear switched systems is addressed that reduces to asymptotic stability of a suitable switched system with guards extracted from it, i.e., switching systems whose discrete state is triggered externally.

Unknown input reconstruction techniques are widely applied to solve control problems of systems with disturbances, for instance in the fault detection problem (see, for example, [2, 17]). Therefore for the best of our knowledge, none of the existing works present the unknown input reconstruction problem acting in nonlinear autonomous switched systems. However, there are few works that deal with the observability of nonlinear switched systems with unknown inputs (see, for example, [13, 27]), which aim at designing unknown input observers for some classes of nonlinear switched systems.

Motivated by these works, under assumption that both the continuous and discrete state are unknown, a robust observer based on high-order sliding modes is proposed that is able to estimate the continuous and discrete state for certain classes of switched systems.

6.2 Problem Statement

Consider the following class of switched systems

$$\begin{aligned} \dot{x} &= f_{\lambda(t,x)}(x) + Fw(t,x), \\ y &= h(x), \end{aligned} \tag{6.1}$$

where $x \in \mathscr{X} \subseteq \Re^n$ is the continuous state vector, $\lambda(t,x) \in \mathscr{Q} = \{1, \ldots, q\}$ is the so-called "discrete state." The discrete state determines the current system dynamics among the possible q "operating modes" $f_1(x), f_2(x), \ldots, f_q(x)$, and this could be dependent on time; i.e., $\lambda(t)$ (exogenous switchings), or dependent on continuous state; i.e., $\lambda(x)$ (autonomous switchings). The system output is $y \in \mathscr{Y} \subseteq \Re^p$, and $w \in \mathscr{W} \subseteq \Re^m$ is the unknown input term. It is assumed that the unknown input term is a piecewise continuous function in t and bounded Lipschitz function in x, i.e., $\|w(t,x_0) - w(t,x_1)\| < M_0(x_0 - x_1),\ \forall t \geq 0$. The vector fields $f_{\lambda(t,x)}(x) : \mathscr{X}_{\lambda(t,x)} \to \Re^n$, are considered smooth, $F \in \Re^{n \times 1}$ is a constant column matrix, and $h(x) : \mathscr{X} \to \mathscr{Y}$ is a smooth function defined on an open set of $\Re^n$.

The goal is to design a finite-time converging observer capable of estimating the continuous and discrete state in spite of the unknown inputs and to establish a method for identifying the unknown input to the system (6.1) using the equivalent output injection.

6.3 Preliminaries

6.3.1 Notation

The following notation is used. The pseudoinverse matrix of $F \in \Re^{n\times m}$ is defined as $F^{+} \in \Re^{m\times n}$. Then, if $\text{rank}(F) = n \rightarrow FF^{+} = I$, and if $\text{rank}(F) = m \rightarrow F^{+}F = I$. With reference to a scalar function h with vector argument x defined in an open set $\Omega \in \Re^n$ such that $h(x) : \Re^n \rightarrow \Re$, denote $dh(x) = \frac{\partial h(x)}{\partial x} = \left[\frac{\partial h(x)}{\partial x_1} \cdots \frac{\partial h(x)}{\partial x_n} \right]$. Denote $\mathscr{U}$ as a subset of a topological space, $P(\mathscr{U})$ denotes the set of all subsets of $\mathscr{U}$, and $T\mathscr{U}$ denotes the tangent bundle of $\mathscr{U}$.

6.3.2 Definitions

Now, some basic definitions for hybrid automaton, hybrid trajectory, execution, and dwell time are recalled in this section.

Definition 6.1 [25] A hybrid automaton $\mathscr{H}$ is a collection $\mathscr{H} = (Q, X, f, \text{Init}, D, E, G, R)$, where Q is the finite set of discrete variables; X is the finite set of continuous variables; $f : \mathscr{Q} \times \mathscr{X} \rightarrow T\mathscr{X}$ is a vector field; $\text{Init} \subseteq \mathscr{Q} \times \mathscr{X}$ is the set of initial states; $D : \mathscr{Q} \rightarrow P(\mathscr{X})$ is a domain; $E \subseteq \mathscr{Q} \times \mathscr{Q}$ is the set of edges; $G : E \rightarrow P(\mathscr{X})$ is the guard condition, and $R : E \times \mathscr{X} \rightarrow P(\mathscr{X})$ is the reset map.

Definition 6.2 [25] A hybrid time trajectory is a finite or infinite sequence of intervals $\tau = \{I\}_{i=0}^{N}$ such that:

- $I_i = [\tau_1, \tau_i']$, for all $i < N$;
- if $N < \infty$, then either $I_N = [\tau_N, \tau_N']$, or $I_N = [\tau_N, \tau_N')$;
- $\tau_i \leq \tau_i' = \tau_{i+1}$ for all i.

In other words, a hybrid time trajectory is a sequence of intervals of the real line, whose endpoints overlaps. For a hybrid time trajectory $\tau = \{I_i\}_{i=0}^{N}$, define $\langle \tau \rangle$ as the set $\{1, \ldots, N\}$ if N is finite, and $\{1, 2, \ldots\}$ if $N = \infty$ and $|\tau| = \sum_{i\in\langle\tau\rangle}(\tau_i' - \tau_i)$.

Definition 6.3 [23] The minimal dwell time is a number $T_\delta > 0$ such that the class of admissible switching signals satisfy the property that the switching times $t_1, t_2, \ldots$ fulfill the inequality $t_{j+1} - t_j \geq T_\delta$ for all j.

In this chapter, we study the systems whose hybrid time trajectories satisfies the minimal dwell time definition. Moreover, it is assumed that the dwell time is sufficiently large or it is possible to estimate it (see [30] for the estimation of the switching times for linear switched systems).

Definition 6.4 [25] An execution of a hybrid automaton $\mathscr{H}$ is a collection $\xi = (\tau, q, x)$, where τ is a hybrid time trajectory, $q : \langle\tau\rangle \to Q$ is a map, and $x = \{x^i : i \in \langle\tau\rangle\}$ is a collection of differentiable maps $x^i : I_i \to \mathscr{X}$ such that

- $(q(0), x^0(0)) \in \mathrm{Init}$;
- for all $t \in [\tau_i, \tau_i')$, $\dot{x}^i(t) = f_{q(i)}(x^i(t))$ and $x^i(t) \in \mathscr{X}$;
- for all $i \in \langle\tau\rangle \backslash \{N\}$, $e = (q(i), q(i+1)) \in E$, $x^i(\tau_i') \in G(e)$, and $x^{i+1}(\tau_{i+1}) \in R(e, x^i(\tau_i'))$.

The execution of a hybrid automaton is a similar concept to the solution of a continuous dynamic systems. Notice that for any infinite execution of the system (6.1), it is necessary that $|\tau| = \infty$. Zeno executions are not allowed. The zeno phenomena can be described by an infinite execution with $|\tau| < \infty$.

The following definition of observability for a Hybrid Automaton is adapted from [20].

Definition 6.5 Consider the system (6.1) and the variable $x = x(t, x)$. Let $x(t, x^1)$ be a trajectory of the automation $\mathscr{H}$ with a hybrid time trajectory T_N and $\langle T_N\rangle$. Suppose for any trajectory (t, x^2) of $\mathscr{H}$ with the same T_N and $\langle T_N\rangle$, the equality $y(t, x^1) = y(t, x^2)$, a.e. in $[t_{\text{ini}}, t_{\text{end}}]$ implies $x(t, x^1) = x(t, x^2)$, a.e. in $[t_{\text{ini}}, t_{\text{end}}]$, then $x = x(t, x)$ is $Z(T_N)$ observable along the trajectory $x(t, x^1)$.

In all this chapter, it is considered that system (6.1) is $Z(T_N)$ observable along any trajectory $x(t, x)$ and for any possible hybrid time trajectory T_N.

In the following three particular problems, derived from the main problem, are described: state observation for the exogenous switchings case, state reconstruction for autonomous switchings case and state estimation for systems with autonomous switchings with unknown inputs.

6.4 State Estimation for Systems with Exogenous Switchings

Consider that the discrete state is unknown and dependent on time, i.e., $\lambda(t)$ (controlled switchings), and there do not exist unknown inputs, i.e.,

$$\begin{aligned} \dot{x} &= f_{\lambda(t)}(x), \\ y &= h(x). \end{aligned} \tag{6.2}$$

The vector fields $f_{\lambda(t)}(x)$ changes abruptly in response to a control command, in this case, an unknown exogenous switching signal $\lambda(t)$. The aim is to design a finite-time converging observer for both the continuous and discrete state by means of the knowledge of the output. The single output case is treated in this section, i.e., $y \in \mathscr{Y} \subseteq \Re$. However, for a multiple output case description, please refer to [14].

6.4.1 Assumptions

The following assumptions over system (6.2) are considered.

Assumption 6.6 Suppose that every vector field $f_{\lambda(t)} : \mathscr{X}_{\lambda(t)} \to T_{\lambda(t)}\mathscr{X}_{\lambda(t)}$ can be extended to each subspace $\mathscr{X}_1, \ldots, \mathscr{X}_q$. Then, there exist smooth vector fields $g(x)$ such that

$$g_j : \mathscr{X}_j \to T_j\mathscr{X}_j;\ g_j = f_1,\ \forall x \in \mathscr{X}_j,\ \forall j = 2, 3, \ldots, q, \tag{6.3}$$

$$g_k : \mathscr{X}_k \to T_k\mathscr{X}_k;\ g_k = f_2,\ \forall x \in \mathscr{X}_k,\ \forall k = 1, 3, \ldots, q, \tag{6.4}$$

$$\vdots$$

$$g_l : \mathscr{X}_l \to T_l\mathscr{X}_l;\ g_l = f_q,\ \forall x \in \mathscr{X}_l,\ \forall l = 1, 2, \ldots, q-1. \tag{6.5}$$

Notice that the previous assumption implies that all vector fields $g(x)$ are extensions of every vector field $f_{\lambda(t)}(x)$ with the smoothness property satisfied for all the domains $D_{\lambda(t)} = \left\{x \mid x \in \mathscr{X}_{\lambda(t)}\right\}$.

Let us define the following mappings:

$$\Phi_{\lambda(t)}(x) = \begin{bmatrix} h(x) \\ L_{f_{\lambda(t)}(x)}h(x) \\ \vdots \\ L^{n-1}_{f_{\lambda(t)}(x)}h(x) \end{bmatrix},\ \forall \lambda = 1, \ldots, q. \tag{6.6}$$

Their corresponding Jacobian matrices (observability mappings) are given by

$$\frac{\partial \Phi_{\lambda(t)}(x)}{\partial x} = \begin{bmatrix} dh(x) \\ dL_{f_{\lambda(t)}(x)}h(x) \\ \vdots \\ dL^{n-1}_{f_{\lambda(t)}(x)}h(x) \end{bmatrix},\ \forall \lambda = 1, \ldots, q. \tag{6.7}$$

Assumption 6.7 Each matrix (6.7) is such that

$$\operatorname{rank}\left(\frac{\partial \Phi_1(x)}{\partial x}\right) = n,\ \forall x \in \mathscr{X}_1,$$

$$\operatorname{rank}\left(\frac{\partial \Phi_2(x)}{\partial x}\right) = n,\ \forall x \in \mathscr{X}_2,$$

$$\vdots$$

$$\operatorname{rank}\left(\frac{\partial \Phi_q(x)}{\partial x}\right) = n,\ \forall x \in \mathscr{X}_q.$$

Moreover, each mapping $\Phi_{\lambda(t)}(x)$ is a local diffeomorphism at least on its corresponding domain $\mathscr{X}_{\lambda(t)}$, $\forall\lambda = 1, \ldots, q$ (see, for example [19]).

6.4.2 Observer Design

Consider the following stack of q state observers

$$\begin{aligned} \dot{\bar{x}}_j &= f_j(\bar{x}_j) + \left(\frac{\partial \Phi_j(\bar{x}_j)}{\partial \bar{x}_j}\right)^{-1} \nu_j, \quad \forall j = 1, \ldots, q, \\ \bar{y}_j &= h(\bar{x}_j), \end{aligned} \tag{6.8}$$

with the estimated state vectors $\bar{x}_j \in \Re^n$, the estimated outputs $\bar{y}_j \in \Re$, and the correction terms $\nu_j \in \Re^n$, that will be designed further on. The solutions of (6.8) are understood in the Filippov sense [18] to provide the possibility to use discontinuous signals in the observer and coincide with the usual solutions, when the right-hand sides are continuous. It is assumed also that all considered correction terms allow the existence and extension of solutions to the whole semi-axis $t \geq 0$.

Remark 6.8 Notice that if the q observability mappings are equal it will be only necessary to design one observer. If $q - 1$ observability mappings are equal, it will be only necessary to design two observers, and so successively.

Remark 6.9 Notice that, due to Assumption 6.7, the q subsystems must be observable only in the corresponding domain $D_{\lambda(t)}$. However, outside their corresponding domains, according to Assumption 6.6, the observers (6.8) are well defined.

Let the following assumption be satisfied:

Assumption 6.10 There are known constants $M_\lambda > 0$, $\forall\lambda = 1, \ldots, q$, such that the following inequalities are satisfied:

$$\left| L^n_{f_1(\bar{x}_1)} h(\bar{x}_1) - L^n_{f_1(x)} h(x) \right| < M_1, \quad \forall x \in \mathscr{X}_1, \tag{6.9}$$

$$\left| L^n_{f_2(\bar{x}_2)} h(\bar{x}_2) - L^n_{f_2(x)} h(x) \right| < M_2, \quad \forall x \in \mathscr{X}_2, \tag{6.10}$$

$$\vdots$$

$$\left| L^n_{f_q(\bar{x}_q)} h(\bar{x}_q) - L^n_{f_q(x)} h(x) \right| < M_q, \quad \forall x \in \mathscr{X}_q. \tag{6.11}$$

Notice that so the above assumption is satisfied, it is necessary, not sufficiently, that the system is bounded-input bounded-state (BIBS) (see, for example, [21]) or that it is possible to stabilize it somehow. Nevertheless, notice that only it is sufficiently that there exist bounds M_λ in the corresponding domains $D_{i(x)}$ and not outside.

The high-order sliding-mode differentiator [22] is used as an auxiliary dynamics in order to design the correction terms. The differentiator has the following form

$$\begin{aligned}
\dot{\vartheta}_{j,1} &= \vartheta_{j,2} - \alpha_{j,1} M_j^{\frac{1}{n}} \left|e_{y_j}\right|^{\frac{n-1}{n}} \operatorname{sign}\left(e_{y_j}\right), \\
\dot{\vartheta}_{j,2} &= \vartheta_{j,3} - \alpha_{j,2} M_j^{\frac{1}{n-1}} \left|\vartheta_{j,2} - \dot{\vartheta}_{j,1}\right|^{\frac{n-2}{n-1}} \operatorname{sign}\left(\vartheta_{j,2} - \dot{\vartheta}_{j,1}\right), \\
&\vdots \\
\dot{\vartheta}_{j,n} &= -\alpha_{j,n} M_j \operatorname{sign}\left(\vartheta_{j,n} - \dot{\vartheta}_{j,n-1}\right),
\end{aligned} \tag{6.12}$$

where $e_{y_\lambda} = \bar{y}_j - y, \forall j = 1, \ldots, q$; are the output errors. The correction terms are taken from (6.12) as:

$$\nu_j = \begin{bmatrix} -\alpha_{j,1} M_j^{\frac{1}{n}} \left|e_{y_j}\right|^{\frac{n-1}{n}} \operatorname{sign}\left(e_{y_j}\right) \\ -\alpha_{j,2} M_j^{\frac{1}{n-1}} \left|\vartheta_{j,2} - \dot{\vartheta}_{j,1}\right|^{\frac{n-2}{n-1}} \operatorname{sign}\left(\vartheta_{j,2} - \dot{\vartheta}_{j,1}\right) \\ \vdots \\ -\alpha_{j,n} M_j \operatorname{sign}\left(\vartheta_{j,n} - \dot{\vartheta}_{j,n-1}\right) \end{bmatrix}. \tag{6.13}$$

The constants $\alpha_{j,k}$ are chosen recursively and sufficiently large. In particular, according to [22], one possible choice is $\alpha_{j,6} = 1.1$, $\alpha_{j,5} = 1.5$, $\alpha_{j,4} = 2$, $\alpha_{j,3} = 3$, $\alpha_{j,2} = 5$, $\alpha_{j,1} = 8$, $\forall j = 1, \ldots, q$, which is enough for the case that $n \leq 6$.

6.4.3 Continuous State Estimation Between Switchings

Let us describe the continuous state estimation during the time interval between switchings. Consider that $\lambda(t) \equiv \lambda^* = \text{const.}$, $\forall t \in [0, t_1)$ with the first switching time $t_1 \geq T_\delta$. The system dynamics on the operating mode j^* is given by

$$\dot{x} = f_{\lambda^*}(x), \quad \forall t \in [0, t_1). \tag{6.14}$$

Due to the Assumption 6.7, only one of the q observers (6.8) can be associated with the corresponding output error $e_{y_j} = \bar{y}_j - y$ and state estimation error $e_{x_j} = \bar{x}_j - x$, in the time interval between switchings.

Taking into account the previous explanations, the following theorem can be established.

Theorem 6.11 *Consider that the* λ^*th *observer* (6.8) *with the correction terms designed according to* (6.13) *is applied to system* (6.14)*, and let Assumption 6.6 and 6.7 be satisfied. Then, provided that constants* $\alpha_{\lambda^*,k}$ *are chosen properly and each* M_{j^*} *is selected satisfying Assumption 6.10, the state estimation error converges to zero in finite time, i.e.,* $e_{x_{\lambda^*}} = \bar{x}_{\lambda^*} - x = 0 \; \forall t \in [t_{\lambda^*}, t_1)$.

Proof See Appendix.

Therefore, when the active dynamics is the i^*th, then the observation error $e_{x_{\lambda^*}}$ of the λ^*th observer tends to zero in finite time.

Remark 6.12 Notice that to design the observers only it is necessary the calculation of the inverse matrices $\left(\frac{\partial \Phi_j(\bar{x}_j)}{\partial \bar{x}_j}\right)^{-1}$, not the inverse transformations $\Phi_j^{-1}(\bar{z}_j)$.

A further result is given consisting of an additional assumption that involve all the possible system dynamics $f_1(x), f_2(x), \ldots, f_q(x)$, guaranteeing that all observers provide the correct estimate of the continuous state irrespectively of the current value of the operating mode.

Assumption 6.13 Let the functions $f_\iota(x)$, $f_\kappa(x)$ and $h(x)$ be such that for each $\iota, \kappa = 1, 2, \ldots, q$, $\iota \neq \kappa$,

$$\begin{bmatrix} L_{\bar{f}_{\iota,\kappa}(x)} h(x) \\ \vdots \\ L_{\bar{f}_{\iota,\kappa}(x)} L_{f_\iota(x)}^{n-2} h(x) \end{bmatrix} = \begin{bmatrix} 0 \\ \vdots \\ 0 \end{bmatrix}, \quad \bar{f}_{\iota,k} = f_\iota - f_\kappa. \tag{6.15}$$

Therefore, if Assumption 6.13 is satisfied, then all observers provide the correct estimate of the continuous state irrespectively of the current value of the operating mode. Notice that the proposed method does not require that the switching parameter satisfies the matching condition (see, for example, [33]). This improvement is a result of the multiobserver approach.

It is important to remark that after each operating mode switches, the observers could loose the correct estimation (due to the discontinuities in the higher order output derivatives) if Assumption 6.13 is not satisfied. However, after a transient, which can be made arbitrarily small by taking sufficiently large values of the correction terms parameters, the correct value of the state is recovered.

Notice that the appropriate selection of the differentiator's gains ensure the convergence of the state estimation error $e_{x_{\lambda^*}}$ to zero in a time smaller than T_δ. It means that, under the presence of jumps in the continuous state, the proposed algorithm ensures the estimation of x after a finite-time transient smaller than T_δ.

The following step to solve the proposed observability problem is to establish a method for reconstructing the operating mode to complete the observer design.

6.4.4 Discrete State Estimation

In this section, the method for reconstructing the switching signal is outlined. In steady state, all entries of vectors e_{x_j} are identically zero, while the terms $\dot{e}_{x_j}$, $i = 1, \ldots, p$, are directly affected by the discontinuous correction terms, i.e., are zero in the "average" sense. Thus, we are in position to exploit one of the main features of sliding-mode observers, the equivalent output injection principle. The expression for $\dot{e}_{x_j}$ is

$$\dot{e}_{x_j} = f_j(\bar{x}_j) + \left(\frac{\partial \Phi_j(\bar{x}_j)}{\partial \bar{x}_j}\right)^{-1} \nu_j - f_{\lambda^*}(x). \tag{6.16}$$

Starting from the moment at which the exact state reconstruction is achieved, (6.16) simplifies as follows

$$\dot{e}_{x_j} = f_j(\bar{x}_j) + \left(\frac{\partial \Phi_j(\bar{x}_j)}{\partial \bar{x}_j}\right)^{-1} \nu_j - f_{\lambda^*}(x) = 0. \tag{6.17}$$

Then, the corrections terms ν_j, will take the value of the equivalent output injection $\nu_{j_{eq}}$, i.e.,

$$\left(\frac{\partial \Phi_j(\bar{x}_j)}{\partial \bar{x}_j}\right)^{-1} \nu_{j_{eq}} = f_{\lambda^*}(x) - f_j(\bar{x}_j), \tag{6.18}$$

which derives from imposing the zeroing of $\dot{e}_{x_j} = 0$ (equivalent control method; see [31]). The above equation implies that among the q observers (6.8), there is only one of them with all the associated equivalent output injections being identically zero according to the following condition of reconstructability of the switching signal:

Condition 6.14 *The discrete state $\lambda(t)$ can be reconstructed by means of equivalent output injection according to*

$$\left(\frac{\partial \Phi_j(\bar{x}_j)}{\partial \bar{x}_j}\right)^{-1} \nu_{j_{eq}} \equiv 0, \quad j = \lambda^*, \tag{6.19}$$

$$\left(\frac{\partial \Phi_j(\bar{x}_j)}{\partial \bar{x}_j}\right)^{-1} \nu_{j_{eq}} \neq 0, \quad \forall j \neq \lambda^*, \tag{6.20}$$

provided that the following set:

$$\mathscr{M} = \left\{x \in \Re^n \,\middle|\, f_{\lambda^*}(x) = f_j(x), \quad \forall \lambda^* \neq j\right\}, \tag{6.21}$$

is a discrete set.

Since ν_j has discontinuous terms, the equivalence $\nu_j = \nu_{j_{eq}}$, holds only in the Filippov sense, so that the recovery of the equivalent output injection $\nu_{j_{eq}}$ from the discontinuous output injection ν_j requires *filtration*. Let us define the following equivalent output injection estimators of $\nu_{j_{eq}}$:

$$\tau_j \dot{\hat{\nu}}_{j_{eq}} = \nu_j - \hat{\nu}_{j_{eq}}, \quad \forall j = 1, \ldots, q, \tag{6.22}$$

where τ_j are designed according to [31], i.e., $\delta \ll \tau_j \ll 1$ with $\tau_j = \delta^{1/2}$, where δ is the sample time. The continuous signals $\hat{\nu}_{j_{eq}}$ must be analyzed in order to extract the information about the current value of the switching signal. Theoretically, a simple threshold would be enough. Indeed it was shown that one and only one of the signals $\hat{\nu}_{j_{eq}}$ becomes identically zero and stays in this value until $\lambda(t)$ changes value. However, all signals $\hat{\nu}_{j_{eq}}$ can occasionally cross the zero value. Therefore, a logic should be implemented that looks for the signal being "closer" to zero over a suitable receding-horizon time interval of finite length. This can be done easily via the numerical method described below.

Let T_s be a small sampling time. The following non-negative quantities are evaluated online at any "sampling instants" $t = kT_s$, $k = 0, 1, 2, \ldots$

$$\mu_j = \textstyle\sum_{i=0}^{\nu} \left|\hat{v}_{j_{eq}}(t - kT_s)\right|, \forall j = 1, 2, \ldots, q, \tag{6.23}$$

The value of j for which μ_j is minimum is evaluated, and this value will be the discrete state estimation $\hat{\lambda}(t)$, as follows

$$\hat{\lambda}(t) = \arg\min_j \mu_j. \tag{6.24}$$

In this way, using the information of the estimated discrete state, the real estimated state $\hat{x}$ is defined as follows:

$$\hat{x}(t) = \bar{x}_{\hat{\lambda}(t)}(t), \quad \forall t > t_{\lambda^*}. \tag{6.25}$$

6.5 State Estimation for Systems with Autonomous Switchings

Consider that the discrete state is unknown and dependent on continuous state, i.e., $\lambda(x)$ (autonomous switchings)

$$\begin{aligned} \dot{x} &= f_{\lambda(x)}(x), \\ y &= h(x), \end{aligned} \tag{6.26}$$

$$\lambda(x) = \begin{cases} 1, & \forall x \mid C(x) \in \mathscr{C}_1, \\ 2, & \forall x \mid C(x) \in \mathscr{C}_2, \\ \vdots & \\ q, & \forall x \mid C(x) \in \mathscr{C}_q, \end{cases} \tag{6.27}$$

where $x \in \mathscr{X} \subseteq \Re^n$ is the continuous state vector, the unknown discrete state $\lambda(x) \in \mathscr{Q} = \{1, \ldots, q\}$ is determined by a smooth continuous scalar function $C(x) : \mathscr{X} \to \mathscr{C} \subseteq \Re$. This function determines which of the q dynamics is acting on the system and the manifolds $\mathscr{C}_i$ are known. The vector fields $f_{\lambda(x)}(x) : \mathscr{X}_{\lambda(x)} \to \Re^n$, are considered smooth, and $h(x) : \mathscr{X} \to \mathscr{Y}$ is a smooth function defined on an open set of $\Re^n$. The single output case is treated in this section. However, the multiple output case is shown in [29].

6.5.1 Assumptions

For the autonomous switching case Assumptions 6.6, 6.7 and 6.10 are also required.

6.5.2 Observer Design

Consider for the autonomous case, the stack of q state observers described in Sect. 6.4.2. The results presented in Sect. 6.4.3, for continuous state estimation, are also applicable for this case under Assumption 6.10 (see, for details [29]).

6.5.3 Discrete State Estimation

The discrete state estimation method described in Sect. 6.4.4 is also applicable for this case. Notice that the proposed method makes use of the equivalent out injection to estimate the discrete state and to know which observer estimate the continuous state correctly. However, due to the structure of the autonomous system (6.26)–(6.27), it is possible to use the output and continuous state information of each observer (6.8) to estimate the discrete state.

Theorem 6.11 establishes that only one of the q observers, if every observability mappings are different, estimates the continuous state correctly but it does not establish how to know which of them makes it. Nevertheless, according to [3], to detect which of the error dynamics converge to zero it is sufficient to verify that the following inequality is satisfied

$$\left|e_{y_j}(t)\right| \leq \gamma_j M_j \delta^n, \quad \forall j = 1, \ldots, q, \quad \forall t \in \left[0, \gamma_{j_t}\delta\right], \tag{6.28}$$

where γ_j and γ_{j_t} are positive constants and δ is the sample time. It is natural to estimate the constants γ_j and γ_{j_t} through simulation. Thus, in this way it is possible to determine when and which of the q observers has converged to the correct continuous state during the time interval $t \in [0, t_1)$.

Remark 6.15 Notice that the inequality (6.28) is used for implementation since due to the presence of the sample time it is impossible to reach the equality $\left|e_{y_j}(t)\right| \equiv 0$ as it is established in the proof of Theorem 6.11.

Once the continuous state is estimated correctly, it is possible to estimate the discrete state with the following discrete state observer:

$$\hat{\lambda}(\hat{x}) = \begin{cases} 1, & \forall \hat{x} \mid C(\hat{x}) \in \mathscr{C}_1, \\ 2, & \forall \hat{x} \mid C(\hat{x}) \in \mathscr{C}_2, \\ \vdots & \\ q, & \forall \hat{x} \mid C(x) \in \mathscr{C}_q, \end{cases} \tag{6.29}$$

where the manifolds $\mathscr{C}_i$ are known. Then, the discrete state $\lambda(x)$ is estimated by means of $\hat{\lambda}(\hat{x})$ since, according to Theorem 6.11, the following equality is satisfied:

$$\hat{x}(t) = x(t), \quad \forall t \in [t_{\lambda^*}, t_1). \tag{6.30}$$

Thus, it is possible to reconstruct the continuous and discrete state before the first switching happens. Now, the following step is to analyze what it happens in the switching times.

6.5.4 Continuous and Discrete State Estimation on Switching Times

Let $t_i{}^+$ be the time instants after the switching times t_i. In order to maintain the state estimation on the switching times the following proposition is done.

Proposition 6.16 *The state estimation of system* (6.26)–(6.27) *is maintained in spite of the switchings if the following reset equations are implemented in the bank of observers* (6.8) *for all* $j \neq \lambda^*$

$$\bar{x}_j(t_i^+) = \hat{x}(t_i^-), \quad \forall i = 1, 2, \ldots. \tag{6.31}$$

Proof See Appendix.

Then, once the discrete state has been identified, it is possible to analyze such signal in order to know when some change takes place. Therefore, the reset equations (6.31) can always be implemented for each observer when the change in the discrete state is detected, i.e., in each switching time.

With the previous explanation, the following identities are obtained:

$$\hat{x}(t) = x(t), \quad \hat{\lambda}(\hat{x}) = \lambda(x), \quad \forall t > t_{\lambda^*}. \tag{6.32}$$

6.6 State Estimation for Systems with Autonomous Switchings and Unknown Inputs

Consider the following autonomous switched system with unknown inputs

$$\begin{aligned} \dot{x} &= f_{\lambda(x)}(x) + Fw(t, x), \\ y &= h(x), \end{aligned} \tag{6.33}$$

$$\lambda(x) = \begin{cases} 1, & \forall x \mid C(x) \in \mathscr{C}_1, \\ 2, & \forall x \mid C(x) \in \mathscr{C}_2, \\ \vdots & \\ q, & \forall x \mid C(x) \in \mathscr{C}_q, \end{cases} \tag{6.34}$$

where $w \in \mathscr{W} \subseteq \Re$ is the unknown input term and the rest of variables was already definite. It is assumed that the unknown input term is a piecewise continuous function in t and bounded Lipschitz function in x, i.e., $|w(t, x_0) - w(t, x_1)| < M_0 (x_0 - x_1), \ \forall t \geq 0$. The multiple unknown input case is shown in [29].

6.6.1 Assumptions

For the autonomous switching case with unknown inputs, Assumption 6.6 and 6.7 are required. For this case, one more assumption is necessary for the observer design, i.e.,

Assumption 6.17 The vector fields $f_{\lambda(x)}$, the matrix F, and the function $h(x)$, are such that $\forall x \in \mathscr{X}_{\lambda(x)}, \forall \lambda(t) = 1, \ldots, q$, it is satisfied

$$\begin{aligned} d(L^k_{f_{\lambda(x)}(x)} h(x))F &= 0, \quad \forall k < n-1, \ \forall \lambda = 1, \ldots, q, \\ d(L^{n-1}_{f_{\lambda(x)}(x)} h(x))F &\neq 0, \quad \forall \lambda = 1, \ldots, q. \end{aligned} \tag{6.35}$$

6.6.2 Observer Design

Consider for the autonomous case with unknown inputs the stack of q state observers described in Sect. 6.4.2. Let the following assumption be satisfied:

Assumption 6.18 There are known constants $M_\lambda > 0, \quad \forall \lambda = 1, \ldots, q$, such that the following inequalities are satisfied:

$$\left| L^n_{f_1(\bar{x}_1)} h(\bar{x}_1) - L^n_{f_1(x)} h(x) - d(L^{n-1}_{f_1(x)} h(x)) F w \right| < M_1, \quad \forall x \in \mathscr{X}_1, \tag{6.36}$$

$$\left| L^n_{f_2(\bar{x}_2)} h(\bar{x}_2) - L^n_{f_2(x)} h(x) - d(L^{n-1}_{f_2(x)} h(x)) F w \right| < M_2, \quad \forall x \in \mathscr{X}_2, \tag{6.37}$$

$$\vdots$$

$$\left| L^n_{f_q(\bar{x}_q)} h(\bar{x}_q) - L^n_{f_q(x)} h(x) - d(L^{n-1}_{f_q(x)} h(x)) F w \right| < M_q, \quad \forall x \in \mathscr{X}_q. \tag{6.38}$$

The results presented in Sect. 6.4.3, for continuous state estimation, are also applicable for this case under Assumption 6.6, 6.7 and 6.18 (see, for details [29]).

6.6.3 Discrete State Estimation

The discrete state estimation method (equivalent output injection) described in Sect. 6.4.4 is not applicable for this case. Notice that the proposed method makes use of the equivalent output injection to estimate the discrete state and to know which observer estimate the continuous state correctly without taking into account unknown inputs. For the autonomous switchings with unknown inputs is not possible to use the equivalent output injection since in such a case this injection is affected for the unknown inputs and is not possible to determine the discrete state from this information. However, due to the structure of the autonomous system (6.33)–(6.34), it is possible to use the output and continuous state information of each observer (6.8) to estimate the discrete state as it was shown in Sect. 6.5.3.

The results presented in Sect. 6.4.3, for continuous state estimation between switchings, are also applicable for this case. Then, the following identities are obtained:

$$\hat{x}(t) = x(t), \quad \hat{\lambda}(\hat{x}) = \lambda(x), \quad \forall t > t_{\lambda^*}. \tag{6.39}$$

In the following section, it is studied the unknown input reconstruction problem.

6.6.4 Unknown Input Reconstruction

In steady state, all entries of state error vector $e_x = \hat{x} - x$, are identically zero. Thus, the expression of $\dot{e}_x$ is

$$\dot{e}_x = f_{\hat{\lambda}(\hat{x})}(\hat{x}) + \left(\frac{\partial \Phi_{\hat{\lambda}(\hat{x})}(\hat{x})}{\partial \hat{x}} \right)^{-1} \nu_{\hat{\lambda}(\hat{x})} - f_{\lambda(x)}(x) - Fw(t,x) = 0. \tag{6.40}$$

Then, the discontinuous corrections terms of $\nu_{\hat{\lambda}(\hat{x})}$ will take the value of the equivalent output injection, i.e.,

$$\left(\frac{\partial \Phi_{\hat{\lambda}(\hat{x})}(\hat{x})}{\partial \hat{x}} \right)^{-1} \nu_{\hat{\lambda}(\hat{x})_{\text{eq}}} = f_{\lambda(x)} - f_{\hat{\lambda}(\hat{x})}(\hat{x}) + Fw(t,x). \tag{6.41}$$

At the moment in which the exact reconstruction of the continuous state is reached, the exact estimation of the discrete state also is reached. In this way, since the continuous and discrete state are perfectly known, it is possible to identify the unknown input term as follows

$$\hat{w}(t,\hat{x}) = F^{+} \left[\left(\frac{\partial \Phi_{\hat{\lambda}(\hat{x})}(\hat{x})}{\partial \hat{x}} \right)^{-1} \nu_{\hat{\lambda}(\hat{x})_{\text{eq}}} \right], \tag{6.42}$$

where $F^+ = \left(F^T F\right)^{-1} F^T$. Therefore, if each $\nu_{\hat{\lambda}(\hat{x})_{\text{eq}}}$ was available, it might extract the necessary information to determine the unknown input on the switched system.

Theoretically, the equivalent output injection is the result of an infinite switching frequency of the discontinuous terms ν_λ. However, the realization of the observer produces finite high switching frequency making necessary the application of a filter.

Thus, in order to estimate the equivalent output injection, estimator (6.22) is used. To maintain the correct reconstruction of the unknown inputs, like in the observation problem, it is necessary to implement the following reset equations on the equivalent output injection estimators (6.22), i.e.,

$$\nu_{\hat{\lambda}(\hat{x})_{\text{eq}}}(t_i^+) = \nu_{\lambda^*_{\text{eq}}}(t_i^-), \quad \forall i = 1, 2, \ldots \tag{6.43}$$

The purpose of the reset equations (6.31) and (6.43) is that switchings have no effect in the continuous state estimation and in the equivalent output injection estimator, respectively. The main idea is to maintain both the state estimation error and the unknown input reconstruction in the "desired surface," i.e., in zero for the continuous state and in the unknown input for the equivalent output injection estimators in spite of the switchings on the system.

6.7 Simulation Example

Consider a simplified particular 2-DOF model of a vertical oilwell drillstring (see [26] for more details). The drillstring torsional behavior is described by a torsional pendulum driven by an electrical motor and the bit-rock contact is described by a dry friction term. The drill pipes are modeled by a linear spring with torsional stiffness k_t and a torsional damping c_t, which connect the inertias J_r and J_b (Fig. 6.1 that illustrate the n-DOF model of the drillstring).

The state vector $x \in \Re^3$ is composed by the angular displacements (φ_r and φ_b) and angular velocities ($\dot{\varphi}_r$ and $\dot{\varphi}_b$) of the top-rotary system and the bit, i.e., $x = \left[\dot{\varphi}_r, \varphi_r - \varphi_b, \dot{\varphi}_b\right]^T$.

A viscous damping torque $c_r x_1$ is considered at the top-drive system. The constant input torque applied by a motor at the surface is $T_m = u$. The torque on the bit is $T_b(x_3) = c_b x_3 + T_{f_b}(x_3)$ where $c_b x_3$ approximating the influence of the mud drilling on the bit behavior and $T_{f_b}(x_3)$ is the friction modeling the bit-rock contact, and $T_{f_b}(x_3) = f_b(x_3)\text{sign}(x_3)$ with:

$$f_b(x_3) = W_{\text{ob}} R_b \left[\mu_{c_b} + \left(\mu_{s_b} - \mu_{c_b}\right) \exp\left(-\frac{\gamma_b}{\nu_f}|x_3|\right)\right], \tag{6.44}$$

where $W_{\text{ob}} > 0$ is the constant weight on the bit, $R_b > 0$ is the bit radius. The static and Coulomb friction coefficients associated with J_b are represented by $\mu_{s_b}, \mu_{c_b} \in$

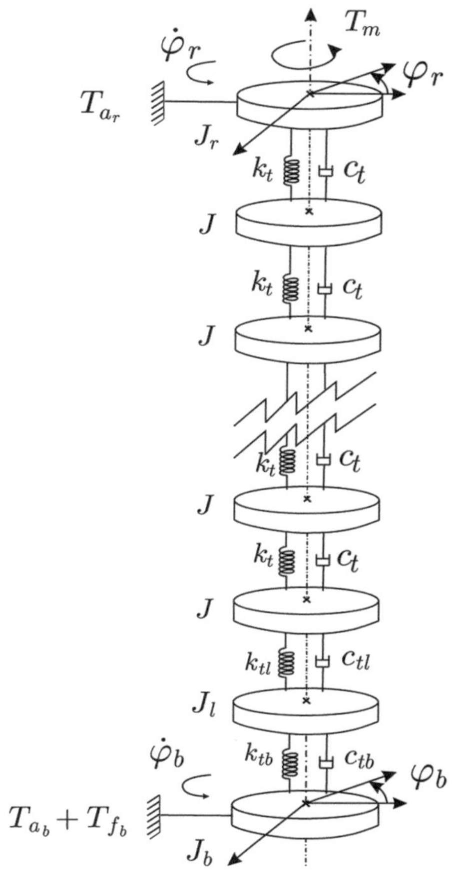

Fig. 6.1 Mechanical model of the torsional behavior of a drillstring

$(1, 0)$, and the positive constants $0 < \gamma_b < 1$ and $v_f > 0$. In addition, the Coulomb and static friction torque are $T_{c_b} = W_{\text{ob}} R_b \mu_{c_b}$ and $T_{s_b} = W_{\text{ob}} R_b \mu_{s_b}$, respectively.

Then, the drillstring switched dynamics with unknown input is given by

$$\Sigma_1 : \left\{ \dot{x} = \begin{bmatrix} \frac{1}{J_r}\left[-(c_t + c_r)x_1 - k_t x_2 + c_t x_3 + u\right] \\ x_1 - x_3 \\ \frac{1}{J_b}\left[c_t x_1 + k_t x_2 - (c_t + c_b)x_3 - T_{f_b}^{+}(x_3)\right] \end{bmatrix} + \begin{bmatrix} 0 \\ k\frac{c_t}{J_r} \\ k\frac{k_t}{J_r} \end{bmatrix} w \right. , \tag{6.45}$$

$$\Sigma_2 : \left\{ \dot{x} = \begin{bmatrix} \frac{1}{J_r}\left[-(c_t + c_r)x_1 - k_t x_2 + c_t x_3 + u\right] \\ x_1 - x_3 \\ \frac{1}{J_b}\left[c_t x_1 + k_t x_2 - (c_t + c_b)x_3 - T_{f_b}^{-}(x_3)\right] \end{bmatrix} + \begin{bmatrix} 0 \\ k\frac{c_t}{J_r} \\ k\frac{k_t}{J_r} \end{bmatrix} w \right. , \tag{6.46}$$

$$\Sigma_3 : \left\{ \dot{x} = \begin{bmatrix} \frac{1}{J_r}\left[-(c_t + c_r)x_1 - k_t x_2 + u\right] \\ x_1 \\ 0 \end{bmatrix} + \begin{bmatrix} 0 \\ k\frac{c_t}{J_r} \\ k\frac{k_t}{J_r} \end{bmatrix} w. \right. \tag{6.47}$$

where $T_{f_b}^{+}$ and $T_{f_b}^{-}$ are $T_{f_b}(x_3)$ for $x_3 > 0$ and $x_3 < 0$, respectively. The output, the discrete state and the unknown input are the following:

$$y = [x_1,\ x_3]^T, \tag{6.48}$$

$$\lambda(x) = \begin{cases} 1, & \forall x \mid C(x) \in (T_{s_b}, \infty) \\ 2, & \forall x \mid C(x) \in (-\infty, -T_{s_b}), \\ 3, & \forall x \mid C(x) \in [-T_{s_b}, T_{s_b}] \end{cases} \tag{6.49}$$

$$w = 2\cos(1.5t) + 6\text{square}(0.5t) + 1, \tag{6.50}$$

where $C(x) = c_t x_1 + k_t x_2$, the function square is a pulse train with angular frequency equal to 0.5, and the unknown input w could represent some nondesired oscillation (parameter k represents a constant coupling of the unknown input). The parameter values of the drillstring dynamics are shown in the Table 6.1.

The system initial conditions are set as $x(0) = \left[-10,\ -50,\ -10\right]^T$. The simulations have been done with the Euler discretization method and sampling time $h = 10e^{-3}$ s.

Then, to design the observers 1 and 2 only will use the output $h_1 = x_1$ to reconstruct the whole state vector (*scalar output case*), while for the observer 3 will use the whole output y to reconstruct the state (*multiple output case*, see [29]). Now, the assumptions are going to be analyzed.

Table 6.1 Drillstring parameters

Parameter	Value	Units
J_r	2122	kgm^2
J_b	471.9698	kgm^2
R_b	0.155575	m
k_t	698.063	Nm/rad
c_t	139.6126	Nms/rad
c_r	425	Nms/rad
c_b	50	Nms/rad
u	6	kNm
W_{ob}	60	kN
T_{s_b}	7.4676	kNm
μ_{c_b}	0.5	–
μ_{s_b}	0.8	–
γ_b	0.9	–
ν_f	1	–

Assumption 6.17, for scalar output case, is satisfied since

$$\begin{aligned} &d(h_1(x))F = 0, \\ &d(L_{f_{\lambda(x)}(x)}h_1(x))F = 0, \quad \forall\lambda = 1, 2, \\ &d(L^2_{f_{\lambda(x)}(x)}h_1(x))F \neq 0, \quad \forall\lambda = 1, 2. \end{aligned} \tag{6.51}$$

Assumption 6.17, for multiple output case, is satisfied since

$$\begin{aligned} &d(h_1(x))F = 0, \\ &d(L_{f_{\lambda(x)}(x)}h_1(x))F = 0, \quad \forall\lambda = 3, \\ &d(h_2(x))F = 0. \end{aligned} \tag{6.52}$$

The matrices in (6.7), for $\lambda(x) = 1, 2$, have the following structure

$$\frac{\partial \Phi_{\lambda(x)}(x)}{\partial x} = \begin{bmatrix} 1 & 0 & 0 \\ -0.26608 & -0.32896 & 6.5795 \times 10^{-2} \\ -0.23870 & 0.18484 & 0.35134 \exp(-0.9\,|x_3|) + 0.285\,02 \end{bmatrix}. \tag{6.53}$$

Notice that in this case the observability mappings are equal. Therefore, it is possible to design only one observer for $\lambda(x) = 1, 2$.

The matrix in (6.7), for multiple output case, with $\lambda(x) = 3$, has the following structure

$$\frac{\partial \Phi_3(x)}{\partial x} = \begin{bmatrix} 1 & 0 & 0 \\ -0.26608 & -0.32896 & 0 \\ 0 & 0 & 1 \end{bmatrix}. \tag{6.54}$$

It could be compute that rank $\left(\frac{\partial \Phi_{\lambda(x)}(x)}{\partial x}\right) = 3, \forall x \in \Re^3$ for $\lambda(x) = 1, 2, 3$. Then, since the Assumption 6.17 is satisfied, it is possible to demonstrate that Assumption 6.7 is satisfied too (see [19]). As for the Definition 6.3, it is considered that the first switching time $t_1 = 1.146\,\text{s} \geq T_\delta$.

Then, the observers are designed as follows, with $j = 1, 3$[1]:

$$\begin{aligned} \dot{\bar{x}}_j &= f_j(\bar{x}_j) + \left(\frac{\partial \Phi_j(\bar{x}_j)}{\partial \bar{x}_j}\right)^{-1} \nu_j, \\ \bar{y}_j &= h(\bar{x}_j), \end{aligned} \tag{6.55}$$

where the correction terms, for $j = 1$, are calculated using the following auxiliary dynamics

[1] Notice that the observer $j = 1$ is working for $\lambda(x) = 1, 2$ since its observability mappings are equals.

$$\begin{aligned}\dot{\vartheta}_{j,1} &= \vartheta_{j,2} - \alpha_{j,1} M_j^{\frac{1}{3}} \left| e_{y_{1j}} \right|^{\frac{2}{3}} \operatorname{sign}\left(e_{y_{1j}}\right), \\ \dot{\vartheta}_{j,2} &= \vartheta_{j,3} - \alpha_{j,2} M_j^{\frac{1}{2}} \left| \vartheta_{j,2} - \dot{\vartheta}_{j,1} \right|^{\frac{1}{2}} \operatorname{sign}\left(\vartheta_{j,2} - \dot{\vartheta}_{j,1}\right), \\ \dot{\vartheta}_{j,3} &= -\alpha_{j,3} M_j \operatorname{sign}\left(\vartheta_{j,3} - \dot{\vartheta}_{j,2}\right).\end{aligned} \tag{6.56}$$

For $j = 3$ the corrections terms are calculated using the following auxiliary dynamics

$$\begin{aligned}\dot{\vartheta}_{1,1_j} &= \vartheta_{1,2_j} - \alpha_{1,1_j} M_{j_1}^{\frac{1}{2}} \left| e_{y_{1_j}} \right|^{\frac{1}{2}} \operatorname{sign}\left(e_{y_{1_j}}\right), \\ \dot{\vartheta}_{1,2_j} &= -\alpha_{1,2_j} M_{j_1} \operatorname{sign}\left(\vartheta_{1,2_j} - \dot{\vartheta}_{1,1_j}\right), \\ \dot{\vartheta}_{2,1_j} &= -\alpha_{2,1_j} M_{2_j} \operatorname{sign}\left(e_{\bar{y}_{2_j}}\right),\end{aligned} \tag{6.57}$$

where $e_{y_{1_j}} = h_1(\bar{x}_j) - h_1(x)$ and $e_{y_{2_j}} = h_2(\bar{x}_j) - h_2(x)$. The parameter values of the corrections terms are shown in the Table 6.2.[2]

Simulations results are shown in the following figures. The output error convergence for every observer is shown in the Fig. 6.2. By means of simulations, it is possible to determine that the minimal error band, defined in inequality (6.28), is approximately 1×10^{-10}. It is possible to see that the output error $e_{y_{11}}$ satisfies the inequality (6.28) after $t = 1.028$ s) and before the real first switching time, which is equal to $t_1 = 1.146$ s, while other output errors not. Then, it is possible to determine that the estimation of the Observer 1 is the correct one and then estimate the correct discrete state.

The real and estimated continuous state are shown in the Fig. 6.3. It is possible to see that the estimated continuous state does not get lost in spite of the switchings on the system and that it shows a finite-time convergence. On the other hand, once the estimation of the continuous state is reached, the reconstruction of the discrete state is immediate by means of the switching signal, as shown in the Fig. 6.4. Notice that there exists an interval of time ($t \in (13, 14)$ s) where the system presents high frequency switchings. Nevertheless, the estimation of the continuous state never gets lost.

Table 6.2 Correction terms parameters

Parameter	$\lambda = 1$	$\lambda = 3$
$\alpha_{\lambda,1}$	2	–
$\alpha_{\lambda,2}$	1.5	–
$\alpha_{\lambda,3}$	1.1	–
M_λ	100	–
$\alpha_{1,1_\lambda}$	–	1.5
$\alpha_{1,2_\lambda}$	–	1.1
$\alpha_{2,1_\lambda}$	–	1.1
M_{λ_1}	–	10
M_{λ_2}	–	10

[2] These parameters provide a minor convergence time to 1.146 s.

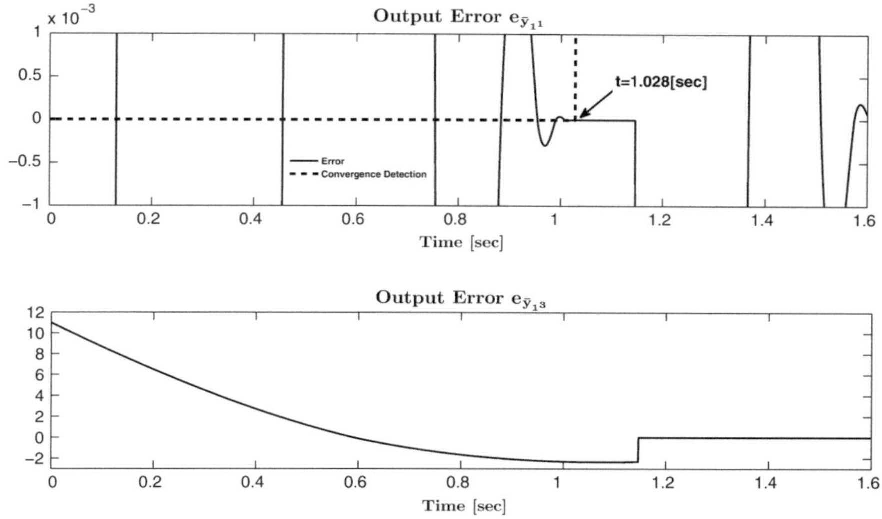

Fig. 6.2 Output error convergence for each observer

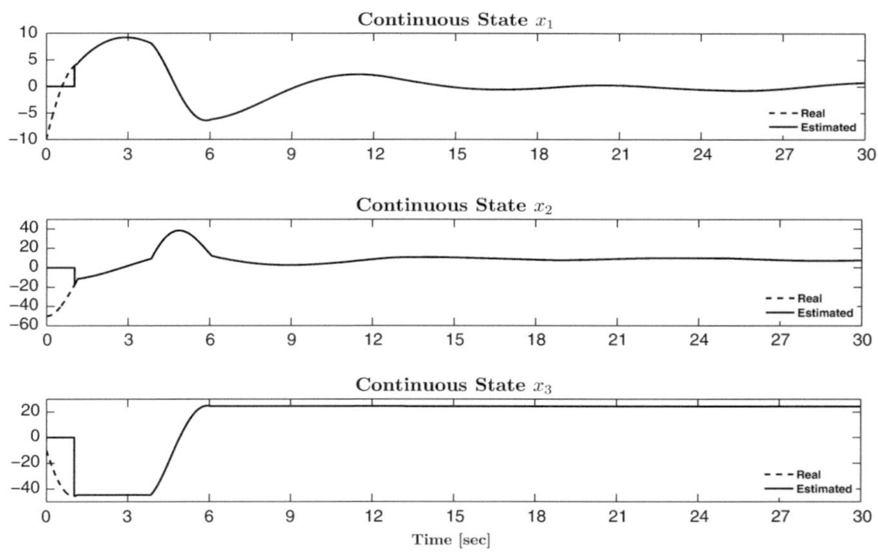

Fig. 6.3 Continuous trajectories of the nonlinear switched system

A comparison between the estimation error convergence with and without reset equation (6.31) is shown in Fig. 6.5. It is clear the effect of the reset equation which maintain the correct estimation of the continuous state on the switching times.

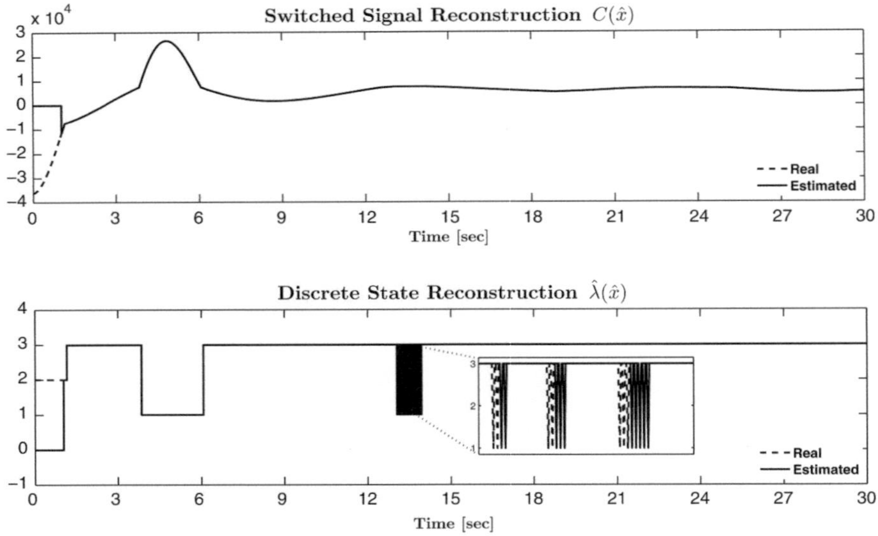

Fig. 6.4 Switching signal reconstruction and discrete state estimation

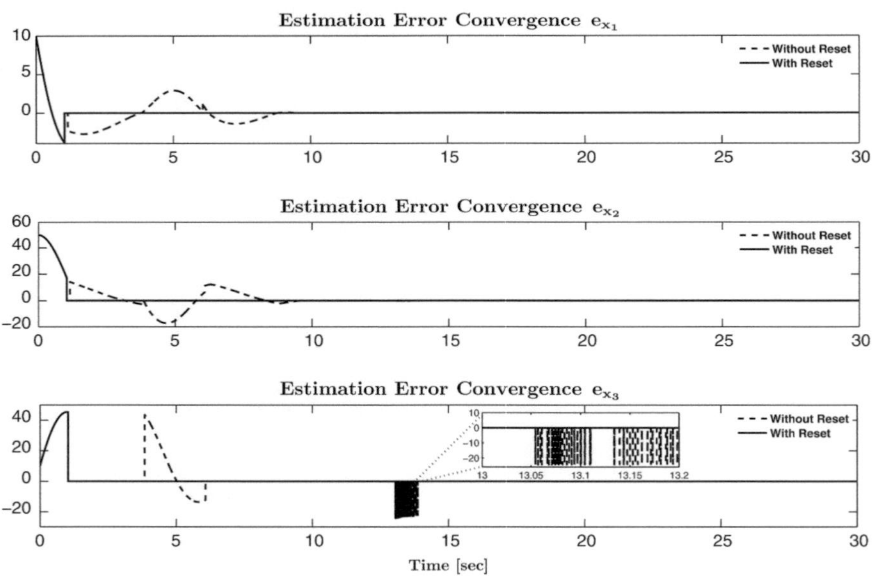

Fig. 6.5 Comparison between estimation errors convergence

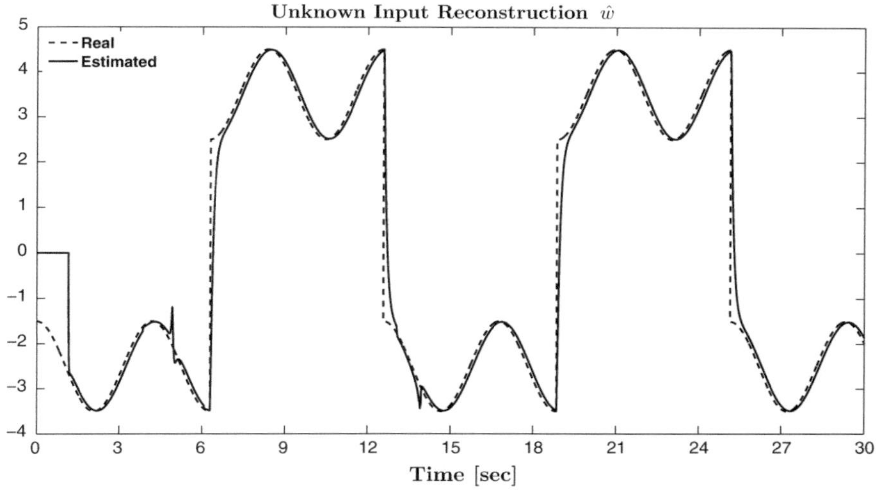

Fig. 6.6 Unknown input reconstruction

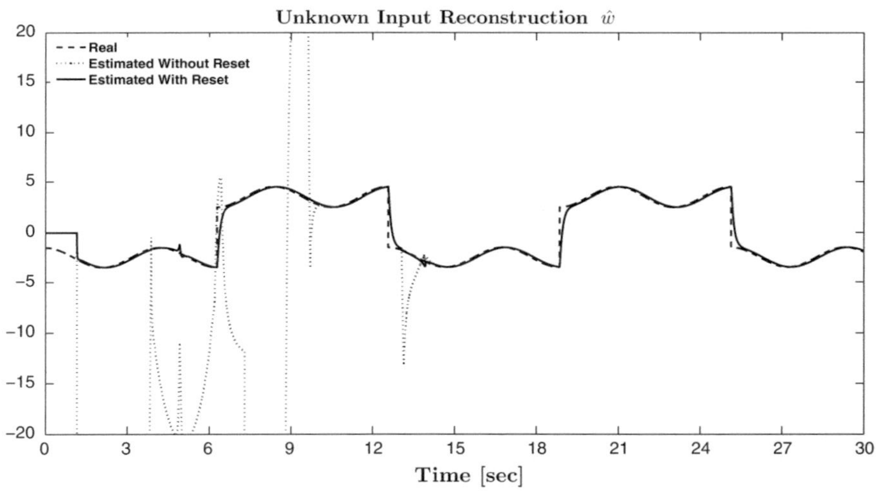

Fig. 6.7 Unknown input reconstruction without reset equations

Finally, in Fig. 6.6 is shown the unknown input reconstruction. The equivalent output injection estimators were designed as in (6.22) with $\tau_j = 0.003162 \; \forall j = 1, 2, 3$. A comparison between the reconstruction with and without reset equation for equivalent output injection estimators (6.43) is shown in Fig. 6.7. It is possible to appreciate the effect of reset estimator equation which maintains the correct unknown input reconstruction on the switching times.

6.8 Conclusions

The problem of continuous and discrete state estimation for switched nonlinear systems is addressed in this chapter. In the first part of the chapter, the solvability of the observation problem, for the continuous and discrete states, is proposed using structural properties of the system with exogenous switchings. This structural properties are exploited to design the observers. The high-order sliding-mode techniques are introduced to guarantee finite time convergence to zero of the estimation error for the continuous state. The discrete state is reconstructed using the information of the equivalent output injection. The continuous state observation, discrete state, and unknown input reconstruction for switched systems are realized in the second part of the chapter. Structural properties of the systems with autonomous switchings are exploited to ensure the observability of the continuous and discrete states. Finite-time convergence to zero of the estimation errors, for continuous and discrete states, are achieved. The information of the equivalent output injection is used for unknown input reconstruction.

Acknowledgments The authors gratefully acknowledge the financial support from CONACyT grant 132125, 151855, and 270504; PAPIIT 113613; SIP-IPN and CDA-IPN.

Appendix

Proof of Theorem 6.1

System (6.14), under Assumption 6.7, can be represented, on new coordinates z, as:

$$\begin{aligned} \dot{z} &= Az + B\varphi_{\lambda^*}(z), \\ y_z &= Cz, \end{aligned} \tag{6.58}$$

where

$$A = \begin{bmatrix} 0 & 1 & \cdots & 0 \\ \vdots & \vdots & \ddots & \vdots \\ 0 & 0 & \cdots & 1 \\ 0 & 0 & \cdots & 0 \end{bmatrix}_{n\times n}, \quad B = \begin{bmatrix} 0 \\ \vdots \\ 0 \\ 1 \end{bmatrix}_{n\times 1}, \quad C = \begin{bmatrix} 1 & 0 & \cdots & 0 \end{bmatrix}_{1\times n}, \tag{6.59}$$

$$\varphi_{\lambda^*}(z) = \left(L^n_{f_{\lambda^*}(x)} h(x) \right) |_{x=\Phi^{-1}_{\lambda^*}(z)}. \tag{6.60}$$

Notice that the functions $\varphi_{i^*}(z)$ contain the unknown input that acts on the system. On the other hand, the observers (6.8), under Assumption 6.7, can be represented as:

$$\begin{aligned}\dot{\hat{z}}_j &= A\hat{z}_j + B\varphi_j(\hat{z}_j) + \nu_j,\\ y_{\hat{z}_j} &= C\hat{z}_j,\end{aligned} \tag{6.61}$$

where $\varphi_j(\hat{z}_j) = \left(L^n_{f(\hat{x}_j)}h(\hat{x}_j)\right)|_{x_j=\Phi_j^{-1}(\hat{z}_j)}$.

Define the state estimation errors as

$$e_{z_j} = \hat{z}_j - z. \tag{6.62}$$

Thus, the state estimation error dynamics takes the following form:

$$\dot{e}_{z_j} = Ae_{z_j} + B\Psi_j(\hat{z}_j, z) + \nu_j, \tag{6.63}$$

where $\Psi_j(\hat{z}_j, z) = \varphi_j(\hat{z}_j) - \varphi_{\lambda^*}(z)$. If it is possible to find appropriate correction terms ν_j, which can steer the vector e_{z_j} to zero, then equality $\hat{z}_j = z$, will be satisfied only for the case when $j = \lambda^*$. Therefore, only one of the observers can be associated with the corresponding state observation error in the time interval $t \in [0, t_1)$.

On the other hand, it is not desirable to design the correction terms in the coordinates $\hat{z}_j$ but in the coordinates $\hat{x}_j$. Therefore, defining the following output error vector:

$$\varepsilon_j = \begin{bmatrix} \varepsilon_{j,1} \\ \vdots \\ \varepsilon_{j,n} \end{bmatrix} = \begin{bmatrix} e_{y_j} \\ \vdots \\ e_{y_j}^{(n-1)} \end{bmatrix}. \tag{6.64}$$

The state observation error dynamics (6.63) turns into output observation error dynamics as follows:

$$\dot{\varepsilon}_j = A\varepsilon_j + B\Psi_j(\Phi_j(\hat{x}_j), \Phi_{\lambda^*}(x)) + \nu_j. \tag{6.65}$$

In an extended structure

$$\begin{aligned}\dot{\varepsilon}_{j,1} &= \varepsilon_{j,2} + \nu_{j,1},\\ \dot{\varepsilon}_{j,2} &= \varepsilon_{j,3} + \nu_{j,2},\\ &\vdots\\ \dot{\varepsilon}_{j,n} &= \Psi_j(\Phi_j(\hat{x}_j), \Phi_{\lambda^*}(x)) + \nu_{j,n}.\end{aligned} \tag{6.66}$$

Notice that the dynamic structures (6.66) are very similar to the high-order sliding-mode differentiator (6.12). Thus, if a variable change is realized in the structure (6.13), it is possible to obtain the following sliding-mode differentiator form:

$$\dot{\varepsilon}_{j,1} = \varepsilon_{j,2} - \alpha_{j,1} M_j^{\frac{1}{n}} \left|\varepsilon_{j,1}\right|^{\frac{n-1}{n}} \operatorname{sign}\left(\varepsilon_{j,1}\right),$$

$$\dot{\varepsilon}_{j,2} = \varepsilon_{j,3} - \alpha_{j,2} M_j^{\frac{1}{n-1}} |\varepsilon_{j,2} - \dot{\varepsilon}_{j,1}|^{\frac{n-2}{n-1}} \operatorname{sign}\left(\varepsilon_{j,2} - \dot{\varepsilon}_{j,1}\right),$$
$$\vdots$$
$$\dot{\varepsilon}_{j,n} = \Psi_j(\Phi_j(\hat{x}_j), \Phi_{\lambda^*}(x)) - \alpha_{j,n} M_j \operatorname{sign}\left(\varepsilon_{j,n} - \dot{\varepsilon}_{j,n-1}\right). \tag{6.67}$$

Let the parameters $\alpha_{j,k}$ be chosen recursively and sufficiently large according to the high-order sliding-mode differentiator properties described in [22]. In view of the Assumption 6.10, the following equality is satisfied in finite time when $j = \lambda^*$:

$$\left[\varepsilon_{\lambda^*,1}, \varepsilon_{\lambda^*,2}, \ldots, \varepsilon_{\lambda^*,n}\right] \equiv [0, 0, \ldots, 0]. \tag{6.68}$$

The condition $\varepsilon_{\lambda^*,1} \equiv 0$, $\forall t \in [t_{\lambda^*}, t_1)$ implies that $\left[\varepsilon_{\lambda^*,2}, \ldots, \varepsilon_{\lambda^*,n}\right] \equiv [0, \ldots, 0]$, $\forall t \in [t_{\lambda^*}, t_1)$ with $t_1 \geq T_\delta$. To prove that, assume that the condition $\varepsilon_{\lambda^*,1} \equiv 0$ is satisfied in a nonzero time interval. This condition implies that $\dot{\varepsilon}_{\lambda^*,1} \equiv 0$ in the same time interval. Thus, from the first row of (6.67) it is obtained that $\varepsilon_{\lambda^*,2} \equiv 0$. Then, since $\varepsilon_{\lambda^*,2} \equiv 0$ and $\dot{\varepsilon}_{\lambda^*,1} \equiv 0$ from the second row of (6.67) it is obtained that $\varepsilon_{\lambda^*,3} \equiv 0$. If the same procedure is iterated the following expressions are obtained

$$\varepsilon_{\lambda^*,k} \equiv 0, \quad \forall k = 1, \ldots, n. \tag{6.69}$$

Given Assumption 6.10, the last row of (6.67) defines the following differential inclusion

$$\dot{\varepsilon}_{\lambda^*,n} \in [-M_{\lambda^*}, M_{\lambda^*}] - \alpha_{\lambda^*,n} M_{\lambda^*} \operatorname{sign}\left(\varepsilon_{\lambda^*,n} - \dot{\varepsilon}_{\lambda^*,n-1}\right), \tag{6.70}$$

where $\Psi_{\lambda^*}(\cdot) \in [-M_{\lambda^*}, M_{\lambda^*}]$. Therefore, according to [22], the dynamics (6.67), for $j = \lambda^*$, converges to zero after a finite time, i.e., $\varepsilon_{\lambda^*} \equiv 0$ $\forall t > t_{\lambda^*}$, and according to Assumption 6.7, it is ensure that the state estimation error $e_{x_{\lambda^*}} = \hat{x}_{\lambda^*} - x$ also converge to zero in finite time in spite of the unknown input. Notice that it is always possible to select the gain M_{λ^*} sufficiently large such that $t_{\lambda^*} < t_1 \geq T_\delta$. In this way, the theorem is proven.

Proof of Proposition 6.1

Let us consider the dynamics (6.67) on the time instants before and after the switching time t_1, i.e.,

$$\dot{\varepsilon}_{\Delta_{\lambda^*\kappa,1}} = \varepsilon_{\Delta_{\lambda^*\kappa,2}} - \alpha_{\Delta_{\lambda^*\kappa,1}} M_{\Delta_{\lambda^*\kappa}}^{\frac{1}{n}} \left|\varepsilon_{\Delta_{\lambda^*\kappa,1}}\right|^{\frac{n-1}{n}} \operatorname{sign}\left(\varepsilon_{\Delta_{\lambda^*\kappa,1}}\right),$$
$$\dot{\varepsilon}_{\Delta_{\lambda^*\kappa,2}} = \varepsilon_{\Delta_{\lambda^*\kappa,3}} - \alpha_{\Delta_{\lambda^*\kappa,2}} M_{\Delta_{\lambda^*\kappa}}^{\frac{1}{n-1}} \left|\varepsilon_{\Delta_{\lambda^*\kappa,2}} - \dot{\varepsilon}_{\Delta_{\lambda^*\kappa,1}}\right|^{\frac{n-2}{n-1}} \operatorname{sign}\left(\varepsilon_{\Delta_{\lambda^*\kappa,2}} - \dot{\varepsilon}_{\Delta_{\lambda^*\kappa,1}}\right),$$
$$\vdots$$
$$\dot{\varepsilon}_{\Delta_{\lambda^*\kappa,n}} = \Psi_{\Delta_{\lambda^*\kappa}}(\cdot) - \alpha_{\Delta_{\lambda^*\kappa,n}} M_{\Delta_{\lambda^*\kappa}} \operatorname{sign}\left(\varepsilon_{\Delta_{\lambda^*\kappa,n}} - \dot{\varepsilon}_{\Delta_{\lambda^*\kappa,n-1}}\right), \tag{6.71}$$

where λ^* is the current operating mode, $\kappa \in \mathscr{Q}$, with $\kappa \neq \lambda^*$, represents the next operating mode, and

$$\varepsilon_{\Delta_{\lambda^*\kappa,k}} = \varepsilon_{\Delta_{\lambda^*,k}}(t_1^-) - \varepsilon_{\Delta_{\kappa,k}}(t_1^+), \; \forall k = 1, \ldots, n,$$
$$\Psi_{\Delta_{\lambda^*\kappa}}(\cdot) = \Psi_{\Delta_{\lambda^*}}(t_1^-) - \Psi_{\Delta_\kappa}(t_1^+).$$

If the reset equations (6.31) are applied on each switching time, then the trajectories of the system (6.71) always remain in the sliding surface, i.e., $\varepsilon_{\Delta_{\lambda^*\kappa,k}} = 0$. Thus, the state estimation is maintained in spite of the switchings. Notice that due to the nature of the system, the transformations $\Phi_{\lambda(x)}$ do not present jumps in the switching times. Nevertheless, the jumps can appear in the trajectories generated by the observers. However, the reset equations (6.31) avoid this happening.

References

1. Alessandri, A., Coletta, P.: Design of observers for switched discretetime linear systems. In: Proceedings of the 2003 American Control Conference, Denver, CO, USA (2003) pp. 2785–2790
2. Alwi, H., Edwards, C.: Fault tolerant control using sliding modes with on-line control allocation. Automatica **44**(7), 1859–1866 (2008)
3. Angulo, M.T., Fridman, L., Levant, A.: Robust exact finite-time output based control using high-order sliding modes. Int. J. Syst. Sci. **42**(11), 1847–1857 (2011)
4. Barbot, J.P., Saadaoui, H., Djemai, M., Manamanni, N.: Nonlinear observer for autonomous switching systems with jumps. Nonlinear Anal.: Hybrid Syst. **1**(4), 537–547 (2007)
5. Bartolini, G., Ferrara, A., Punta, E., Usai, E.: Chattering elimination in the hybrid control of constrained manipulators via first/second order sliding modes. Dyn. Control **9**(2), 99–124 (1999)
6. Bejarano, F.J., Pisano, A.: Switched observers for switched linear systems with unknown inputs. IEEE Trans. Autom. Control **56**(3), 681–686 (2011)
7. Bejarano, F.J., Pisano, A., Usai, E.: Finite-time converging jump observer for switched linear systems with unknown inputs. Nonlinear Anal.: Hybrid Syst. **5**(2), 174–188 (2011)
8. Bemporad, A., Ferrari-Trecate, G., Morari, M.: Observability and controllability of piecewise affine and hybrid systems. IEEE Trans. Autom. Control **45**(10), 1864–1876 (2000)
9. Bobtsov, A.A., Efimov, D.V., Pyrkin, A.: Hybrid adaptive observers for locally lipschitz systems with application to mechanical oscillators. Control Applications (CCA) and Intelligent Control, (ISIC), pp. 619–624. St. Petersburg, Russia (2009)
10. Boukhobza, T., Hamelin, F.: Observability of switching structured linear systems with unknown input. A graph theoretic approach. Automatica **47**, 395–402 (2011)
11. Branicky, M.S.: Multiple lyapunov functions and other analysis tools for switched and hybrid systems. IEEE Trans. Autom. Control **43**(5), 475–482 (1998)
12. Chaib, S., Boutat, D., Benali, A., Barbot, J.P.: Observability of the discrete state for dynamical piecewise hybrid systems. Nonlinear Anal. **63**, 423–438 (2005)
13. Chaib, S., Boutat, D., Benali, A., Kratz, F.: Failure detection and reconstruction in switched nonlinear systems. Nonlinear Anal.: Hybrid Syst. **3**, 225–238 (2009)
14. Davila, J., Ríos, H., Fridman, L.: State observation for nonlinear switched systems using non-homogeneous high-order sliding mode observers. Asian J. Control **14**(4), 911–923 (2012)
15. De-Santis, E., Di-Benedetto, M.D., Pola, G.: A structural approach to detectability for a class of hybrid systems. Automatica **45**, 1202–1206 (2009)

16. Defoort, M., Djemai, M., Floquet, T., Perruquetti, W.: Robust finite time observer design for multicellular converters. Int. J. Syst. Sci. **38**(10), 773–791 (2011)
17. Efimov, D., Zolghadri, A., Raïssi, T.: Actuator fault detection and compensation under feedback control. Automatica **47**(8), 1699–1705 (2011)
18. Filippov, A.F.: Differential Equations with Discontinuos Right-Hand Sides. Kluwer Academic, Dordrecht (1988)
19. Isidori, A.: Nonlinear Control Syst. Springer, London (1996)
20. Kang, W., Barbot, J.P., Xu, L.: On the observability of nonlinear and switched systems. In: Ghosh, B., Martin, C., Zhou, Y. (eds.) Emergent Problems in Nonlinear Systems and Control. Lecture Notes in Control and Information Science, pp. 199–216. Springer, Berlin (2009)
21. Khalil, H.: Nonlinear Systems. Prentice Hall, New Jersey (2002)
22. Levant, A.: High-order sliding modes: differentiation and output-feedback control. Int. J. Control **76**(9–10), 924–941 (2003)
23. Liberzon, D.: Switching in Systems and Control. Systems and Control: Foundations and Applications. Birkhuser, Boston (2003)
24. Lin, H., Antsaklis, P.: Stability and stabilizability of switched linear systems: a survey of recent results. IEEE Trans. Autom. Control **54**(2), 308–322 (2009)
25. Lygeros, J., Johansson, K.H., Simić, S.N., Zhang, J., Sastry, S.S.: Dynamical properties of hybrid automata. IEEE Trans. Autom. Control **48**(1), 2–17 (2003)
26. Navarro-Lopez, E.M., Carter, R.: Hybrid automata: an insight into the discrete abstraction of discontinuous systems. Int. J. Syst. Sci. **42**(11), 1883–1898 (2011)
27. Orani, N., Pisano, A., Franceschelli, M., Giua, A., Usai, E.: Robust reconstruction of the discrete state for a class of nonlinear uncertain switched systems. Nonlinear Anal.: Hybrid Syst. **5**(2), 220–232 (2011)
28. Pina, L., Botto, M.A.: Simultaneous state and input estimation of hybrid systems with unknown inputs. Automatica **42**, 1755–1762 (2006)
29. Ríos, H., Davila, J., Fridman, L.: High-order sliding mode observers for nonlinear autonomous switched systems with unknown inputs. J. Franklin Inst. **349**(10), 2975–3002 (2012)
30. Tian, Y., Floquet, T., Belkoura, L., Perruquetti, W.: Switching time estimation for linear switched systems: an algebraic approach. In: Proceedings of the 48th IEEE Conference Decision Control and 28th Chinese Control Conference, Shangai, P.R. China (2009) pp. 3909–3913
31. Utkin, V., Guldner, J., Shi, J.: Sliding Modes in Electromechanical Systems. Taylor and Francis, London (1999)
32. Vu, L., Liberzon, D.: Invertibility of switched linear systems. Automatica **44**, 949–958 (2008)
33. Yu, L., Barbot, J.P., Boutat, D., Benmerzouk, D.: Observability forms for switched systems with zeno phenomenon or high switching frequency. IEEE Trans. Autom. Control **56**(2), 436–441 (2011)

Chapter 7
Observer Design for Switched Linear Systems with State Jumps

Aneel Tanwani, Hyungbo Shim and Daniel Liberzon

Abstract An observer design for switched linear systems with state resets is proposed based on the geometric conditions for large-time observability from our recent work. Without assuming the observability of individual subsystems, the basic idea is to combine the maximal information available from each mode to obtain a good estimate of the state after a certain time interval (over which the switched system is observable) has passed. We first study systems where state reset maps at switching instants are invertible, in which case it is possible to collect all the observable and unobservable information separately at one time instant. One can then annihilate the unobservable component of all the modes and obtain an estimate of the state by introducing an error correction map at that time instant. However, for the systems with non-invertible jump maps, this approach needs to be modified and a recursion-based error correction scheme is proposed. In both approaches, the criterion for choosing the output injection matrices is given, which leads to the asymptotic recovery of the system state.

7.1 Introduction

State estimation in dynamical systems is one of the classical control-theoretic problems that relates to constructing estimates of the state of the system using the measurements of the inputs and outputs. This chapter studies the problem of state

A. Tanwani (✉)
Department of Mathematics, University of Kaiserslautern, Kaiserslautern, Germany
e-mail: tanwani@mathematik.uni-kl.de

H. Shim
ASRI School of Electrical Engineering, Seoul National University, Seoul, South Korea
e-mail: hshim@snu.ac.kr

D. Liberzon
Coordinated Science Laboratory, University of Illinois at Urbana-Champaign, Urbana, IL, USA
e-mail: liberzon@illinois.edu

M. Djemai and M. Defoort (eds.), *Hybrid Dynamical Systems*, Lecture Notes in Control and Information Sciences 457, DOI 10.1007/978-3-319-10795-0_7

estimation for a class of hybrid systems characterized by linear continuous-time dynamics, switching vector fields, and state jumps, which are described as

$$\dot{x}(t) = A_q x(t) + B_q u(t), \qquad t \in [t_{q-1}, t_q), \tag{7.1a}$$

$$x(t_q) = G_q x(t_q^-) + H_q v_q, \tag{7.1b}$$

$$y(t) = C_q x(t) + D_q u(t), \qquad t \in [t_{q-1}, t_q) \tag{7.1c}$$

where $x : [t_0, \infty) \to \mathbb{R}^{\mathbf{n}}$ is the state, $y : [t_0, \infty) \to \mathbb{R}^{\mathbf{y}}$ is the output, $v_i \in \mathbb{R}^{\mathbf{v}}$ and $u : [t_0, \infty) \to \mathbb{R}^{\mathbf{u}}$ are the inputs, and u is a measurable function. The index $q \in \mathbb{N}$ determines the active subsystem over the interval $[t_{q-1}, t_q)$ and the system trajectories are right-continuous. It is assumed that there are a finite number of switching times in any finite time interval, thus we rule out the Zeno phenomenon in our problem formulation. The switching mode $q \in \mathbb{N}$ and the switching times $\{t_q\}$ may be governed by a supervisory logic controller, or determined internally depending on the system state, or considered as an external input. In any case, it is assumed in this paper that the active subsystem and the switching times $\{t_q\}$ as well are known. For estimation of the active subsystem, one may be referred to, e.g., [4, 6, 8, 18, 25, 26].

Over the past decade and a half, the structural properties of switched systems have been investigated by many researchers and observability along with observer construction has been one of them. For switched systems, observability can be studied from various perspectives. If we allow for the use of the differential operator in the observer, then it may be desirable to determine the state of the system instantaneously from the measured output. This in turn requires each subsystem to be observable; however, the problem becomes nontrivial when the switching signal is treated as a discrete state and simultaneous recovery of the discrete and continuous state is required for observability. Some results on this problem are published in [2, 5, 25].

On the other hand, with the knowledge of switching signal, even though the subsystems at individual modes are not observable, it is possible to recover the initial state $x(t_0)$ when the output is observed over an interval $[t_0, T)$ that involves multiple switching instants. This phenomenon is of particular interest for switched systems as the notion of instantaneous observability and observability over an interval[1] coincide for linear time invariant systems. This variant of the observability in switched systems has been studied most notably by [4, 9, 10, 17, 27], and we refer the reader to Chap. 8 for more references on different notions of observability. The observer design has also received some attention in the literature [1, 3, 13], where authors have assumed that each mode in the system is in fact observable, hence admitting a state observer, and have treated the switching as a source of perturbation effect. This approach immediately incurs the need of a common Lyapunov function for the switched error dynamics, or a fixed amount of dwell-time between switching instants, because it is intrinsically a stability problem of the error dynamics.

---

[1] See Definition 7.1 for precise meaning.

The approach adopted for observer design in our work is based on the notion of observability over an interval, and is conceptually similar to the work of [4]. This relaxed notion of observability for switched systems does not require observability of individual subsystems in the classical sense. As a result, one *cannot* simply take the Luenberger observer for individual subsystem and work out the stability of the error dynamics using slow switching, or common Lyapunov function approach. Since the amount of observable information coming from different modes may vary, the interesting aspect of this approach is to design an algorithm that recovers the maximal possible information available from each mode and combines this information in an appropriate manner that results in an asymptotically converging state estimate. This is the fundamental idea behind our recent papers on observer design for switched linear systems [19, 20], switched nonlinear systems [14, 15], and systems with switched linear differential-algebraic equations [22].

In this chapter, we address the problem of observer design in the context of switched systems with linear ordinary differential equations, and the technical content is primarily based on our papers [19–21]. We focus only on the linear case because this relatively simpler class of systems brings out our design methodology in the most transparent manner. The construction of the observer is based on the necessary and sufficient conditions of *forward observability* (see Definition 7.1), or what is also called determinability in [17], and final-state observability in [16]. The detailed treatment of this notion of observability is considered in Chap. 8 of this book, and here in Sect. 7.2, for the sake of completeness, we will only recall the definition and the related formulae that set-up the ground work for observer construction.

Section 7.3 then considers the construction of the observers. The key idea is to combine the partial information available from each mode and collect them at one instant of time to get the estimate of the state at that time. We show that under mild assumptions, such estimates converges to the actual state of the plant. More emphasis will be given to the case when the individual modes of the system (7.1) are not observable (in the classical sense of linear time-invariant systems theory) since it is obvious that the system becomes immediately observable when the system is switched to the observable mode. The distinct feature of our observer design is that we do not inject error in the continuous dynamics of the proposed observer, but rather apply the error correction at discrete switching instants. This way the state estimation error may grow in between two consecutive switches, but the error correction terms are designed in a manner such that the error eventually converges to zero. One can already see that, in contrast to slow switching, our approach would work only if the switching is persistent and the estimation gets better if the frequency of switching is high since we can apply error corrections more often in that case. We basically treat two different cases: in the first case [19, 20], we assume that all the jump maps G_q are invertible, because it is relatively easier to do computations for this case. The second case [21] allows for jump maps G_q to be non-invertible, but the calculations are more involved in this case. As a result the error correction term in the later case is computed using a recursion-based algorithm instead of a direct formula.

In the end, we give some concluding remarks on how the ideas presented in this paper have been applied to more general classes of switched systems, and where the proposed observer has been applied in practice.

Notation: For a matrix A, $\mathscr{R}(A)$ denotes the column space (range space) of A. The sum of two subspaces $\mathscr{V}_1$ and $\mathscr{V}_2$ is defined as $\mathscr{V}_1 + \mathscr{V}_2 := \{v_1 + v_2 : v_1 \in \mathscr{V}_1, v_2 \in \mathscr{V}_2\}$. For a possibly non-invertible matrix A, the pre-image of a subspace $\mathscr{V}$ under A is given by $A^{-1}\mathscr{V} = \{x : Ax \in \mathscr{V}\}$. Let $\ker A := A^{-1}\{0\}$; then it is seen that $A^{-1} \ker C = \ker(CA)$ for a matrix C. For convenience of notation, let $A^{-\top}\mathscr{V} := (A^\top)^{-1}\mathscr{V}$ where $A^\top$ is the transpose of A, and it is understood that $A_2^{-1}A_1^{-1}\mathscr{V} = A_2^{-1}(A_1^{-1}\mathscr{V})$. Also, we denote the products of matrices A_i as $\prod_{i=j}^k A_i := A_j A_{j+1} \dots A_k$ when $j < k$, and $\prod_{i=j}^k A_i := A_j A_{j-1} \dots A_k$ when $j > k$. The notation $\mathrm{col}(A_1, \dots, A_k)$ means the vertical stack of matrices $A_1, \dots$, A_k, that is, $[A_1^\top, \dots, A_k^\top]^\top$.

7.2 Preliminaries: Observability Notion

As mentioned earlier, our observer design is based on geometric conditions for forward observability. The related observability notions are treated in detail in this chapter. For the sake of completeness, here we recall the definition, and the related formulae that will be used in the design of observer. The notion of observability that we consider is formulated as follows:

Definition 7.1 (*Forward observability*) The system (7.1) is said to be *forward observable* if, and only if, for every pair of solutions (u^1, v^1, y^1, x^1), (u^2, v^2, y^2, x^2), there exists $T > t_0$, such that the following implication holds:

$$(u^1, v^1, y^1) = (u^2, v^2, y^2) \quad \Rightarrow \quad x^1_{[T,\infty)} = x^2_{[T,\infty)}$$

where $x^i_{[T,\infty)}$, $i = 1, 2$, denotes the restriction of x^i over the interval $[T, \infty)$.

Since the value of the state at time T, $x(T)$, and the inputs (u, v) uniquely determine x on $[T, \infty)$ through the Eq. (7.1), forward observability is achieved if and only if $x(T)$, for some $T > t_0$, is uniquely determined by the inputs and the output. In case, all the jump maps are invertible, one can, in theory, recover $x(t_0)$ from the knowledge of $x(T)$ and in that case forward observability implies global (in time) observability. However, if the state reset maps G_q are not invertible, then the two notions may not coincide. We refer the reader to Chap. 8 for more details and examples related to this issue.

7.2.1 Characterization of Forward Observability

Roughly speaking, the switched system (7.1) is forward observable in the sense of Definition 7.1 if there exists $m \in \mathbb{N}$ such that $x(t_m)$ could be determined from the knowledge of external signals (u, v, y) measured over the interval $[t_0, t_{m+1})$. Because $x(t_{m+1}^-) = e^{A_{m+1}(t_{m+1}-t_m)}x(t_m)$, uncertainty in the knowledge of $x(t_m)$ and $x(t_{m+1}^-)$ is the same, so that, recovering $x(t_m)$ is equivalent to recovering $x(t_{m+1}^-)$. We now proceed toward quantifying the unknown information about the state using the measurements of (u, v, y) over a certain interval. Since Definition 7.1 does not require individual subsystems to be observable, the basic idea in formulating the geometric conditions that quantify the unknown information is to characterize how much information could be extracted from each subsystem about the state by measuring the output over a certain interval. To do so, it is seen that system (7.1) is an LTI system between two consecutive switching times, so that its unobservable subspace on the interval $[t_{q-1}, t_q)$ is simply given by the largest A_q-invariant subspace contained in $\ker C_q$, i.e., $\ker O_q$ where

$$O_q := \mathrm{col}(C_q, C_q A_q, \ldots, C_q A_q^{n-1}).$$

For system (7.1), let $\mathcal{N}_q^m$ be the subspace such that $x(t_m^-)$ is determined modulo $\mathcal{N}_q^m$ using the knowledge of external signals (u, v, y) over the interval $[t_{q-1}, t_m)$. We call $\mathcal{N}_q^m$ the *unobservable subspace for* $[t_{q-1}, t_m)$ and compute it recursively as follows, for $q \geq 1$:

$$\begin{aligned} \mathcal{N}_q^q &:= \ker O_q \\ \mathcal{N}_q^k &:= \ker O_k \cap G_{k-1} e^{A_{k-1}\tau_{k-1}} \mathcal{N}_q^{k-1}, \quad q+1 \leq k \leq m, \end{aligned} \tag{7.2}$$

where $\tau_k := t_k - t_{k-1}$.

An alternative dual characterization of forward observability is possible by inspecting whether the complete state information is available. This is achieved in terms of the *observable subspace* $\mathcal{Q}_q^m$, defined in this chapter as the orthogonal complement of $\mathcal{N}_q^m$. It is noted that a recursive expression for $\mathcal{Q}_q^m$ is given by

$$\begin{aligned} \mathcal{Q}_q^q &= \mathcal{R}(O_q^\top) \\ \mathcal{Q}_q^k &= G_{k-1}^{-\top} e^{-A_{k-1}^\top \tau_{k-1}} \mathcal{Q}_q^{k-1} + \mathcal{R}(O_k^\top), \quad q+1 \leq k \leq m. \end{aligned} \tag{7.3}$$

We characterize the observability of system (7.1) using these subspaces in the following result, which is essentially a restatement of [20, Theorem 2]:

Theorem 7.2 (Forward observability characterization) *Consider the switched system (7.1) with $(u, v) \equiv 0$. Then, $\mathcal{N}_q^m$ for some $m \geq q \geq 1$ characterizes the unobservable space in the following sense:*

$$y_{[t_{q-1}, t_m)} \equiv 0 \quad \Leftrightarrow \quad x(t_{m-1}) \in \mathcal{N}_q^m.$$

In particular, if there exists $m \geq q$ such that $\mathcal{N}_q^m = \{0\}$, or equivalently $\mathcal{Q}_q^m = \mathbb{R}^n$, then the state $x(t_{m-1})$ (and hence the complete future trajectory) can be determined from the knowledge of (u, v, y) on the interval $[t_{q-1}, t_m)$.

We are often interested in deriving a direct formula for $\mathcal{Q}_q^m$ instead of the recursive one given in (7.2). For that, let us consider the matrix

$$\Phi_j^k = G_{k-1}e^{A_{k-1}\tau_{k-1}} \dots G_j e^{A_j \tau_j}, \quad k > j$$

which defines the flow of system (7.1) with zero inputs from t_{j-1} to t_{k-1}, and assume that, for $k \geq q+2$, $i = 1, 2, \dots, k-q-1$, $q \in \mathbb{N}$, the following condition holds[2]:

$$\Phi_{k-i}^k(\ker O_{k-i} \cap \Phi_{k-i-1}^{k-i} \mathcal{N}_q^{k-i-1}) = \Phi_{k-i}^k \ker O_{k-i} \cap \Phi_{k-i-1}^k \mathcal{N}_q^{k-i-1}. \quad (7.4)$$

It is readily checked that, if (7.4) holds, then the sequential definition (7.2) leads to another equivalent expression for $\mathcal{N}_q^m$, $m \geq q \geq 1$, given by:

$$\begin{aligned} \mathcal{N}_q^m &= \bigcap_{j=m,\dots,q} \Phi_j^m \ker O_j \\ &= \ker O_m \cap G_{m-1} \ker(O_{m-1}) \cap \left(\bigcap_{i=q}^{m-2} \prod_{l=m-1}^{i+1} G_l e^{A_l \tau_l} G_i \ker O_i \right), \end{aligned} \quad (7.5)$$

where Φ_k^k denotes the identity matrix, and we used the fact that $e^{A_j \tau_j} \ker G_j = \ker G_j$. Condition (7.4) indeed holds when each of the matrix $G_q, q \in \mathbb{N}$, is invertible because in that case the mapping Φ_j^k, for all $j, k \in \mathbb{N}$, $k > j$, is invertible.

Similarly, when (7.4) holds, $\mathcal{Q}_q^m$ in (7.3) is equivalently expressed as:

$$\begin{aligned} \mathcal{Q}_q^m &= (\mathcal{N}_q^m)^\perp \\ &= \sum_{i=q}^{m-2} \prod_{l=m-1}^{i+1} G_l^{-\top} e^{-A_l^\top \tau_l} G_i^{-\top} \mathcal{R}(O_i^\top) + G_{m-1}^{-\top} \mathcal{R}(O_{m-1}^\top) + \mathcal{R}(O_m^\top). \end{aligned} \quad (7.6)$$

7.3 Observer Design

Using the geometric conditions for forward observability stated in the previous section, we now proceed to design an observer. Our proposed observer is given by:

$$\dot{\hat{x}}(t) = A_q \hat{x}(t) + B_q u(t), \qquad t \in [t_{q-1}, t_q), \quad (7.7a)$$

$$\hat{x}(t_q) = G_q(\hat{x}(t_q^-) - \xi_q) + H_q v_q, \quad (7.7b)$$

---

[2] Note that, $A(\mathcal{V}_1 \cap \mathcal{V}_1) \subset A\mathcal{V}_1 \cap A\mathcal{V}_2$, and the equality does not hold in general. The necessary and sufficient condition for equality to hold is that $(\mathcal{V}_1 + \mathcal{V}_2) \cap \ker A = \mathcal{V}_1 \cap \ker A + \mathcal{V}_2 \cap \ker A$, which is the case when A is invertible. For systems with non-invertible jump maps, the flow matrix Φ_i^j is not necessarily invertible and (7.4) does not hold in general.

with an arbitrary initial condition $\hat{x}(t_0) \in \mathbb{R}^n$ and the expression for ξ_q will be computed in the sequel. It is seen that the observer consists of a system copy and unlike classical methods where the continuous dynamics of the estimate are driven by an error injection term, the observer (7.7) updates the state estimate only at discrete switching instants by an error correction vector ξ_q. If for some $q \in \mathbb{N}$, ξ_q equals the state estimation error $\hat{x}(t_q^-) - x(t_q^-)$, then the Eq. (7.7) gives $\hat{x}(t_q) = x(t_q)$, and from then onward we can recover the exact value of the trajectory x. However, in practice, where we do not use the derivatives of the output, it is not easy to recover the exact value of the state estimation error. Thus, our goal is to compute ξ_q, for each $q \in \mathbb{N}$, such that it *approximates* the value of state estimation error at time t_q^-, which will result in $\hat{x}(t)$ converging to $x(t)$ as t increases.

With this motivation, we introduce the state estimation error $\tilde{x} := \hat{x} - x$, and the error dynamics are given by

$$\dot{\tilde{x}}(t) = A_q \tilde{x}(t), \qquad t \in [t_{q-1}, t_q), \tag{7.8a}$$

$$\tilde{x}(t_q) = G_q(\tilde{x}(t_q^-) - \xi_q). \tag{7.8b}$$

The corresponding output error is defined as

$$\tilde{y}(t) := C_q \hat{x}(t) + D_q u(t) - y(t) = C_q \tilde{x}(t), \quad t \in [t_{q-1}, t_q).$$

The basic idea in computing ξ_q is to

- Firstly, identify the observable components of the individual subsystems that can be estimated using classical state-estimation techniques. For subsystem $q \in \mathbb{N}$, let $z_q \in \mathscr{R}(O_q^\top)$ denote the vector of such observable component.
- Secondly, derive an equation for $\tilde{x}(t_q^-)$ of the form[3]

$$\tilde{x}(t_q^-) = \Xi_q(z_q, z_{q-1}, \ldots, z_{q-\mathbf{m}^*}, \xi_{q-1}, \ldots, \xi_{q-\mathbf{m}^*}) \tag{7.9}$$

  for some $\mathbf{m}^* \in \mathbb{N}$.
- Finally, let

$$\xi_q = \Xi_q(\hat{z}_q, \hat{z}_{q-1}, \ldots, \hat{z}_{q-\mathbf{m}^*}, \xi_{q-1}, \ldots, \xi_{q-\mathbf{m}^*}). \tag{7.10}$$

We will develop calculations for each of the aforementioned steps in detail and arrive at a formal statement on error convergence that results from the observer. To do that, we need to introduce some assumptions that allow us to follow this proposed line of thought.

[3] With slight abuse of notation, the vectors z_j in (7.9) will be replaced by $z_j(t_j^-)$, so that the notation z_j will be used to denote a function.

The identification of observable components in the first step could be achieved easily by Kalman-like decomposition without imposing any constraints on system structure. For the second step, however, where we want to write $\tilde{x}(t_q^-)$, for each $q \in \mathbb{N}$, in terms of the observable components of the currently active mode and some past modes, we need the following assumption on the switching signal and system dynamics:

Assumption 7.3 The switched system (7.1) is persistently forward observable in the sense that there exists an $\mathtt{m}^* \in \mathbb{N}$ such that

$$\dim \mathscr{Q}^q_{q-\mathtt{m}^*} = \mathtt{n}, \qquad \forall q \geq \mathtt{m}^* + 1. \tag{7.11}$$

The integer $\mathtt{m}^*$ in Assumption 7.3 is interpreted as the minimal number of switches required to gain forward observability.

For the third step, it is seen that if $\hat{z}_q$ closely approximates z_q, and Ξ_q is globally Lipschitz (in our calculations, it will be linear), then it follows from (7.8) that the norm of the state estimation error at switching instants $\tilde{x}(t_q)$ becomes small. Since there is no error correction between the switching instants, it is important to update the estimate repeatedly for asymptotic convergence and also make sure that the error does not get arbitrarily large between the two switching instants. This motivates us to introduce the following assumptions for our observer design:

Assumption 7.4 The switching is persistent in the sense that a switch occurs at least once in any time interval of length D; that is,

$$t_q - t_{q-1} < D, \qquad \forall q \in \mathbb{N}. \tag{7.12}$$

Assumption 7.5 The induced matrix norms $\|A_q\|$ and $\|G_q\|$ are uniformly bounded for all $q \in \mathbb{N}$.

Note that Assumption 7.5 holds when A_q, G_q belong to a set of finite elements. Assumption 7.4 is in contrast to the conditions proposed for observer designs in [1] in the sense that we require the switching to be sufficiently fast and not too slow. This is not surprising because the works like [1] assume the observability of individual modes, so that the resulting error dynamics are stable for each subsystem, and a result on stability of switched systems with slow switching could be invoked [7] to show error convergence. In our work, however, since the individual subsystems are not assumed to observable, so that the resulting error dynamics for a particular mode are not necessarily stable, we need to switch fast enough between the unstable (or, partly stable) switched systems to obtain error convergence.

In the remainder of this section, the above thought process is formalized by following the steps outlined earlier to compute the correction vector ξ_q. The identification of observable components for individual subsystems is carried out in Sect. 7.3.1. For the computation of the map Ξ_q, we will discuss two cases separately: in the first case, we assume that all the jump maps G_q, $q \in \mathbb{N}$, are invertible and in the second case, we allow for non-invertible jump maps. For each of these cases, we show that the resulting state estimation error converges to zero (Theorems 7.6 and 7.7, respectively) if the observable components are estimated accurately enough. Followed by the theorem statements, we will arrive at a specific criteria for obtaining the estimates by carrying out the error analysis for each case. Our computations are summed up in Algorithm 7.1 (for invertible jump maps) and Algorithm 7.2 (for non-invertible jump maps).

7.3.1 Observability Decomposition of Error Dynamics

As a first step in computing ξ_q, $q \in \mathbb{N}$, we want to write $\tilde{x}$ in terms of observable components of individual subsystems. To do that, we first find a coordinate change for each mode, similar to the Kalman decomposition. For each $q \in \mathbb{N}$, choose a matrix Z_q such that its columns are an orthonormal basis of $\mathscr{R}(O_q^\top)$, so that $\mathscr{R}(Z_q) = \mathscr{R}(O_q^\top)$. Similarly, choose a matrix W_q such that its columns are an orthonormal basis of $\ker O_q$. From the construction, there are matrices $S_q \in \mathbb{R}^{r_q \times r_q}$ and $R_q \in \mathbb{R}^{\mathbf{y} \times r_q}$, where $r_q = \operatorname{rank} O_q$, such that $Z_q^\top A_q = S_q Z_q^\top$ and $C_q = R_q Z_q^\top$, and that the pair (S_q, R_q) is observable. Let $z_q := Z_q^\top \tilde{x} \in \mathbb{R}^{r_q}$ and $w_q := W_q^\top \tilde{x} \in \mathbb{R}^{\mathbf{n}-r_q}$. So, for the interval $[t_{q-1}, t_q)$, we obtain,

$$\dot{z}_q = Z_q^\top A_q \tilde{x} = S_q z_q, \quad \tilde{y} = C_q \tilde{x} = R_q z_q, \tag{7.13a}$$

$$z_q(t_{q-1}) = Z_q^\top \tilde{x}(t_{q-1}). \tag{7.13b}$$

Since z_q is observable over the interval $[t_{q-1}, t_q)$, a standard Luenberger observer is designed as

$$\dot{\hat{z}}_q = S_q \hat{z}_q + L_q(\tilde{y} - R_q \hat{z}_q), \quad t \in [t_{q-1}, t_q), \tag{7.14a}$$

$$\hat{z}_q(t_{q-1}) = 0, \tag{7.14b}$$

whose role is to estimate $z_q(t_q^-)$ at the end of the interval. Note that we have fixed the initial condition of the estimator to be zero for each interval. A consequence of introducing the observable and unobservable components is that the vector $\tilde{x}(t_q^-)$ can be written as,

$$\tilde{x}(t_q^-) = \begin{bmatrix} Z_q^\top \\ W_q^\top \end{bmatrix}^{-1} \begin{bmatrix} z_q(t_q^-) \\ w_q(t_q^-) \end{bmatrix} = Z_q z_q(t_q^-) + W_q w_q(t_q^-), \tag{7.15}$$

where $w_q(t_q^-)$ on the right-hand side remains unknown. Our objective now is to write $\tilde{x}(t_q^-)$ only in terms of known or recoverable quantities, that is, only as a function of the vectors $z_j(t_j^-)$, $j = q, q-1, \ldots, q-\mathbf{m}^*$, and ξ_k, $k = q-1, \ldots, q-\mathbf{m}^*$, for $\mathbf{m}^*$ given in Assumption 7.3. The calculations for arriving at such a formula for the general case are given in Sect. 7.3.3, but for the case when all jumps are invertible, one can derive a simpler formula. Because of simplicity, and to give an intuition about the calculations leading up to the computable expression for ξ_q, we choose to treat the case with invertible jumps first in the following section.

7.3.2 Error Correction with Invertible State Reset Maps

The goal of this subsection is to derive an expression of the form (7.9) when the jump maps G_q are invertible. For that, we first define the state-flow matrix Ψ_p^q, $p, q \in \mathbb{N}$, $p < q$, as

$$\Psi_p^q := e^{A_q \tau_q} G_q e^{A_{q-1}\tau_{q-1}} G_{q-1}, \ldots, e^{A_{p+1}\tau_{p+1}} G_p,$$

which transports $\tilde{x}(t_p^-)$ to $\tilde{x}(t_q^-)$ along (7.8) by

$$\tilde{x}(t_q^-) = \Psi_p^q \tilde{x}(t_p^-) - \sum_{k=p}^{q-1} \Psi_k^q \xi_k, \tag{7.16}$$

where for convenience, we let Ψ_q^q to be the identity matrix. For $q > \mathbf{m}^*$, we now have the following series of equivalent expressions for $\tilde{x}(t_q^-)$:

$$\begin{aligned}
\tilde{x}(t_q^-) &= Z_q z_q(t_q^-) + W_q w_q(t_q^-) \\
&= \Psi_{q-1}^q Z_{q-1} z_{q-1}(t_{q-1}^-) + \Psi_{q-1}^q W_{q-1} w_{q-1}(t_{q-1}^-) - \Psi_{q-1}^q \xi_{q-1} \\
&= \Psi_{q-2}^q Z_{q-2} z_{q-2}(t_{q-2}^-) + \Psi_{q-2}^q W_{q-2} w_{q-2}(t_{q-2}^-) - \sum_{k=1}^{2} \Psi_{q-k}^q \xi_{q-k} \\
&\;\;\vdots \\
&= \Psi_{q-\mathbf{m}^*}^q Z_{q-\mathbf{m}^*} z_{q-\mathbf{m}^*}(t_{q-\mathbf{m}^*}^-) + \Psi_{q-\mathbf{m}^*}^q W_{q-\mathbf{m}^*} w_{q-\mathbf{m}^*}(t_{q-\mathbf{m}^*}^-) - \sum_{k=q-\mathbf{m}^*}^{q-1} \Psi_k^q \xi_k.
\end{aligned} \tag{7.17}$$

In these equations, the vector $\tilde{x}(t_q^-)$ is not only expressed in terms of the observable and unobservable components of mode $q \in \mathbb{N}$, but also those of previously active $q - \mathbf{m}^*$ modes. In other words, for each $q - \mathbf{m}^* \le k \le q$, the term $\Psi_k^q Z_k z_k(t_k^-)$ transports the observable information of the kth mode from the interval $[t_{k-1}, t_k)$ to the time instant t_q^-. However, in each equation, there is an added unknown term $w_k(t_k^-)$. Thus, in order to obtain an explicit expression for $\tilde{x}(t_q^-)$ in terms of z_k, $k = q, q-1, \ldots, q - \mathbf{m}^*$, we must

- first, eliminate $w_k(t_k^-)$ from each equation in (7.17), and

- secondly, make sure that the resulting set of equations is not under-determined with $\tilde{x}(t_q^-)$ as the unknown.

To achieve the first objective, we introduce the matrices Θ_k^q whose columns form the basis of the subspace $\mathscr{R}(\Psi_k^q W_k)^\perp$; that is,

$$\mathscr{R}(\Theta_k^q) = \mathscr{R}(\Psi_k^q W_k)^\perp, \quad k = q - \mathbf{m}^*, \ldots, q.$$

Then, for each equality in (7.17), we obtain the relation

$$\Theta_k^{q\top} \tilde{x}(t_q^-) = \Theta_k^{q\top} \left(\Psi_k^q Z_k z_k(t_k^-) - \sum_{j=k}^{q-1} \Psi_j^q \xi_j \right), \quad k = q - \mathbf{m}^*, \ldots, q. \quad (7.18)$$

It now follows that if the matrix

$$\Theta_q^\top := \mathrm{col}(\Theta_q^{q\top}, \Theta_{q-1}^{q\top}, \ldots, \Theta_{q-\mathbf{m}^*}^{q\top})$$

has full column rank equal to $\mathbf{n}$, then the set of equations (7.18) can be solved for $\tilde{x}(t_q^-)$. It can be shown that the matrix Θ_q, $q > \mathbf{m}^*$, has rank $\mathbf{n}$ if, and only if, Assumption 7.3 holds. Indeed, $\Theta_q^\top$ has full column rank $\mathbf{n}$ if, and only if, $\ker(\Theta_q^\top) = \{0\}$, or equivalently,

$$\mathscr{R}(\Theta_q^q) + \mathscr{R}(\Theta_{q-1}^q) + \cdots + \mathscr{R}(\Theta_{q-\mathbf{m}^*}^q) = \mathbb{R}^\mathbf{n}.$$

Using the fact that $\mathscr{R}(W_k)^\perp = (\ker O_k)^\perp = \mathscr{R}(O_k^\top)$, $e^{-A_k^\top \tau_k} \mathscr{R}(O_k^\top) = \mathscr{R}(O_k^\top)$, and the expression (7.6), it follows under Assumption 7.3 that

$$\begin{aligned}
&\mathscr{R}(W_q)^\perp + \mathscr{R}(\Psi_{q-1}^q W_{q-1})^\perp + \cdots + \mathscr{R}(\Psi_{q-\mathbf{m}^*}^q W_{q-\mathbf{m}^*})^\perp \\
&= e^{-A_q^\top \tau_q} \left(\mathscr{R}(O_q^\top) + G_{q-1}^{-\top} \mathscr{R}(O_{q-1}^\top) + \sum_{i=q-\mathbf{m}^*}^{q-2} \prod_{l=q-1}^{i+1} G_l^{-\top} e^{-A_l^\top \tau_l} G_i^{-\top} \mathscr{R}(O_i^\top) \right) \\
&= e^{-A_q^\top \tau_q} \mathscr{Q}_{q-\mathbf{m}^*}^q = \mathbb{R}^\mathbf{n},
\end{aligned} \quad (7.19)$$

where we recall from Sect. 7.2.1 that the second to last equality only holds when the jump maps G_q are invertible. Thus, the matrix $\Theta_q^\top$ is left-invertible, so that $(\Theta_q^\top)^\dagger = (\Theta_q \Theta_q^\top)^{-1} \Theta_q$, where $\dagger$ denotes the left-pseudo-inverse. Let us now introduce the matrix

$$\Omega_{q-\mathbf{m}^*}^q := \begin{bmatrix} \Theta_q^{q\top} \Psi_q^q Z_q z_q(t_q^-) \\ \vdots \\ \Theta_{q-\mathbf{m}^*}^{q\top} \left(\Psi_{q-\mathbf{m}^*}^q Z_{q-\mathbf{m}^*} z_{q-\mathbf{m}^*}(t_{q-\mathbf{m}^*}^-) - \sum_{l=q-\mathbf{m}^*}^{q-1} \Psi_l^q \xi_l \right) \end{bmatrix},$$

so that the arguments of the matrix $\Omega^q_{q-\mathbf{m}^*}$ are $z_q(t_q^-), z_{q-1}(t_{q-1}^-), \ldots, z_{q-\mathbf{m}^*}(t_{q-\mathbf{m}^*}^-)$ and $\xi_{q-1}, \ldots, \xi_{q-\mathbf{m}^*}$. It then follows that

$$\begin{aligned}\tilde{x}(t_q^-) &= (\Theta_q^\top)^\dagger \Omega^q_{q-\mathbf{m}^*}(z_q(t_q^-), \ldots, z_{q-\mathbf{m}^*}(t_{q-\mathbf{m}^*}^-), \xi_{q-1}, \ldots, \xi_{q-\mathbf{m}^*}) \\ &=: \Xi_q(z_q(t_q^-), \ldots, z_{q-\mathbf{m}^*}(t_{q-\mathbf{m}^*}^-), \xi_{q-1}, \ldots, \xi_{q-\mathbf{m}^*}). \end{aligned} \tag{7.20}$$

Once again, it is seen from (7.20) that, if we can estimate $z_k(t_k^-)$, $k = q - \mathbf{m}^*$, $\ldots, q$, without error, then by (7.20) the plant state $x(t_q^-)$ is exactly recovered because $x(t_q^-) = \hat{x}(t_q^-) - \tilde{x}(t_q^-)$, and both entities on the right side of the equation are known. However, since this is not the case, we set ξ_q to be an estimate of $\tilde{x}(t_q^-)$ through

$$\boxed{\xi_q = \Xi_q(\hat{z}_q(t_q^-), \hat{z}_{q-1}(t_{q-1}^-), \ldots, \hat{z}_{q-\mathbf{m}^*}(t_{q-\mathbf{m}^*}^-), \xi_{q-1}, \ldots, \xi_{q-\mathbf{m}^*}), \quad q > \mathbf{m}^*} \tag{7.21}$$

and for $1 \le q \le \mathbf{m}^*$, we let $\xi_q = 0$. The following theorem now states that if the estimates used in (7.21) are good enough, then the resulting estimate converges to the actual state asymptotically.

Theorem 7.6 *Consider the observer proposed in* (7.7) *under Assumptions 7.3–7.5 and also assume that the jump maps G_q, $q \in \mathbb{N}$, are invertible. If the error correction vector ξ_q is computed using* (7.21) *in which $\hat{z}_j$, $j = q, \ldots, q - \mathbf{m}^*$ are obtained from the Luenberger observers* (7.14), *then the output injection gains L_j in* (7.14) *can be chosen such that*

$$\lim_{t\to\infty} |\hat{x}(t) - x(t)| = 0.$$

To complete the design procedure, we need to choose the gain matrices $L_q, q \in \mathbb{N}$. This is done in Sect. 7.3.2.1, where we analyze the state estimation error resulting from injecting the expression for ξ_q from (7.21) in Eq. (7.8).

7.3.2.1 Error Analysis and Gain Criterion

The gain matrices $L_q, q \in \mathbb{N}$, are basically chosen such that $\tilde{x}(t_q)$ converges to zero as q increases, because it follows from (7.8) and Assumptions 7.4 and 7.5 that the estimation error $\tilde{x}(t)$ for the interval $[t_q, t_{q+1})$ is bounded by

$$|\tilde{x}(t)| = |e^{A_{q+1}t}\tilde{x}(t_q)| \le e^{a(t-t_q)}|\tilde{x}(t_q)|$$

with constant a such that $\|A_q\| \le a$, and thus,

$$|\tilde{x}(t)| \le e^{aD}|\tilde{x}(t_q)|.$$

Therefore, if $|\tilde{x}(t_q)| \to 0$ as $q \to \infty$, then we achieve that

$$\lim_{t\to\infty} |\tilde{x}(t)| = 0. \tag{7.22}$$

Since the operator $\Omega^q_{q-\mathbf{m}^*}$ is linear in its arguments, it is noted that,

$$\tilde{x}(t_q) = G_q(\tilde{x}(t_q^-) - \xi_q) \tag{7.23a}$$

$$= G_q\Big(\Xi_q(z_q(t_q^-), \ldots, z_{q-\mathbf{m}^*}(t^-_{q-\mathbf{m}^*}), \xi_{q-1}, \ldots, \xi_{q-\mathbf{m}^*})$$
$$- \Xi_q(\hat{z}_q(t_q^-), \ldots, \hat{z}_{q-\mathbf{m}^*}(t^-_{q-\mathbf{m}^*}), \xi_{q-1}, \ldots, \xi^-_{q-\mathbf{m}^*})\Big) \tag{7.23b}$$

$$= -G_q(\Theta_q^\top)^\dagger \Omega^q_{q-\mathbf{m}^*}(\tilde{z}_q(t_q^-), \ldots, \tilde{z}_{q-\mathbf{m}^*}(t^-_{q-\mathbf{m}^*}), 0, \ldots, 0) \tag{7.23c}$$

where $\tilde{z}_j := \hat{z}_j - z_j$, $j = q, \ldots, q - \mathbf{m}^*$. It follows from (7.13) and (7.14) that

$$\tilde{z}_j(t_{j-1}) = \hat{z}_j(t_{j-1}) - z_j(t_{j-1}) = 0 - Z_j^\top \tilde{x}(t_{j-1}),$$

and

$$\tilde{z}_j(t_j^-) = e^{(S_j - L_j R_j)\tau_j} \tilde{z}_j(t_{j-1}) = -e^{(S_j - L_j R_j)\tau_j} Z_j^\top \tilde{x}(t_{j-1}).$$

Plugging this expression in (7.23), and introducing the matrices M_j^q, for $j = q, q-1, \ldots, q - \mathbf{m}^*$, as follows:

$$[M_q^q, M_{q-1}^q, \ldots, M^q_{q-\mathbf{m}^*}] := G_q(\Theta_q^\top)^\dagger \times$$
$$\text{blockdiag}\left(\Theta_q^{q^\top}\Psi_q^q, \Theta_{q-1}^{q^\top}\Psi_{q-1}^q, \ldots, \Theta_{q-\mathbf{m}^*}^{q^\top}\Psi^q_{q-\mathbf{m}^*}\right), \tag{7.24}$$

we obtain

$$\tilde{x}(t_q) = \sum_{j=q-\mathbf{m}^*}^{q} M_j^q Z_j e^{(S_j - L_j R_j)\tau_j} Z_j^\top \tilde{x}(t_{j-1}), \tag{7.25}$$

where we note that the argument of every $\mathbf{n} \times \mathbf{n}$ matrix M_j^q, $j = q - \mathbf{m}^*, \cdots, q$, is the vector $(\tau_q, \ldots, \tau_{q-\mathbf{m}^*+1})$ because of the matrices Ψ_j^q in the definition.

In order to bound the norm of $\tilde{x}(t_q)$, one can always find the constants $\alpha_j, \gamma_j > 0$ such that $\|Z_j e^{(S_j - L_j R_j)\tau_j} Z_j^\top\| \le \alpha_j e^{-\gamma_j \tau_j}$. With constants $\lambda_j^q > 0$ denoting the induced norm of M_j^q, we get

$$|\tilde{x}(t_q)| \le \sum_{j=q-\mathbf{m}^*}^{q} \lambda_j^q \alpha_j e^{-\gamma_j \tau_j} |\tilde{x}(t_{j-1})|. \tag{7.26}$$

If for each $q > \mathbf{m}^*$, and $j = q - \mathbf{m}^*, \ldots, q$, the gains L_j are chosen such that

$$\lambda_j^q \alpha_j e^{-\gamma_j \tau_j} \le c < \frac{1}{\mathbf{m}^* + 1}$$

(such a choice is always feasible [12, Lemma 1]), then

$$|\tilde{x}(t_q)| < c \sum_{j=q-\mathbf{m}^*}^{q} |\tilde{x}(t_{j-1})|.$$

One can now use [19, Lemma 1] to conclude that $\lim_{q\to\infty} |\tilde{x}(t_q)| = 0$.

Algorithm 7.1 Implementation of hybrid observer for invertible jump maps

Require: u, v, y
Ensure: Run (7.7) for $t \in [t_0, t_{\mathbf{m}^*+1})$ with some $\hat{x}(t_0)$
for all $q \geq \mathbf{m}^* + 1$ **do**
  for $j = q - \mathbf{m}^*$ **to** q **do**
    Compute the injection gain L_j such that

$$\|M_j^q Z_j e^{(S_j - L_j R_j)\tau_j} Z_j^\top\| \leq c < \frac{1}{\mathbf{m}^* + 1}. \tag{7.27}$$

    Obtain $\hat{z}_j(t_j^-)$ by running the individual observer (7.14) for the j-th mode.
  end for
  Compute ξ_q from (7.21) , to implement (7.7).
  Compute $\hat{x}(t_q)$ using (7.7) and run (7.7) over the interval $[t_q, t_{q+1})$.
end for

7.3.3 Error Correction for Non-invertible State Reset Maps

The formula for ξ_q computed in Sect. 7.3.2 is only valid for the case of invertible jump maps. To derive a more general formula, which is also valid for the case of non-invertible jump maps, we basically follow the same procedure but the details are slightly more involved.

For $p, q \in \mathbb{N}$ with $p < q$, let Q_p^q and N_p^q be matrices such that their columns are an orthonormal basis of $e^{-A_q^\top \tau_q} \mathcal{Q}_p^q$ and $e^{A_q \tau_q} \mathcal{N}_p^q$, respectively. The corresponding projections of $\tilde{x}(t_q^-)$ onto these subspaces are defined by letting $\chi_p^q := Q_p^{q\top} \tilde{x}(t_q^-)$ and $\nu_p^q := N_p^{q\top} \tilde{x}(t_q^-)$. Thus, it is seen that in addition to (7.15), another way of expressing $\tilde{x}(t_q^-)$ is:

$$\tilde{x}(t_q^-) = \begin{bmatrix} Q_p^{q\top} \\ N_p^{q\top} \end{bmatrix}^{-1} \begin{bmatrix} \chi_p^q \\ \nu_p^q \end{bmatrix} = Q_p^q \chi_p^q + N_p^q \nu_p^q. \tag{7.28}$$

The definition of χ_p^q implies that it contains the information of the error $\tilde{x}(t_q^-)$, which we are able to extract from the output on the interval $[t_{p-1}, t_q)$ as given by the observability space $\mathcal{Q}_p^q$. For $q > \mathbf{m}^*$, the forward observability assumption ensures that $\chi^q_{q-\mathbf{m}^*}$ contains all information of $\tilde{x}(t_q^-)$; in fact $Q^q_{q-\mathbf{m}^*}$ is then an invertible matrix, and hence the equation $\chi^q_{q-\mathbf{m}^*} = Q^{q\top}_{q-\mathbf{m}^*}\tilde{x}(t_q^-)$ is uniquely solvable for $\tilde{x}(t_q^-)$.

Thus, once again we are interested in representing $\tilde{x}(t_q^-)$ only in terms of the known vectors χ^q_j, and eliminate its dependency over the terms involving ν^q_j, $j = q, q-1, \ldots, q-\mathbf{m}^*$. For that, we once again introduce the matrix Θ^q_p whose columns form the basis of the subspace $\mathscr{R}(e^{A_{q+1}\tau_{q+1}}G_q N^q_p)^\perp$; that is,

$$\Theta^{q\top}_p e^{A_{q+1}\tau_{q+1}} G_q N^q_p = 0. \tag{7.29}$$

Compared to the case treated earlier, the key difference is that we do not transport the observable components of the individual subsystems to one time instant through the state-transition matrix. Instead, we gather all the observable information for $\tilde{x}(t_{q-1}^-)$ over the interval $[t_{p-1}, t_{q-1})$ into the vector χ^{q-1}_p, $p < q$, and combine it with the local observability information $z_q(t_q^-)$ for $\tilde{x}(t_q^-)$ obtained on the interval $[t_{q-1}, t_q)$ in order to recover more information for $\tilde{x}(t_q^-)$, represented by χ^q_p. For that, the following relationship between $\tilde{x}(t_q^-)$ and χ^{q-1}_p, $p < q$, is crucial:

$$\begin{aligned}\tilde{x}(t_q^-) &= e^{A_q\tau_q} G_{q-1}(\tilde{x}(t_{q-1}^-) - \xi_{q-1}) \\ &= e^{A_q\tau_q} G_{q-1}\left(Q^{q-1}_p \chi^{q-1}_p + N^{q-1}_p \nu^{q-1}_p - \xi_{q-1}\right).\end{aligned} \tag{7.30}$$

Combining this with (7.15) we obtain

$$\begin{pmatrix} Z_q^\top \\ \Theta^{q-1\top}_p \end{pmatrix} \tilde{x}(t_q^-) = \begin{pmatrix} z_q(t_q^-) \\ \Theta^{q-1\top}_p \left(e^{A_q\tau_q} G_{q-1}\left(Q^{q-1}_p \chi^{q-1}_p - \xi_{q-1}\right)\right)\end{pmatrix},$$

and hence we have more information about $\tilde{x}(t_q^-)$ by combining $z_q(t_q^-)$ and χ^{q-1}_p accordingly. Now, consider a full column rank matrix U^q_p such that

$$[Z_q, \Theta^{q-1}_p] U^q_p = Q^q_p.$$

Such a matrix always exists because from the definition of $\mathscr{Q}^q_p$ and Z_q, we have

$$\mathscr{R}(Q^q_p) = \mathscr{R}([Z_q, \Theta^{q-1}_p]).$$

From $\chi^q_p = Q^{q\top}_p \tilde{x}(t_q^-)$, it now follows that

$$\chi^q_p = U^{q\top}_p \begin{bmatrix} Z_q^\top \\ \Theta^{q-1\top}_p \end{bmatrix} \tilde{x}(t_q^-)$$

$$= U_p^{q\top} \begin{pmatrix} z_q(t_q^-) \\ \Theta_p^{q-1\top} \left(e^{A_q \tau_q} G_{q-1} \left(Q_p^{q-1} \chi_p^{q-1} - \xi_{q-1} \right) \right) \end{pmatrix} \tag{7.31}$$

$$= U_p^{q\top} \left[\begin{array}{c|c} Z_q^\top & 0 \\ \hline 0 & \Theta_p^{q-1\top} e^{A_q \tau_q} G_{q-1} \end{array} \right] \begin{pmatrix} Z_q z_q(t_q^-) \\ Q_p^{q-1} \chi_p^{q-1} - \xi_{q-1} \end{pmatrix} \tag{7.32}$$

$$\triangleq J_p^q Z_q z_q + K_p^q \left(Q_p^{q-1} \chi_p^{q-1} - \xi_{q-1} \right). \tag{7.33}$$

Note that (7.32) expresses the vector χ_p^q recursively in terms of χ_p^{q-1}. Recall that $\mathscr{Q}_p^p = \mathcal{R}(O_p^\top)^\perp = \mathscr{R}(Z_p)$, hence we can assume $Q_p^p = Z_p$ and we have the "initial value" for the recursion (7.32) given by $\chi_p^p = z_p$.

If $z_{q-\mathbf{m}^*}, \dots, z_q$ were known, then we would be able to compute the error $\tilde{x}(t_q^-)$ exactly and would pick $\xi_q = \tilde{x}(t_q^-)$. Since this is not the case, we work with the estimates $\hat{z}_{q-\mathbf{m}^*}, \dots, \hat{z}_q$ to compute ξ_q.

In summary, having introduced the matrices Z_q as in (7.15), Q_p^q as in (7.28), and Θ_p^q as in (7.29), for $q \in \mathbb{N}$, we let

$$\boxed{\xi_q = \begin{cases} 0, & 1 \le q \le \mathbf{m}^* \\ Q_{q-\mathbf{m}^*}^q \hat{\chi}_{q-\mathbf{m}^*}^q, & q \ge \mathbf{m}^* + 1 \end{cases}} \tag{7.34}$$

where $\hat{\chi}_{q-\mathbf{m}^*}^{q-k}$, for $k = \mathbf{m}^* - 1, \dots, 0$, is computed recursively as follows:

$$\boxed{\begin{aligned} \hat{\chi}_{q-\mathbf{m}^*}^{q-\mathbf{m}^*} &= \hat{z}_{q-\mathbf{m}^*} \\ \hat{\chi}_{q-\mathbf{m}^*}^{q-k} &= J_{q-\mathbf{m}^*}^{q-k} Z_{q-k} \hat{z}_{q-k} + K_{q-\mathbf{m}^*}^{q-k} \left(Q_{q-\mathbf{m}^*}^{q-k-1} \hat{\chi}_{q-\mathbf{m}^*}^{q-k-1} - \xi_{q-k-1} \right), \end{aligned}} \tag{7.35}$$

and

$$[J_{q-\mathbf{m}^*}^{q-k}, K_{q-\mathbf{m}^*}^{q-k}] := U_{q-\mathbf{m}^*}^{q-k\ \top} \begin{bmatrix} Z_{q-k}^\top & 0 \\ 0 & \Theta_{q-\mathbf{m}^*}^{q-k-1\top} e^{A_{q-k} \tau_{q-k}} G_{q-k-1} \end{bmatrix}. \tag{7.36}$$

Using the formula for ξ_q in (7.34), we can again state a result very similar to Theorem 7.6, but this time we do not place any constraints on the jump maps.

Theorem 7.7 *Consider observer* (7.7) *under Assumptions 7.3–7.5. If the error correction vector ξ_q is computed using* (7.34) *in which $\hat{\chi}_{q-\mathbf{m}^*}^q$ is computed in a recursive manner from Eqs.* (7.35) *and* (7.36)*, then the output injection gains L_j in* (7.14) *can be chosen such that*

$$\lim_{t \to \infty} |\hat{x}(t) - x(t)| = 0.$$

It just remains to show how well $\hat{\chi}^q_{q-\mathbf{m}^*}$ should approximate $\chi^q_{q-\mathbf{m}^*}$ by appropriate choice of gains L_j, $j \in \mathbb{N}$. Once again we motivate the gain criterion by analyzing the error.

7.3.3.1 Error Analysis and Gain Criterion

As in Sect. 7.3.2.1, we want to derive a gain criterion for L_j, $j \in \mathbb{N}$, such that $\lim_{q\to\infty} |\tilde{x}(t_q)| = 0$. It is noted that, for $q > \mathbf{m}^*$:

$$\begin{aligned}\tilde{x}(t_q) &= G_q(\tilde{x}(t_q^-) - \xi_q) = G_q Q^q_{q-\mathbf{m}^*}(\chi^q_{q-\mathbf{m}^*} - \hat{\chi}^q_{q-\mathbf{m}^*}) \\ &= -G_q Q^q_{q-\mathbf{m}^*}\tilde{\chi}^q_{q-\mathbf{m}^*},\end{aligned} \tag{7.37}$$

where $\tilde{\chi}^q_{q-\mathbf{m}^*} := \hat{\chi}^q_{q-\mathbf{m}^*} - \chi^q_{q-\mathbf{m}^*}$. In the sequel, we will derive an expression for $\tilde{\chi}^q_{q-\mathbf{m}^*}$ for a fixed $q > \mathbf{m}^*$ and plug it in (7.37) to show that $|\tilde{x}(t_q)|$ converges to zero as q increases for appropriate choice of the matrices L_j, $j \in \mathbb{N}$.

Toward this end, we first compute the difference $\tilde{z}_q := \hat{z}_q - z_q$, for $q \in \mathbb{N}$ as follows:

$$\tilde{z}_q(t_q^-) = \hat{z}_q(t_q^-) - z_q(t_q^-) = e^{(S_q - L_q R_q)\tau_q}\tilde{z}_q(t_{q-1}) = -\Lambda_q {Z_q}^\top \tilde{x}(t_{q-1}),$$

where we define $\Lambda_q := e^{(S_q - L_q R_q)\tau_q}$. As a first step in arriving at the expression for $\tilde{\chi}^q_{q-\mathbf{m}^*}$, we observe that $\tilde{\chi}^{q-\mathbf{m}^*}_{q-\mathbf{m}^*} = \tilde{z}_{q-\mathbf{m}^*}(t^-_{q-\mathbf{m}^*})$ and we compute $\tilde{\chi}^{q-\mathbf{m}^*+1}_{q-\mathbf{m}^*}$ as follows:

$$\begin{aligned}\tilde{\chi}^{q-\mathbf{m}^*+1}_{q-\mathbf{m}^*} &= \hat{\chi}^{q-\mathbf{m}^*+1}_{q-\mathbf{m}^*} - \chi^{q-\mathbf{m}^*+1}_{q-\mathbf{m}^*} \\ &= J^{q-\mathbf{m}^*+1}_{q-\mathbf{m}^*} Z_{q-\mathbf{m}^*+1}\tilde{z}_{q-\mathbf{m}^*+1}(t^-_{q-\mathbf{m}^*+1}) + K^{q-\mathbf{m}^*+1}_{q-\mathbf{m}^*} Z_{q-\mathbf{m}^*}\tilde{z}_{q-\mathbf{m}^*}(t^-_{q-\mathbf{m}^*}) \\ &= -\sum_{i=0}^{1}\left(V^{q-\mathbf{m}^*+1}_{q-\mathbf{m}^*,q-\mathbf{m}^*+i} Z_{q-\mathbf{m}^*+i}\Lambda_{q-\mathbf{m}^*+i} Z^\top_{q-\mathbf{m}^*+i}\tilde{x}(t_{q-\mathbf{m}^*+i-1})\right),\end{aligned}$$

where

$$V^{q-\mathbf{m}^*+1}_{q-\mathbf{m}^*,q-\mathbf{m}^*} := K^{q-\mathbf{m}^*+1}_{q-\mathbf{m}^*} \tag{7.38a}$$

$$V^{q-\mathbf{m}^*+1}_{q-\mathbf{m}^*,q-\mathbf{m}^*+1} := J^{q-\mathbf{m}^*+1}_{q-\mathbf{m}^*}. \tag{7.38b}$$

Finally, by introducing the matrices, $q > \mathbf{m}^*$, $k = \mathbf{m}^* - 2, \ldots, 0$ and $i = 0, \ldots, \mathbf{m}^* - k - 1$

$$V^{q-k}_{q-\mathbf{m}^*,q-k} := J^{q-k}_{q-\mathbf{m}^*} \tag{7.38c}$$

$$V^{q-k}_{q-\mathbf{m}^*,q-\mathbf{m}^*+i} := K^{q-k}_{q-\mathbf{m}^*} Q^{q-k-1}_{q-\mathbf{m}^*} V^{q-k-1}_{q-\mathbf{m}^*,q-\mathbf{m}^*+i} \tag{7.38d}$$

the expression for $\tilde{\chi}_{q-\mathbf{m}^*}^{q-k}$, $k = \mathbf{m}^* - 1, \ldots, 0$, is derived recursively:

$$\begin{aligned}
\tilde{\chi}_{q-\mathbf{m}^*}^{q-k} &= \hat{\chi}_{q-\mathbf{m}^*}^{q-k} - \chi_{q-\mathbf{m}^*}^{q-k} \\
&= J_{q-\mathbf{m}^*}^{q-k} Z_{q-k} \tilde{z}_{q-k}(t_{q-k}^-) + K_{q-\mathbf{m}^*}^{q-k} Q_{q-\mathbf{m}^*}^{q-k-1} \tilde{\chi}_{q-\mathbf{m}^*}^{q-k-1} \\
&= -\sum_{i=0}^{\mathbf{m}^*-k} V_{q-\mathbf{m}^*, q-\mathbf{m}^*+i}^{q-k} Z_{q-\mathbf{m}^*+i} \Lambda_{q-\mathbf{m}^*+i} Z_{q-\mathbf{m}^*+i}^\top \tilde{x}(t_{q-\mathbf{m}^*+i-1}).
\end{aligned}$$

Plugging this expression for $\tilde{\chi}_{q-\mathbf{m}^*}^{q}$ in (7.37), we now obtain

$$\tilde{x}(t_q) = G_q Q_{q-\mathbf{m}^*}^{q} \sum_{i=q-\mathbf{m}^*}^{q} V_{q-\mathbf{m}^*, i}^{q} Z_i \Lambda_i Z_i^\top \tilde{x}(t_{i-1}). \tag{7.39}$$

If, for each $k = 0, \ldots, \mathbf{m}^*$, and $q > \mathbf{m}^*$, the output injection matrices L_{q-k} are chosen to minimize the norm of Λ_{q-k} such that

$$\|G_q Q_{q-\mathbf{m}^*}^{q} V_{q-\mathbf{m}^*, q-k}^{q} Z_{q-k} \Lambda_{q-k} Z_{q-k}^\top\| \le c < \frac{1}{\mathbf{m}^* + 1}, \tag{7.40}$$

then it follows that

$$|\tilde{x}(t_q)| \le c \sum_{i=q-\mathbf{m}^*}^{q} |\tilde{x}(t_{i-1})|.$$

We can again invoke Lemma 1 from [19] to obtain $\lim_{q\to\infty} |\tilde{x}(t_q)| = 0$, which proves the desired result.

Algorithm 7.2 Hybrid observer for systems with non-invertible jump maps

Require: u, v, y
Ensure: Run (7.7) for $t \in [t_0, t_{\mathbf{m}^*+1})$ with some $\hat{x}(t_0)$
1: **for all** $q \ge \mathbf{m}^* + 1$ **do**
2:   **for** $j = q - \mathbf{m}^*$ **to** q **do**
3:     Compute the injection gain L_j such that (7.40) holds.
4:     Obtain $\hat{z}_j(t_j^-)$ by running the individual observer (7.14) for the j-th mode.
5:   **end for**
6:   Compute ξ_q from (7.34), (7.35), and (7.36), to implement (7.7).
7:   Compute $\hat{x}(t_q)$ using (7.7) and run (7.7) over the interval $[t_q, t_{q+1})$.
8: **end for**

7.4 Illustrative Examples

We will now apply our results for two academic examples. The first one considers the case without any state resets, and in the second case, we consider switching dynamics with non-invertible state reset maps.

7.4.1 Invertible State Reset Maps

Consider a switched system given by:

$$A_{2k-1} = \begin{bmatrix} 0 & 0 \\ 0 & 0 \end{bmatrix}, \qquad A_{2k} = \begin{bmatrix} 0 & 1 \\ -1 & 0 \end{bmatrix}, \qquad k \geq 1$$

$$C_{2k-1} = \begin{bmatrix} 1 & 0 \end{bmatrix}, \qquad C_{2k} = \begin{bmatrix} 0 & 0 \end{bmatrix}, \qquad k \geq 1.$$

with $G_k = I$, $H_k = 0$, $B_k = 0$, and $D_k = 0$ for each $k \geq 1$.

We assume that each mode is activated for τ seconds and $\tau \neq \kappa\pi$ for any $\kappa \in \mathbb{N}$. For simplicity, let us call $[(2k-2)\tau, (2k-1)\tau), k \in \mathbb{N}$, an *odd* interval, and the mode active on the odd intervals as the *odd* mode. Similarly, the intervals $[(2k-1)\tau, 2k\tau)$, $k \in \mathbb{N}$ are called *even* intervals, and the mode active on these intervals is called the *even* mode. We also use the notation q_o, k_o for odd positive integers and q_e, k_e for even positive integers. It can be verified that the system is forward observable over a time interval that involves the mode sequence *odd* → *even* → *odd*. Hence, we pick $\mathbf{m}^* = 3$ so that Assumption 7.3 holds. With an arbitrary initial condition $\hat{x}(0)$, the observer to be implemented is:

$$\left.\begin{aligned} \dot{\hat{x}}(t) &= A_{2k-1}\hat{x}(t) \\ \hat{y}(t) &= C_{2k-1}\hat{x}(t) \end{aligned}\right\}, \quad t \in [(2k-2)\tau, (2k-1)\tau), \tag{7.41a}$$

$$\left.\begin{aligned} \dot{\hat{x}}(t) &= A_{2k}\hat{x}(t) \\ \hat{y}(t) &= C_{2k}\hat{x}(t) \end{aligned}\right\}, \quad t \in [(2k-1)\tau, 2k\tau), \tag{7.41b}$$

$$\hat{x}(t_k) = \hat{x}(t_k^-) - \xi_k, \quad k \in \mathbb{N}. \tag{7.41c}$$

In order to determine the value of ξ_k, we start off with the estimators for the observable part of each subsystem, denoted by z_q in (7.13). Note that the *odd* mode has a one-dimensional unobservable subspace, whereas for *even* mode, the unobservable subspace is $\mathbb{R}^2$. Let z_{q_o} represent the partial information obtained from the *odd* mode, and z_{q_e} be a null vector as no information is gathered from the *even* mode. So the one-dimensional partial observer in (7.14) is implemented only for odd intervals. Also, for the *odd* mode, we obtain:

$$O_{q_\text{o}} = \begin{bmatrix} 1 & 0 \\ 0 & 0 \end{bmatrix}, \; W_{q_\text{o}} = \begin{bmatrix} 0 \\ 1 \end{bmatrix}, \; Z_{q_\text{o}} = \begin{bmatrix} 1 \\ 0 \end{bmatrix},$$

so that $S_{q_o} = 0$ and $R_{q_o} = 1$, which yields the observer in (7.14) as

$$\dot{\hat{z}}_{q_o} = -l_{q_o}\hat{z}_{q_o} + l_{q_o}\tilde{y}, \qquad t \in [(q_o - 1)\tau, q_o\tau),$$

with the initial condition $\hat{z}_{q_o}((q_o - 1)\tau) = 0$, and $\tilde{y}$ being the difference between the measured output and the estimated output of (7.41). The gain l_{q_o} will be chosen later (see (7.42)). For the *even* mode, we get $W_{q_e} = I_{2\times 2}$, and $O_{q_e} = 0_{2\times 2}$, so that Z_{q_e}, S_{q_e}, and R_{q_e} are null-matrices.

The next step is to use the value of $\hat{z}_{q_o}(t_{q_o}^-)$ to compute ξ_k, $k \in \mathbb{N}$, using the procedure outlined in Sect. 7.3.2. The matrices appearing in the computation of ξ_k are given as follows. For every $q_e > 3$:

$$\begin{aligned}
\Psi_{q_e-3}^{q_e} &= \begin{bmatrix} \cos 2\tau & \sin 2\tau \\ -\sin 2\tau & \cos 2\tau \end{bmatrix} \Rightarrow \Theta_{q_e-3}^{q_e} = \begin{bmatrix} \cos 2\tau \\ -\sin 2\tau \end{bmatrix}, \\
\Psi_{q_e-2}^{q_e} &= \begin{bmatrix} \cos \tau & \sin \tau \\ -\sin \tau & \cos \tau \end{bmatrix} \Rightarrow \Theta_{q_e-2}^{q_e} = \text{null}, \\
\Psi_{q_e-1}^{q_e} &= \begin{bmatrix} \cos \tau & \sin \tau \\ -\sin \tau & \cos \tau \end{bmatrix} \Rightarrow \Theta_{q_e-1}^{q_e} = \begin{bmatrix} \cos \tau \\ -\sin \tau \end{bmatrix}, \\
\Psi_{q_e}^{q_e} &= I_{2\times 2} \Rightarrow \Theta_{q_e}^{q_e} = \text{null}
\end{aligned}$$

where, as a convention, we have taken $\Theta_j^{q_e}$ as a null matrix whenever $\mathscr{R}(\Psi_j^{q_e} W_j)^{\perp} = \{0\}$. Using the matrices $\Theta_j^{q_e}$, $j = q_e - 3, \ldots, q_e$, we obtain for every $q_e > 3$:

$$\Theta_{q_e} = \begin{bmatrix} \Theta_{q_e-1}^{q_e} & \Theta_{q_e-3}^{q_e} \end{bmatrix} = \begin{bmatrix} \cos \tau & \cos 2\tau \\ -\sin \tau & -\sin 2\tau \end{bmatrix}.$$

Thus, for every $q_e > 3$, the error correction term ξ_{q_e} can be computed by the formula:

$$\xi_{q_e} = \Theta_{q_e}^{-\top} \begin{bmatrix} \hat{z}_{q_e-1}(t_{q_e-1}^-) - \xi_{q_e-1}(1) \\ \hat{z}_{q_e-3}(t_{q_e-3}^-) - \xi_{q_e-3}(1) - [\cos \tau \;\; -\sin \tau](\xi_{q_e-2} + \xi_{q_e-1}) \end{bmatrix},$$

where we use the notation $\xi_q(j)$ to denote the jth component of the vector ξ_q. Next, for every $q_o > 3$, we repeat the same calculations and obtain

$$\Theta_{q_o} = \begin{bmatrix} \Theta_{q_o}^{q_o} & \Theta_{q_o-2}^{q_o} \end{bmatrix} = \begin{bmatrix} 1 & \cos \tau \\ 0 & -\sin \tau \end{bmatrix}$$

which further gives

$$\xi_{q_o} = \Theta_{q_o}^{-\top} \begin{bmatrix} \hat{z}_{q_o}(t_{q_o}^-) \\ \hat{z}_{q_o-2}(t_{q_o-2}^-) - \xi_{q_o-2}(1) - [\cos \tau \;\; -\sin \tau]\, \xi_{q_o-1} \end{bmatrix}.$$

To compute the gain l_{q_o}, we note that $M^{q_e}_{q_e}$, $M^{q_e}_{q_e-2}$ are null matrices, and

$$M^{q_e}_{q_e-1} = \begin{bmatrix} \frac{\sin 2\tau}{\sin \tau} & 0 \\ \frac{\cos 2\tau}{\sin \tau} & 0 \end{bmatrix} \quad \text{and} \quad M^{q_e}_{q_e-3} = \begin{bmatrix} -1 & 0 \\ -\frac{\cos \tau}{\sin \tau} & 0 \end{bmatrix}.$$

Also, for $q_o > 3$, $M^{q_o}_{q_o-1}$ and $M^{q_o}_{q_o-3}$ are null matrices, and

$$M^{q_o}_{q_o} = \begin{bmatrix} 1 & 0 \\ \frac{\cos \tau}{\sin \tau} & 0 \end{bmatrix} \quad \text{and} \quad M^{q_o}_{q_o-2} = \begin{bmatrix} 0 & 0 \\ -\frac{1}{\sin \tau} & 0 \end{bmatrix}.$$

By taking l_{q_o} equal to l for each q_o, and computing the induced 2-norm of the matrix, it is seen that, $\max_{q-3 \leq j \leq q, j:\text{odd}, q>3} \|M^q_j Z^j e^{(S_j - lR_j)\tau} Z^{j\top}\| = e^{-l\tau}/|\sin \tau|$. So, the lower bound for the gain l, is obtained as follows:

$$\frac{e^{-l\tau}}{|\sin \tau|} < \frac{1}{\mathbf{m}^* + 1} = \frac{1}{4} \quad \Rightarrow \quad l > \frac{1}{\tau} \ln \frac{4}{|\sin \tau|}. \tag{7.42}$$

It can be seen that the singularity occurs when τ is an integer multiple of π. Moreover, if τ approaches this singularity, then the gain required for convergence gets arbitrarily large. This shows that even though the condition $\sin \tau \neq 0$ guarantees observability, it may cause some difficulty in practice if $\sin \tau \approx 0$. This also explains why the knowledge of the switching signal is required in general to compute the observer gains.

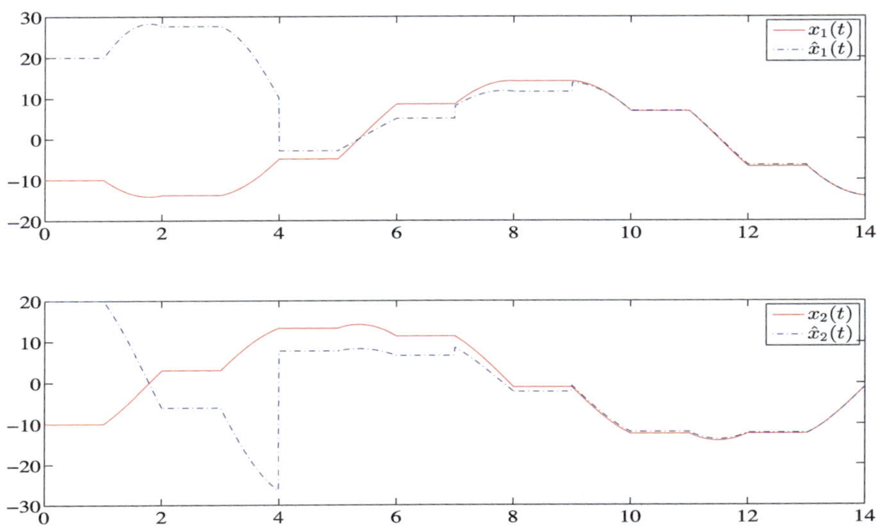

Fig. 7.1 State estimation error in x_1 and x_2

The results of simulations with $\tau = 1$, and $l = 2$, are illustrated in Fig. 7.1. The error initially evolves according to the (marginally stable) system dynamics as no correction is applied till t_4. When the error correction is applied, there is a jump in the state estimation error, which highlights the hybrid nature of the proposed observer.

7.4.2 Non-invertible State Reset Maps

We next consider an academic example of a third order ($\mathbf{n} = 3$) switched system with three modes where $A_q, B_q, H_q, D_q, q \in \mathbb{N}$, are zero matrices of appropriate dimensions. The output measurements are given by:

$$C_{3k-2} = [1\ 0\ 0], \quad C_{3k-1} = [0\ 1\ 0], \quad C_{3k} = [0\ 0\ 1], \quad k \geq 1,$$

and the state reset maps are:

$$G_{3k-2} = G_{3k} = I_{3\times 3}, \quad G_{3k-1} = \begin{bmatrix} 1 & 1 & 0 \\ 1 & 1 & 0 \\ 0 & 0 & 1 \end{bmatrix}, \quad k \geq 1.$$

For this system, it can be checked that the Assumption 7.3 indeed holds, that is, $\dim \mathcal{Q}^q_{q-\mathbf{m}^*} = 3$, for each $q > 2$, where we take $\mathbf{m}^* = 2$. The observer (7.7) is now implemented to obtain the state estimate in which we let $\xi_1 = \xi_2 = 0$.

For $q \geq 3$, the following expressions are obtained for the vector ξ_q using the calculations in Sect. 7.3.3:

$$\xi_{3k} = \begin{pmatrix} \hat{z}_{3k-2} + \hat{z}_{3k-1} \\ \hat{z}_{3k-2} + \hat{z}_{3k-1} \\ \hat{z}_{3k} \end{pmatrix} - \begin{pmatrix} \xi_{3k-2}(1) + \xi_{3k-1}(1) + \xi_{3k-1}(2) \\ \xi_{3k-2}(1) + \xi_{3k-1}(1) + \xi_{3k-1}(2) \\ 0 \end{pmatrix}, \quad k \geq 1,$$

$$\xi_{3k+1} = \begin{pmatrix} \hat{z}_{3k+1} \\ 0 \\ \hat{z}_{3k} \end{pmatrix} - \begin{pmatrix} \frac{1}{\sqrt{2}}\xi_{3k}(1) - \frac{1}{\sqrt{2}}\xi_{3k}(2) \\ \frac{1}{\sqrt{2}}\xi_{3k}(2) - \frac{1}{\sqrt{2}}\xi_{3k}(1) \\ \xi_{3k}(3) \end{pmatrix}, \quad k \geq 1,$$

$$\xi_{3k+2} = \begin{pmatrix} \hat{z}_{3k+1} \\ \hat{z}_{3k+2} \\ \hat{z}_{3k} \end{pmatrix} - \begin{pmatrix} \xi_{3k+1}(1) \\ 0 \\ \xi_{3k+1}(3) + \xi_{3k}(3) \end{pmatrix}, \quad k \geq 1,$$

where the short-hand $\hat{z}_q$ is used to denote $\hat{z}_q(t_q^-)$, which for each $q \in \mathbb{N}$ is obtained from the following equation:

$$\dot{\hat{z}}_q(t) = -l_q \tilde{y}(t), \quad t \in [t_{q-1}, t_q), \qquad \hat{z}(t_{q-1}) = 0.$$

For simplicity, if we let $l_q = l$, and $\tau_q = \tau$ for some $l, \tau > 0$ and each $q \in \mathbb{N}$, then the condition (7.40) boils down to:

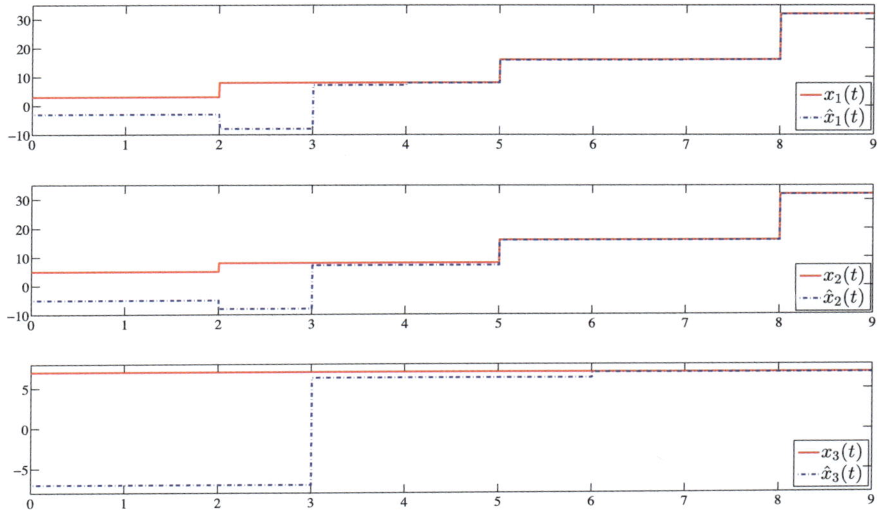

Fig. 7.2 The plot shows the state estimates $\hat{x}_i$, $i = 1, 2, 3$ (*dashed lines* in *blue*) converging to the actual states of the plant x_i, $i = 1, 2, 3$ (*solid lines* in *red*)

$$\sqrt{2} \cdot e^{-l\tau} < \frac{1}{3} \quad \Leftrightarrow \quad l > \frac{\log 3\sqrt{2}}{\tau}.$$

For $\tau = 1$, the simulation results are shown in Fig. 7.2. The plot shows the continuous and discrete nature of the error dynamics where the estimate does not improve between the two switching instants and only when the correction ξ_q is applied, the estimate gets closer to the actual state value.

7.5 Conclusions

This chapter has addressed the problems of observer design for switched linear systems with state reset maps based on a notion of observability, which does not necessarily require the observability of individual subsystems in the classical sense. The proposed state estimators apply error correction only at discrete switching instants and are inherently hybrid in nature. The examples considered in this chapter are purely academic, but it is not difficult to encounter practical systems where such techniques could be applied. First and foremost application that comes to mind are the electrical circuits: Multicellular converters could be modeled as switched systems where each mode is not observable. Our observer design has been used to study diagnostic problems in such systems [24]. Another instance of the utility of our observer design in a power converter has been reported in [11].

Although, we only consider the linear systems with ordinary differential equations in this chapter, the ideas presented in this chapter have been applied to a more general class of systems. The first of these extensions has been studied for the case when the dynamics of individual subsystems are represented by differential-algebraic equations [22]. Such systems have more structure because the solution only evolves in the consistency space determined by the algebraic constraints of individual modes. The state jumps in these systems are also determined by the algebraic constraints, and moreover the solutions of such systems may contain derivatives of jumps for which we adopt the distributional framework proposed in [23].

In another related work, we have used similar ideas to study the problem of observer design in switched nonlinear systems [14, 15]. The major difficulty in dealing with nonlinear systems is that one cannot explicitly solve the system equations to transport the observable information from one time instant to another and neither this map is expected to be linear. Thus, we have to introduce some additional assumptions on the dynamics of individual subsystems that allow for previously recovered information (or part of it) to flow through the unobservable manifold of the following subsystems without being perturbed by the unknown variables. This approach leads to a sufficient condition for forward observability, and the observer design based on this approach has somewhat different structure than the one proposed in this chapter, as one would expect it to be the case when making transition from linear to nonlinear systems.

References

1. Alessandri, A., Coletta, P.: Switching observers for continuous-time and discrete-time linear systems. Proc. Am. Control Conf. **2001**, 2516–2521 (2001)
2. Babaali, M., Pappas, G.J.: Observability of switched linear systems in continuous time. Hybrid Systems: Computation and Control. Lecture Notes in Computer Science, vol. 3414, pp. 103–117. Springer, Berlin (2005)
3. Balluchi, A., Benvenuti, L., Di Benedetto, M., Sangiovanni-Vincentelli, A.: Design of observers for hybrid systems. Hybrid Systems: Computation and Control. Lecture Notes in Computer Science, vol. 2289, pp. 76–89. Springer, Berlin (2002)
4. Balluchi, A., Benvenuti, L., Di Benedetto, M., Sangiovanni-Vincentelli, A.: Observability for hybrid systems. In: Proceedings of the 42nd IEEE Conference on Decision and Control, Hawaii, USA, vol. 2, pp. 1159–1164 (2003). doi:10.1109/CDC.2003.1272764
5. Collins, P., van Schuppen, J.H.: Observability of piecewise-affine hybrid systems. In: Alur, R., Pappas, G.J. (eds.) Hybrid Systems: Computation and Control. Lecture Notes in Computer Science, vol. 2993, pp. 265–279. Springer, Heidelberg (2004)
6. De Santis, E.: On location observability notions for switching systems. Syst. Control Lett. **60**(10), 807–814 (2011). http://dx.doi.org/10.1016/j.sysconle.2011.06.004
7. Hespanha, J.P., Morse, A.S.: Stability of switched systems with average dwell-time. In: Proceedings of the 38th IEEE Conference on Decision and Control, pp. 2655–2660 (1999)
8. Lee, C., Ping, Z., Shim, H.: Real time switching signal estimation of switched linear systems with measurement noise. In: Proceedings of the 12th European Control Conference 2013, Zurich, Switzerland, pp. 2180–2185 (2013)
9. Medina, E.A., Lawrence, D.A.: Reachability and observability of linear impulsive systems. Automatica **44**, 1304–1309 (2008)

10. Medina, E.A., Lawrence, D.A.: State estimation for linear impulsive systems. Proc. Am. Control Conf. **2009**, 1183–1188 (2009)
11. Meghnous, A.R., Pham, M.T., Lin-Shi, X.: A hybrid observer for a class of DC-DC power converters. Proc. Am. Control Conf. **2013**, 6225–6230 (2013)
12. Pait, F.M., Morse, A.S.: A cyclic switching strategy for parameter-adaptive control. IEEE Trans. Autom. Control **6**(39), 1172–1183 (1994)
13. Pettersson, S.: Designing switched observers for switched systems using multiple Lyapunov functions and dwell-time switching. In: IFAC Conference on Analysis and Design of Hybrid Systems, pp. 18–23 (2006)
14. Shim, H., Tanwani, A.: On a sufficient condition for observability of switched nonlinear systems and observer design strategy. Proc. Am. Control Conf. **2011**, 1206–1211 (2011)
15. Shim, H., Tanwani, A.: Hybrid-type observer design based on a sufficient condition for observability in switched nonlinear systems. Int. J. Robust Nonlinear Control **24**(6), 1064–1089 (2014). doi:10.1002/rnc.2901
16. Sontag, E.D.: Mathematical Control Theory: Deterministic Finite Dimensional Systems, 2nd edn. Springer, New York (1998)
17. Sun, Z., Ge, S.S.: Switched linear systems. Communications and Control Engineering. Springer, London (2005). doi:10.1007/1-84628-131-8
18. Tanwani, A., Liberzon, D.: Robust invertibility of switched linear systems. In: Proceedings of the 50th IEEE Conference on Decision and Control and European Control Conference ECC 2011, Orlando, USA, pp. 441–446 (2011)
19. Tanwani, A., Shim, H., Liberzon, D.: Observability implies observer design for switched linear systems. In: Proceedings of the ACM Conference on Hybrid Systems: Computation and Control, pp. 3–12 (2011)
20. Tanwani, A., Shim, H., Liberzon, D.: Observability for switched linear systems: Characterization and observer design. IEEE Trans. Autom. Control **58**(4), 891–904 (2013). doi:10.1109/TAC.2012.2224257
21. Tanwani, A., Shim, H., Liberzon, D.: Comments on "Observability of switched linear systems: characterization and observer design" (2014). Submitted for publication; Preprint available on authors' websites
22. Tanwani, A., Trenn, S.: An observer for switched differential-algebraic equations based on geometric characterization of observability. In: Proceedings of the 52nd IEEE Conference on Decision and Control, Florence, Italy, pp. 5981–5986 (2013)
23. Trenn, S.: Regularity of distributional differential algebraic equations. Math. Control Signals Syst. **21**(3), 229–264 (2009). doi:10.1007/s00498-009-0045-4
24. Van Gorp, J., Defoort, M., Djemai, M., Manamanni, N.: Hybrid observer for the multicellular converter. In: 4th IFAC Conference on Analysis and Design of Hybrid Systems (2012)
25. Vidal, R., Chiuso, A., Soatto, S., Sastry, S.: Observability of linear hybrid systems. Hybrid Systems: Computation and Control. Lecture Notes in Computer Science, vol. 2623, pp. 526–539. Springer, Berlin (2003)
26. Vu, L., Liberzon, D.: Invertibility of switched linear systems. Automatica **44**(4), 949–958 (2008)
27. Xie, G., Wang, L.: Necessary and sufficient conditions for controllability and observability of switched impulsive control systems. IEEE Trans. Autom. Control **49**(6), 960–966 (2004). doi:10.1109/TAC.2004.829656

Chapter 8
Observability of Switched Linear Systems

Mihaly Petreczky, Aneel Tanwani and Stephan Trenn

Abstract Observability of switched linear systems has been well studied during the past decade and depending on the notion of observability, several criteria have appeared in the literature. The main difference in these approaches is how the switching signal is viewed: Is it a fixed and known function of time, is it an unknown external signal, is it the result of a discrete dynamical system (an automaton) or is it controlled and is therefore an input? We will focus on the recently introduced geometric characterization of observability, which assumes knowledge of the switching signal. These geometric conditions depend on computing the exponential of the matrix and require the exact knowledge of switching times. To relieve the computational burden, some relaxed conditions that do not rely on the switching times are given; this also allows for a direct comparison of the different observability notions. Furthermore, the generalization of the geometric approach to linear switched differential algebraic systems is possible and presented as well.

8.1 Introduction

In the context of dynamical systems, *observability* is a fundamental property that plays an important role in realization theory, state estimation, output feedback controller design, and even diagnosis and fault monitoring. Roughly speaking, observability concerns extracting information about the internal variables, called *states*, of the system using the external signals consisting of *output* measurements and the inputs. Hence (for a given input), observability is related to the study of the

M. Petreczky
Department of Automatic Control and Computer Science,
Ecole des Mines de Douai, 941, rue Charles Bourseul, 59508 Douai, France
e-mail: mihaly.petreczky@mines-douai.fr

S. Trenn (✉) · A. Tanwani
Department of Mathematics, University of Kaiserslautern, Kaiserslautern, Germany
e-mail: trenn@mathematik.uni-kl.de

A. Tanwani
e-mail: tanwani@mathematik.uni-kl.de

M. Djemai and M. Defoort (eds.), *Hybrid Dynamical Systems*, Lecture Notes in Control and Information Sciences 457, DOI 10.1007/978-3-319-10795-0_8

mapping from the set of state trajectories to the set of outputs, and in general a system is *observable* if this mapping is injective.

This chapter is concerned with observability of switched linear systems and we consider models given by ordinary differential equations (ODEs) with jumps or given by differential algebraic equations (DAEs).

Various notions for observability of switched systems have appeared in the literature. From the point of view of hybrid systems, the switching signal may be treated either as an unknown discrete state or as a known external signal. In the former case, observability relates to simultaneous recovery of the discrete and continuous state. Some results in this direction for continuous-time hybrid systems appear in [3, 9, 11, 13, 16, 20, 27, 43], and for discrete-time hybrid systems in [2, 4, 28, 42]. Also related is the problem of the reconstruction of the discrete mode without imposing conditions for recovery of the continuous state, and for references in this direction, see for example, [10, 12, 21, 32, 44].

In this chapter, we will treat the other case, i.e., we will view the discrete mode as a known switching signal. In this case, even though the individual subsystems are not observable, it is still possible to recover the state trajectory by appropriately processing the measured signals over a time interval that involves multiple switching instants. This phenomenon is of particular interest for switched systems or systems with state jumps as the notion of instantaneous observability and observability over an interval coincide for nonswitched linear time invariant systems. This variant of the observability problem in switched systems has been studied most notably by [5, 25, 31, 36, 48] for switched linear systems, [18, 19] for linear impulsive systems, [26, 30, 49] for switched nonlinear systems, and [33–35] for switched linear differential algebraic equations. In the work of [17, 24–26, 31, 48], the authors derive conditions in terms of system data under which there exists a switching signal that makes state trajectories distinguishable. In addition, in [24–26, 31] algorithms for observability reduction (i.e., transforming a state space representation to an observable one while preserving input-output behavior) were proposed. Unfortunately, the observability concept used in the papers mentioned above does not guarantee existence of an observer. In contrast, the authors in [30, 33–36] study the observability of the underlying system for a fixed switching signal and use the geometric conditions for designing observers as well.

In this chapter, we present different notions of observability depending on what role the switching signal plays in reconstruction of the state trajectories, i.e., whether the switching signal is fixed, or whether it is viewed as an input. We then derive geometric conditions for each of these notions and in the process draw connections with the existing work. The adopted approach has the advantage that it allows us to treat ordinary differential equations and differential algebraic equations with similar tools, although the later class of systems requires more sophisticated treatment because of a nonstandard solution framework.

The outline of this chapter is as follows: We first present the system classes we consider and give different definitions of observability. After a detailed discussion of these definitions, we first study the single switch case in Sect. 8.4; the switched ODE case and the switched DAE case are treated separately. The latter includes a short discussion about the implicit jump rule (the consistency projector) and the distributional solution framework. We illustrate the observability characterization for switched DAEs with a single switch by a detailed example. In Sect. 8.5, we build on the single switch results to obtain a characterization of observability for a general switching signal. Furthermore, we formulate necessary conditions and sufficient conditions for observability which do not depend on the specific switching times but only on the mode sequence. Finally, we present conditions for observability, where the switching signal is not fixed a priori and compare these results with the former observability conditions.

The following notational conventions are used within this chapter. For a function $f : \mathbb{R} \to \mathbb{R}$, we denote the restriction to an interval I by $f_I : \mathbb{R} \to \mathbb{R}$ given by $f_I(t) = f(t)$ if $t \in I$ and $f_I(t) = 0$, otherwise; the same notation is also used for distributions (see Appendix 8.6.3). If for a function $f : \mathbb{R} \to \mathbb{R}$, the right- and left-sided limit at some $t \in \mathbb{R}$ exist, we denote these by $f(t^+)$ and $f(t^-)$; furthermore, we assume in general right-continuity, i.e., $f(t^+) = f(t)$. For a matrix M, we denote the null space as $\ker M$ and the linear space spanned by the columns of M by $\operatorname{im} M$. For two matrices M and N with the same number of columns, we denote by $[M/N]$ the matrix resulting from stacking M over N. For a switched system, $\mathscr{P}$ denotes the set of possible parameters and we assume that the switching signal $\sigma : \mathbb{R} \to \mathscr{P}$ is right-continuous and the left-sided limit $\sigma(t^-)$ exists for all $t \in \mathbb{R}$ (i.e., we exclude Zeno-behavior). The switching times of σ are denoted by t_i, $i \in \mathbb{N}$ and the duration of the ith mode is $\tau_i := t_{i+1} - t_i$.

8.2 System Classes

We consider switched systems with possible jumps, i.e., we consider systems of the form

$$\dot{x}(t) = A_{\sigma(t)}x(t) + B_{\sigma(t)}u(t), \qquad t \notin \{\, t_i \mid i \in \mathbb{N} \,\}, \tag{8.1a}$$

$$x(t_i^+) = G_{\sigma(t_i^+)}x(t_i^-) + H_{\sigma(t_i^+)}v_i, \qquad i \in \mathbb{N}, \tag{8.1b}$$

$$y(t) = C_{\sigma(t)}x(t) + D_{\sigma(t)}u(t), \qquad t \in \mathbb{R}, \tag{8.1c}$$

where $\sigma : \mathbb{R} \to \mathscr{P} := \{0, 1, 2, \dots, \mathtt{p}\}$, $\mathtt{p} \in \mathbb{N} \cup \{\infty\}$, is the switching signal with nonaccumulating switching times $t_1 < t_2 < t_3 < \dots$; $A_p, B_p, C_p, D_p, G_p, H_p$, $p \in \mathscr{P}$, are matrices of appropriate size describing the dynamics and the jump at each mode, $u : \mathbb{R} \to \mathbb{R}^{\mathtt{u}}$ is the input to the continuous dynamics, $v : \mathbb{N} \to \mathbb{R}^{\mathtt{v}}$ is the input to the discrete dynamics, $x : \mathbb{R} \to \mathbb{R}^{\mathtt{n}}$ is the state trajectory, and $y : \mathbb{R} \to \mathbb{R}^{\mathtt{y}}$ is the measured output. Note that we assume the state space dimension to be the same for all modes; allowing different state space dimensions as in, e.g., [25] is partly a topic for future research.

In the case that $G_p = I$ and $H_p = 0$ for all $p \in \mathscr{P}$, we call (8.1) a *classical* switched system. We will also study switched systems of the form

$$E_\sigma \dot{x} = A_\sigma x + B_\sigma u \tag{8.2a}$$

$$y = C_\sigma x + D_\sigma u \tag{8.2b}$$

which we call *switched differential algebraic equations (switched DAEs)*. In this context, we also call (8.1) a switched ordinary differential equation (switched ODE) with or without jumps. Note that although a switched DAE does not have an explicit jump rule, the solutions nevertheless have jumps in general (for details see Sect. 8.4.2.1). In general, one should be aware, that the solution theory of (8.2) is a bit more involved compared to the switched systems of the form (8.1) (see Sect. 8.4.2.2); nevertheless the observability properties of (8.2) and (8.1) are very much alike (see Sect. 8.5, where the main results hold in an identical form for both system classes).

Remark 8.1 Usually, in order to avoid notational inconveniences, it is assumed that the switching signal σ has infinitely many switching times. For classical switched systems or switched DAEs, this is not a problem as arbitrarily many artificial switching times can be added where the modes do not change. However for systems of the form (8.1), this is not possible in general because any introduction of an additional switching time introduces a jump in the state even if the mode does not change. In particular, the usual semigroup property of switched systems does not hold any more. If $\mathscr{P}$ is finite, one could add for each $p \in \mathscr{P}$ one new mode $\mathtt{p} + 1 + p$ given by $A_{\mathtt{p}+1+p} = A_p$, $B_{\mathtt{p}+1+p} = B_p$, $C_{\mathtt{p}+1+p} = C_p$, $D_{\mathtt{p}+1+p} = D_p$ and $G_{\mathtt{p}+1+p} = I$, $H_{\mathtt{p}+1+p} = 0$. Then, it is always possible to introduce arbitrarily many switching times; however, the new switched system now also allows trajectories which do *not* jump when a switch occurs and hence is not equivalent to the original system (8.1) (unless one makes restrictions on the allowed switching sequences). Another way out is to consider instead of $G_{\sigma(t_i)}$ jump maps $G_{\sigma(t_i-),\sigma(t_i)}$ which also depend on the mode before the switch. In this framework, one could simply set $G_{pp} = I$ and $G_{pq} = G_q$ for all $p \neq q$. Furthermore, this framework has also the advantage that different state space dimensions can be handled, but as mentioned above this is not in the scope of this chapter. ◁

Finally, it should be noted that it makes a conceptionally important difference whether the time-dependent switching signal is seen as a (fixed, but arbitrary) part of the system description or is seen as an additional input to system (8.1). In the former case, (8.1) is a linear—albeit time-varying—system, whereas in the latter case, the system is nonlinear. In particular, when one speaks of observability of (8.1), it is important to distinguish these two viewpoints clearly; resulting in the different notions of "observability for a specific switching signal" and "controlled observability" (see Definition 8.2 for precise meaning). In this chapter, the focus is mainly on the first viewpoint.

8.3 Observability Definitions

We first review the observability notions for nonswitched linear systems

$$\dot{x} = Ax + Bu, \quad y = Cx + Du. \tag{8.3}$$

The system (8.3) is called observable if, and only if, knowledge of the external signals (u, y) implies knowledge of the internal signal x. More precisely, if (x^1, u^1, y^1) and (x^2, u^2, y^2) are two solutions of (8.3) and $(u^1, y^1) = (u^2, y^2)$, then it must follow that $x^1 = x^2$. Note that the equality is global (in time), however, it is well-known that this is equivalent to local observability where one only considers an arbitrarily small interval (with nonempty interior). Furthermore, knowledge of $x(t_0)$ for some $t_0 \in \mathbb{R}$ implies knowledge of x on the whole time axis with fixed inputs, hence observability for (8.3) can often be reduced to the question: Is it possible to reconstruct the initial value $x(0)$ via the (local) knowledge of u and y?

It is easily seen that, due to linearity, observability of (8.3) does not depend on the input u. Hence, the matrices B and D play no role and the question of observability can be further reduced to the question of zero-distinguishability, i.e., does $y \equiv 0$ imply $x \equiv 0$? Taking derivatives of the output, it is easily seen that

$$y \equiv 0 \quad \Leftrightarrow \quad y^{(i)}(0) = 0 \;\; \forall i \in \mathbb{N} \quad \Leftrightarrow \quad x(0) \in \ker O_{(A,C)}$$

where $O_{(A,C)} := [C/CA/CA^2/\cdots/CA^{n-1}]$ is the Kalman observability matrix. In particular, (8.3) is observable if, and only if, $\ker O_{(A,C)} = \{0\}$.

When considering observability for switched systems, it is first obvious that local and global observability need not coincide anymore. Furthermore, in the presence of jumps, it is not true in general that knowledge of $x(0)$ implies knowledge of x on the whole time axis. Hence, we arrive at different notions of observability:

Definition 8.2 (*Observability*) Consider the switched system (8.1) (resp. (8.2)).

- We call (8.1) (resp. (8.2)) *strongly observable* if, and only if, for all solutions $(\sigma^1, u^1, v^1, y^1, x^1)$, $(\sigma^2, u^2, v^2, y^2, x^2)$ the following implication holds:

$$(\sigma^1, u^1, v^1, y^1) \equiv (\sigma^2, u^2, v^2, y^2) \quad \Rightarrow \quad x^1 \equiv x^2. \tag{8.4}$$

- We call (8.1) (resp. (8.2)) *observable for the switching signal* σ if, and only if, the implication (8.4) holds for all solutions with $\sigma \equiv \sigma^1 \equiv \sigma^2$.
- We call (8.1) (resp. (8.2)) *controlled observable* if, and only if, there exist (σ, u, v) such that for all solutions (σ, u, v, y^1, x^1), (σ, u, v, y^2, x^2) the following implication holds

$$y^1 \equiv y^2 \quad \Rightarrow \quad x^1 \equiv x^2.$$

- We call (8.1) (resp. (8.2)) *weakly controlled observable* if, and only if, any two distinct initial states $x_0^1, x_0^2 \in \mathbb{R}^n$ are distinguishable for some σ, i.e., for all $x_0^1, x_0^2 \in \mathbb{R}^n$ there exists (σ, u, v) such that the corresponding solutions (σ, u, v, y^1, x^1) and (σ, u, v, y^2, x^2) with $x^1(0^-) = x_0^1$ and $x^2(0^-) = x_0^2$ satisfy

$$y^1 \not\equiv y^2.$$

- We call (8.1) (resp. (8.2)) *forward observable for the switching signal* σ if, and only if, for all solutions $(\sigma, u^1, v^1, y^1, x^1)$, $(\sigma, u^2, v^2, y^2, x^2)$ there exists $T \in \mathbb{R}$ such that the following implication holds

$$(u^1, v^1, y^1) \equiv (u^2, v^2, y^2) \quad \Rightarrow \quad x^1{}_{(T,\infty)} \equiv x^2{}_{(T,\infty)}. \qquad \triangleleft$$

Remark 8.3 Some remarks on the different concepts of observability follow.

(i) Clearly, strong observability of (8.1) (resp. (8.2)) implies observability for each individual switching signal. In particular, it implies observability of each mode (just chose the constant switching signal). It is easily seen that observability of each mode also implies strong observability of the switched system. Hence, we already obtained the following equivalence: (8.1) (resp. (8.2)) is strongly observable if, and only if, each mode is observable in the classical sense. For this reason, strong observability is not such an interesting concept for switched systems and not considered here in detail.

(ii) Due to linearity, the observability notions do not depend on u and v (c.f. the forthcoming Proposition 8.4), hence observability for a particular switching signal implies controlled observability. On the other hand, controlled observability implies observability for the corresponding switching signal.

(iii) Controlled observability implies existence of a single switching signal, which distinguishes any two initial values; whereas weakly controlled observability implies that for any pair of initial values, there exists a switching signal (that may depend on these specific initial values) which distinguishes these initial values. Hence, controlled observability implies weakly controlled observability, but the converse cannot be expected in general. However, for a classical switched system which is weakly controlled observable, one can construct a single switching signal which is able to distinguish any two initial values, hence controlled observability and weakly controlled observability are equivalent in this case, see Theorem 8.29. For switched ODEs with jumps or for switched DAEs, this issue is not resolved yet. To the best of our knowledge, the term controlled observability was first used in [17].

(iv) If any mode of the switched system is observable, then it follows that the switched system is controlled observable (just take the corresponding constant switching signal). The converse is not true, for counter examples see [16, 23].

(v) It does not make much sense to define forward observability without fixing the switching signal, because on the one hand the time T from which onwards the state can be reconstructed depends on σ. For example, switching just faster will in general make T arbitrarily small. On the other hand, forward observability for all switching signals implies forward observability of each mode.

For nonswitched systems, forward observability is the same as observability. Since observability of each mode implies strong observability of the switched system (8.1) (resp. (8.2)), it follows that forward observability without fixing a switching signal and strong observability are identical.
(vi) If the jump matrices G_p in (8.1) are invertible for all $p \in \mathscr{P}$, then solutions can be extended uniquely also back in time, hence in this case forward observability and observability (for a fixed switching signal) are equivalent. ◁

As already mentioned in the above remark, the notion of observability does not depend on the specific inputs u and v (but it does depend on the "input" σ), hence in the following, we will only consider (8.1) with $B_p = 0$ and $H_p = 0$ for all $p \in \mathscr{P}$. Furthermore, also due to linearity it suffices to consider the output $y \equiv 0$ and check whether this output can be produced by a nonzero state. These observations are summarized in the following proposition.

Proposition 8.4 (Observability independent of inputs, c.f. [36]) *Consider the switched system* (8.1) *without inputs:*

$$\left.\begin{aligned} \dot{x}(t) &= A_{\sigma(t)}x(t), && t \notin \{\, t_i \mid i \in \mathbb{N} \,\}, \\ x(t_i) &= G_{\sigma(t_i)}x(t_i^-), && i \in \mathbb{N}, \\ y(t) &= C_{\sigma(t)}x(t), && t \in \mathbb{R}, \end{aligned}\right\} \tag{8.5}$$

- (8.1) *is observable for the switching signal* σ *if, and only if, for all corresponding solutions* (x, y) *of* (8.5) *it holds that*

$$y \equiv 0 \quad \Rightarrow \quad x \equiv 0. \tag{8.6}$$

- (8.1) *is controlled observable if, and only if, there exists* σ *such that* (8.1) *is observable for* σ.
- (8.1) *is weakly controlled observable if, and only if, for every* $x_0 \in \mathbb{R}^n \setminus \{0\}$ *there exists a switching signal* σ *such that for the corresponding solution* (x, y) *of* (8.5) *with* $x(0^-) = x_0$ *it hold that*

$$y \not\equiv 0.$$

- (8.1) *is forward observable for the switching signal* σ *if, and only if, for all solutions* (x, y) *of* (8.5) *the following implication holds for some* $T \in \mathbb{R}$

$$y \equiv 0 \quad \Rightarrow \quad x_{(T,\infty)} \equiv 0.$$

Corresponding results concerning the switched DAE without an input:

$$E_\sigma \dot{x} = A_\sigma x, \quad y = C_\sigma x \tag{8.7}$$

hold as well (even when considering distributional inputs, c.f. [35, Prop. 7]*).*

From a practical point of view, we only consider switching signals which are constant prior to some time (say $t = 0$); in fact, we already made this assumption implicitly when we denoted the switching times of σ by $t_1 < t_2 < t_3 < \ldots$ earlier. Hence, the trajectories of the switched system (8.1) or (8.5) on $(-\infty, 0)$ are similar to a system without switches. In particular, if the inputs σ, u and v are known, then $x(0^-)$ uniquely determines the whole trajectory x. Therefore, it make sense to define the unobservability space as follows.

Definition 8.5 (*Unobservable space*) Consider (8.5) (resp. (8.7)). For a switching signal σ being constant on $(-\infty, 0)$, let $\mathcal{M}^\sigma_{0-} \subseteq \mathbb{R}^n$ be such that for all solutions (x, y) of (8.5) it holds that

$$y \equiv 0 \quad \Leftrightarrow \quad x(0^-) \in \mathcal{M}^\sigma_{0-}. \qquad \lhd$$

Definition 8.6 (*T-unobservable space*) Consider (8.5) (resp. (8.7)). For some $T \geq 0$, let $\mathcal{N}^\sigma_{T+} \subseteq \mathbb{R}^n$ be the smallest set for which the following implication holds for all solutions (x, y) of (8.5) (resp. (8.7))

$$y \equiv 0 \quad \Rightarrow \quad x(T^+) \in \mathcal{N}^\sigma_{T+}. \qquad \lhd$$

It is easily seen, that $\mathcal{M}^\sigma_{0-}$ and $\mathcal{N}^\sigma_{T+}$ are linear subspaces of $\mathbb{R}^n$. The different observability notions can now be characterized in terms of the unobservable spaces as follows:

Corollary 8.7 (Observability and the unobservable space) *Consider* (8.5) *(resp.* (8.7)*) and let*

$$\Sigma := \left\{ \sigma : \mathbb{R} \to \mathcal{P} \;\middle|\; \begin{array}{l} \sigma \textit{ is constant on } (-\infty, 0) \textit{ and has} \\ \textit{finitely many switches in every finite interval} \end{array} \right\}$$

denote the space of feasible switching signals.

- (8.5) *(resp.* (8.7)*) is observable for $\sigma \in \Sigma$ if, and only if, $\mathcal{M}^\sigma_{0-} = \{0\}$.*
- (8.5) *(resp.* (8.7)*) is controlled observable if, and only if, there exists $\sigma \in \Sigma$ such that $\mathcal{M}^\sigma_{0-} = \{0\}$.*
- (8.5) *(resp.* (8.7)*) is weakly controlled observable if, and only if, $\bigcap_{\sigma \in \Sigma} \mathcal{M}^\sigma_{0-} = \{0\}$.*
- (8.5) *(resp.* (8.7)*) is forward observable for σ if, and only if, $\mathcal{N}^\sigma_{T+} = \{0\}$ for some $T \geq 0$.*

For nonswitched ODEs (8.3) we have already established that the unobservable space $\mathcal{M}$ is given by

$$\mathcal{M} = \ker O_{(A,C)} = \ker[C/CA/\cdots/CA^{n-1}].$$

Since we have also established that (8.1) is strongly observable if, and only if, each mode is observable, we have the following characterization.

Corollary 8.8 *The switched system* (8.1) *is strongly observable if, and only if,* $\ker O_{(A_p,C_p)} = \{0\}$ *for all* $p \in \mathscr{P}$.

Characterizing the other observability notions is more complicated and a crucial step towards this characterization is the consideration of the simplest nontrivial switching signal first.

8.4 Characterization of Observability: The Single Switch Case

In this section we restrict ourselves to the following simple switching signal

$$\sigma_1 : \mathbb{R} \to \{0, 1\}, \quad t \mapsto \begin{cases} 0, & t < 0, \\ 1, & t \geq 0, \end{cases} \tag{8.8}$$

i.e., we consider switched systems (8.1) with exactly one switch (which occurs at $t = 0$).

The characterization of observability with this switching signal cannot be used to characterize (weakly) controlled observability as even in the case of two modes more than one switch might be necessary to achieve observability (see the forthcoming Example 8.22). Note furthermore, that for the single switch case the observability characterization significantly differs between switched ODEs and switched DAEs; this is the reason these two cases are treated separately. However, when studying the case of general switching signals in the forthcoming Sect. 8.5 the approach can again be unified by introducing the notion of local unobservable spaces which are based on the analysis in this section.

8.4.1 Observability for the Single Switch Case for (8.1)

For notational convenience, let $O_p := O_{(A_p,C_p)}$, $p \in \mathscr{P}$. It is clear that for all solutions (x, y) of the switched systems (8.5) with switching signal (8.8) the following two implications hold:

$$\begin{aligned} y_{(-\infty,0)} \equiv 0 \quad &\Leftrightarrow \quad x(0^-) \in \ker O_0, \\ y_{(0,\infty)} \equiv 0 \quad &\Leftrightarrow \quad x(0^+) \in \ker O_1. \end{aligned}$$

Since $x(0^+) = G_1 x(0^-)$ the above two equivalences become

$$y \equiv 0 \quad \Leftrightarrow \quad x(0^-) \in \ker O_0 \cap G_1^{-1}(\ker O_1) = \ker O_0 \cap \ker O_1 G_1$$

Hence, we have arrived at the following characterization of observability.

Lemma 8.9 (Observability for σ_1: ODE case) *Consider the switched system* (8.1) *with switching signal* σ_1 *given in* (8.8). *Then the unobservable space for* (8.5) *is*

$$\boxed{\mathcal{M}_{0-}^{\sigma_1} = \ker O_0 \cap \ker O_1 G_1.}$$

In particular, (8.1) *is observable for* σ_1 *if, and only if,*

$$\ker O_0 \cap \ker O_1 G_1 = \{0\}. \tag{8.9}$$

Remark 8.10 (Order of switching relevant) If there is no jump at the switch, i.e., $G_1 = I$ then the switched system (8.1) with a single switch is observable if, and only if, the intersection of the individual unobservable subspaces is trivial. In this case the order of the switching sequence doesn't matter, c.f. [43]. In the presence of jumps, it is however important whether the system jumps from mode 0 to mode 1 or vice versa. $\lhd$

Since there is no switch after time zero, it is easily seen that, for $T \geq 0$, $\mathcal{N}_{T+}^{\sigma_1} = e^{A_1 T}\mathcal{N}_{0+}^{\sigma_1}$, hence forward observability is characterized by $\mathcal{N}_{0+}^{\sigma_1} = \{0\}$. Clearly (taking into account some basic facts from linear algebra, see Appendix 8.6.1)

$$y \equiv 0 \quad \Rightarrow \quad x(0^+) \in G_1 \mathcal{M}_{0-}^{\sigma_1} = G_1(\ker O_0 \cap \ker O_1 G_1) = G_1 \ker O_0 \cap \ker O_1,$$

hence

$$\mathcal{N}_{0+}^{\sigma_1} \subseteq G_1 \ker O_0 \cap \ker O_1.$$

In fact, we have equality as for every $x_0^+ \in G_1 \ker O_0 \cap \ker O_1 = G_1 \mathcal{M}_{0-}^{\sigma_1}$ there exists at least one $x_0^- \in \mathcal{M}_{0-}^{\sigma_1}$ with $x_0^+ = G_1 x_0^-$ which shows that $\mathcal{N}_{0+}^{\sigma_1}$ cannot be chosen smaller. So we have arrived at the following characterization for forward observability for the single switch case.

Lemma 8.11 (Forward observability for σ_1: ODE case) *Consider the switched system* (8.1) with switching signal σ_1 given in (8.8) and $T \geq 0$. Then the T-unobservable space for (8.5) is given by

$$\boxed{\mathcal{N}_{T+}^{\sigma_1} = e^{A_1 T}\mathcal{N}_{0+}^{\sigma_1} = e^{A_1 T}(G_1 \ker O_0 \cap \ker O_1).}$$

In particular, (8.1) is forward observable if, and only if,

$$G_1 \ker O_0 \cap \ker O_1 = \{0\}. \tag{8.10}$$

8.4.2 The Single Switch Result for Switched DAEs

We would like to generalize the above results to switched DAEs (8.2). Due to Proposition 8.4 we can restrict our attention to the homogeneous switched DAE (8.7) and the observability question: Does an observed zero output implies a zero initial state?

Before continuing the discussion on observability of switched DAEs we have to first clarify what we mean by a solution of (8.2) or (8.7). Since each mode is given by a DAE the description of the mode itself contains algebraic constraints, enforcing the solutions to evolve within a certain subspace of $\mathbb{R}^n$. In particular, not for all initial values a solution exists. This is a problem at any switching time, because the value of the state just before the switch is in general not *consistent* with the algebraic constraints of the mode after the switch. So even when there is no explicit jump map given, the solutions have to jump (or there is no solution at all). However these jumps are not arbitrary as the following derivation shows:

8.4.2.1 The Consistency Projector

Consider a single DAE

$$E\dot{x} = Ax \tag{8.11}$$

where the matrix pair (E, A) is regular, i.e., $\det(sE - A) \not\equiv 0$. It is a classical result [14, 45] that regularity of a matrix pair is characterized by the existence of the Weierstrass canonical form, i.e., there exist invertible matrices S and T such that

$$(SET, SAT) = \left(\begin{bmatrix} I & 0 \\ 0 & N \end{bmatrix}, \begin{bmatrix} J & 0 \\ 0 & I \end{bmatrix} \right) \tag{8.12}$$

where J and N are in Jordan canonical form and N is nilpotent with nilpotency index $\nu \in \mathbb{N}$. The latter is called the *index* of the matrix pair (E, A). In the following it is not necessary to assume that J and N are in Jordan canonical form, following [6] we therefore call (8.12) the *quasi Weierstrass form (QWF)* of the matrix pair (E, A). An easy way to obtain the transformation matrices S and T via the Wong-sequences is given in the Appendix 8.6.2.

The relevance of the QWF for the solutions of the DAE (8.11) is stated as follows:

$$x \text{ solves (8.11)} \quad \Leftrightarrow \quad \begin{pmatrix} v \\ w \end{pmatrix} = T^{-1}x \text{ solves } \begin{cases} \dot{v} = Jv \\ N\dot{w} = w \end{cases}$$

i.e., the DAE (8.11) decouples into the standard ODE $\dot{v} = Jv$ and a pure DAE $N\dot{w} = w$. For the ODE any initial value is consistent. The pure DAE on the other hand only has the trivial solution:

$$N\dot{w} = w \quad \Rightarrow \quad N^2\ddot{w} = N\dot{w} = w \quad \Rightarrow \ldots \Rightarrow \quad 0 = N^{\nu}w^{(\nu)} = \ldots = N\dot{w} = w.$$

In particular, the only consistent initial value is zero.

Now assume that the matrix pair (E, A) is in QWF (8.12) and the corresponding DAE is switched on at time $t = 0$ with initial values $v(0^-) = v_0$ and $w(0^-) = w_0$ prior to the switch. The initial value v_0 for the ODE $\dot{v} = Jv$ is consistent, hence no jump occurs in this component. The initial value w_0 for the pure DAE $N\dot{w} = w$ is in general not consistent and since zero is the only solution of $N\dot{w} = w$ there will be a jump from any w_0 to zero. Altogether the *only plausible* jump in the QWF-coordinates is given by the map

$$\begin{pmatrix} v(0) \\ w(0) \end{pmatrix} = \begin{pmatrix} v_0 \\ 0 \end{pmatrix} = \begin{pmatrix} I & 0 \\ 0 & 0 \end{pmatrix} \begin{pmatrix} v(0^-) \\ w(0^-) \end{pmatrix}.$$

Translating this back to the original coordinates via $x = T\binom{v}{w}$ we arrive at the jump rule

$$x(0) = T \begin{bmatrix} I & 0 \\ 0 & 0 \end{bmatrix} T^{-1} x(0^-)$$

and the corresponding *consistency projector*

$$\Pi_{(E,A)} := T \begin{bmatrix} I & 0 \\ 0 & 0 \end{bmatrix} T^{-1}.$$

It is not difficult to see that the consistency projector does not depend on the specific choice of the (nonunique) transformation matrix T. In view of the Wong-sequences approach (see Appendix 8.6.2) with limits $\mathcal{V}^*$ and $\mathcal{W}^*$, it can be seen that $\Pi_{(E,A)}$ is a projection on $\mathcal{V}^*$ along $\mathcal{W}^*$. The above analysis also shows that the *consistency space*

$$\mathfrak{C}_{(E,A)} := \{x(t) \mid t \in \mathbb{R}, x \text{ is a classical (i.e., differentiable) solution of (8.11)}\}$$

is exactly the image of the consistency projector.

8.4.2.2 Distributional Solutions of a Switched DAE

The presence of jumps in the solution of the switched DAE (8.2) results in mathematical problems as these jumps are differentiated in the expression $E_\sigma \dot{x}$. One way out is to consider the switched DAE only on the open intervals between the switching times (in the case of a single switch, the switched DAE is supposed to hold only on the intervals $(-\infty, 0)$ and $(0, \infty)$). The forthcoming Corollary 8.15 shows that in this case it is possible to rewrite the homogeneous switched DAE (8.7) as a homogeneous switched ODE with jumps (8.5), c.f. [41]; and the characterizations of observability for (8.1) carry over without change to the switched DAE case.

However, simple examples based on electrical circuits (see e.g., [40, Example 2]) show that derivatives of the jumps play an important role and it will turn out that they will also play a crucial role in the observability characterization.

The derivative of a jump is not well-defined for usual functions, but when considering *distributions* (or generalized functions) the derivative of a jump is the well known *Dirac impulse* (or Dirac delta). In the Appendix 8.6.3 we give a short introduction to the theory of distributions, but it should be noted that enlarging the solution space of (8.2) to the space of distributions does *not* resolve the problems without further adjustment. The problem is that the multiplication of a piecewise-constant coefficient matrix ($E_{\sigma(\cdot)}$ or $A_{\sigma(\cdot)}$) with a distribution ($\dot{x}$ or x) is not well-defined. Even worse, it can be shown (see e.g., [39]) that it is in fact *impossible* to define such a product for general distributions. A possible way out of this dilemma is the consideration of the smaller space of *piecewise-smooth distributions* denoted by $\mathbb{D}_{\mathrm{pw}\mathscr{C}^\infty}$ as introduced in [37, 38]. The formal definition of $\mathbb{D}_{\mathrm{pw}\mathscr{C}^\infty}$ is given in the Appendix 8.6.3; for the understanding of the following it suffices to keep in mind that any piecewise-smooth distribution $D \in \mathbb{D}_{\mathrm{pw}\mathscr{C}^\infty}$ can be decomposed into a piecewise-smooth function f and a purely impulsive part, denoted by $D[\cdot]$, i.e.,

$$D \in \mathbb{D}_{\mathrm{pw}\mathscr{C}^\infty} \quad \Leftrightarrow \quad D = f_{\mathbb{D}} + D[\cdot],$$

where $f_{\mathbb{D}}$ denotes the distribution induced by the function f. In particular, it is possible to evaluate a piecewise-smooth distribution at a certain point $t \in \mathbb{R}$ in the three following ways:

- Right-evaluation: $D(t^+) := f(t^+) = f(t)$,
- Left-evaluation: $D(t^-) := f(t^-)$,
- Impulse-evaluation: $D[t]$.

For example, for the Dirac impulse δ we have for all $t \in \mathbb{R} \setminus \{0\}$

$$\delta(t-) = 0 = \delta(t+) \quad \text{and} \quad \delta[t] = 0$$

and for $t = 0$

$$\delta(0-) = 0 = \delta(0+) \quad \text{and} \quad \delta[0] = \delta.$$

Definition 8.12 (*Distributional solutions*) We call (x, u, y) (or just x if the context is clear) a *distributional solution* of the switched DAE (8.1) if, and only if, $(x, u, y) \in (\mathbb{D}_{\mathrm{pw}\mathscr{C}^\infty})^{\mathtt{n}} \times (\mathbb{D}_{\mathrm{pw}\mathscr{C}^\infty})^{\mathtt{u}} \times (\mathbb{D}_{\mathrm{pw}\mathscr{C}^\infty})^{\mathtt{y}}$ and (8.1) holds as an equation within $\mathbb{D}_{\mathrm{pw}\mathscr{C}^\infty}$. ◁

For classical (i.e., piecewise differentiable) solutions of (8.1) we have used the following equivalence to obtain observability characterizations for the single switch case:

$$y \equiv 0 \quad \Leftrightarrow \quad y_{(-\infty,0)} \equiv 0 \ \wedge \ y_{(0,\infty)} \equiv 0.$$

But this equivalence is not true anymore for switched DAEs (8.1) in a distributional framework and must be replaced by the following equivalence:

$$y \equiv 0 \quad \Leftrightarrow \quad y_{(-\infty,0)} \equiv 0 \ \wedge \ y[0] = 0 \ \wedge \ y_{(0,\infty)} \equiv 0, \tag{8.13}$$

which takes into account the impulsive part of y as well.

8.4.2.3 The Differential and Impulse Projectors

In order to utilize the equivalence (8.13) we need to find convenient representations of the three terms on the right-hand side of (8.13). To this end we define the so called differential and impulsive "projectors" as follows.

Definition 8.13 (*Differential and impulse projector,* [35]) Consider a regular matrix pair (E, A) with QWF (8.12) and corresponding transformation matrices S and T. The *differential projector* is

$$\Pi^{\text{diff}}_{(E,A)} := T \begin{bmatrix} I & 0 \\ 0 & 0 \end{bmatrix} S$$

and the *impulse projector* is

$$\Pi^{\text{imp}}_{(E,A)} := T \begin{bmatrix} 0 & 0 \\ 0 & I \end{bmatrix} S,$$

where the block sizes correspond to the block sizes in the QWF. Furthermore, define

$$A^{\text{diff}} := \Pi^{\text{diff}}_{(E,A)} A \quad \text{and} \quad E^{\text{imp}} := \Pi^{\text{imp}}_{(E,A)} E. \qquad \triangleleft$$

Note that the differential and impulse projector are not projectors in the usual sense as they are not idempotent in general, but they play a similar role as projectors in the explicit solution formula for inhomogeneous DAEs [39].

The significance of the matrix A^{diff} becomes clear in the following lemma:

Lemma 8.14 (Role of A^{diff}, [35]) *Consider a regular matrix pair (E, A) with corresponding matrix A^{diff}. Then every solution x of the DAE $E\dot{x} = Ax$ also solves*

$$\dot{x} = A^{\text{diff}} x.$$

The converse is also true, but only if $x(0) \in \operatorname{im}\Pi_{(E,A)}$, i.e., $x(0)$ is consistent.

A simple consequence of this result is the following result which establishes the strong similarity of a switched DAE with a switched ODE with jumps:

Corollary 8.15 (Switched DAE interpreted as switched ODE with jumps, [41]) *Consider a switched DAE* (8.2) *with switching signal $\sigma \in \Sigma$ and with switching times $t_1 < t_2 < t_3 \ldots$, where each matrix pair (E_p, A_p), $p \in \mathscr{P}$, is regular with corresponding consistency projector Π_p and matrix A_p^{diff}. Then the impulse-free part $x - x[\cdot]$ of any (distributional) solution x of* (8.2) *is given by the solution of the switched ODE with jumps:*

$$\begin{aligned} \dot{x} &= A^{\text{diff}}_{\sigma(t)} x, & t &\notin \{\, t_i \mid i \in \mathbb{N} \,\} \\ x(t_i^+) &= \Pi_{\sigma(t_i)} x(t_i^-), & i &\in \mathbb{N} \\ x(t_1^-) &\in \operatorname{im}\Pi_{\sigma(t_1^-)}. \end{aligned}$$

Note that $A_p^{\text{diff}} \Pi_p = A_p^{\text{diff}} = \Pi_p A_p^{\text{diff}}$ for all p, hence Remark 8.1 is not relevant here, because arbitrarily many "trivial" switches can be introduced without altering the solutions.

To understand the role of the impulse projector it is first necessary to consider an initial trajectory problem (ITP) for the DAE given by (E, A):

$$\begin{aligned} x_{(-\infty,0)} &= x^0_{(-\infty,0)}, \\ (E\dot{x})_{[0,\infty)} &= (Ax)_{[0,\infty)}, \end{aligned} \tag{8.14}$$

where $x^0 \in (\mathbb{D}_{\text{pw}\mathscr{C}^\infty})^n$ is an arbitrary initial trajectory. In [37] it is shown that for regular matrix pairs (E, A) there exists a unique solution $x \in (\mathbb{D}_{\text{pw}\mathscr{C}^\infty})^n$ for any ITP (8.14). In particular,

$$x(0^+) = \Pi_{(E,A)} x^0(0^-)$$

and the impulsive part is uniquely given as follows:

Lemma 8.16 (Impulsive part of ITP solution, [35]) *Consider the ITP* (8.14) *for a regular matrix pair (E, A) and corresponding matrix E^{imp}. Then the unique solution $x \in (\mathbb{D}_{pw\mathscr{C}^\infty})^n$ fulfills*

$$x[0] = -\sum_{i=0}^{n-2} (E^{\text{imp}})^{i+1} x^0(0^-) \delta^{(i)},$$

where $\delta^{(i)}$ denotes the ith derivative of the Dirac impulse δ.

Remark 8.17 In [35] the formula for $x[0]$ is a bit more complicated, but is identical to the one presented here, once it is realized that $E^{\text{imp}} \Pi_{(E,A)} = 0$. Furthermore, the upper limit of the sum can be reduced to $\nu - 2 \leq n - 2$, where ν is the index of

(E, A), because $(E^{\text{imp}})^k = 0$ for all $k \geq \nu$. However, in the context of switched DAE we use the formula with $n-2$ instead of $\nu - 2$ because the dimension n of x does not depend on the specific mode whereas the index ν might depend on the mode. ◁

8.4.2.4 Observability Condition

We are now ready to state the generalization of Lemmas 8.9 and 8.11 to switched DAEs.

Theorem 8.18 (Observability for σ_1: DAE case, [35]) *Consider the switched DAE* (8.1) *with regular matrix pairs* (E_p, A_p), $p \in \mathscr{P}$, *and corresponding consistency projectors* Π_p *and matrices* A_p^{diff}, E_p^{imp}. *Let*

$$O_p^{\text{diff}} := [C_p / C_p A_p^{\text{diff}} / C_p (A_p^{\text{diff}})^2 / \cdots / C_p (A_p^{\text{diff}})^{n-1}],$$
$$O_p^{\text{imp}} := [C_p E_p^{\text{imp}} / C_p (E_p^{\text{imp}})^2 / \cdots / C_p (E_p^{\text{imp}})^{n-2}],$$

then for the single switch switching signal σ_1 *given by* (8.8) *we have*

$$\boxed{\mathscr{M}_{0^-}^{\sigma_1} = \operatorname{im}\Pi_0 \cap \ker O_0^{\text{diff}} \cap \ker O_1^{\text{diff}} \Pi_1 \cap \ker O_1^{\text{imp}}}$$

and, for any $T \geq 0$,

$$\boxed{\mathscr{N}_{T+}^{\sigma_1} = e^{A_1^{\text{diff}} T} \mathscr{N}_{0+}^{\sigma_1} = e^{A_1^{\text{diff}} T} \Pi_1 \mathscr{M}_{0^-}^{\sigma_1},}$$

where $\mathscr{M}_{0^-}^{\sigma_1}$ *and* $\mathscr{N}_{T+}^{\sigma_1}$ *are as in Definitions 8.5 and 8.6, respectively. In particular, the switched DAE* (8.1) *is observable for* σ_1 *if, and only if,*

$$\operatorname{im}\Pi_0 \cap \ker O_0^{\text{diff}} \cap \ker O_1^{\text{diff}} \Pi_1 \cap \ker O_1^{\text{imp}} = \{0\}$$

and forward observable if, and only if,

$$\Pi_1 (\operatorname{im}\Pi_0 \cap \ker O_0^{\text{diff}} \cap \ker O_1^{\text{diff}} \Pi_1 \cap \ker O_1^{\text{imp}}) = \{0\}.$$

Remark 8.19 Each of the four subspaces involved in the intersection defining $\mathscr{M}_{0^-}^{\sigma_1}$ has an intuitive meaning. Recall that $\mathscr{M}_{0^-}^{\sigma_1}$ consists of all initial values just before $t = 0$ yielding a zero output. Hence, we can derive the following inclusions:

- $\mathscr{M}_{0^-}^{\sigma_1} \subseteq \operatorname{im}\Pi_0$ because $x(0^-)$ must be consistent with the mode before the switch.
- $\mathscr{M}_{0^-}^{\sigma_1} \subseteq \ker O_0^{\text{diff}}$ because $y_{(-\infty,0)} \equiv 0$ implies, invoking Lemma 8.14, that $x(0^-) \in \ker O_0^{\text{diff}}$.
- $\mathscr{M}_{0^-}^{\sigma_1} \subseteq \ker O_1^{\text{diff}}\Pi_1$ because $y_{(0,\infty)} \equiv 0$ implies that $x(0^+) = \Pi_1 x(0^-) \in \ker O_1^{\text{diff}}$, which in turn implies that $x(0^-) \in \Pi_1^{-1}(\ker O_1^{\text{diff}}) = \ker O_1^{\text{diff}}\Pi_1$.
- $\mathscr{M}_{0^-}^{\sigma_1} \subseteq \ker O_1^{\text{imp}}$ because $y[0] = 0$ implies, due to Lemma 8.16, that $x(0^-) \in \ker O_1^{\text{imp}}$.

This already shows that $\mathscr{M}_{0^-}^{\sigma_1} \subseteq \operatorname{im}\Pi_0 \cap \ker O_0^{\text{diff}} \cap \ker O_1^{\text{diff}}\Pi_1 \cap \ker O_1^{\text{imp}}$. The converse inclusion follows from the observation that for any x_0 in the intersection, there exists a unique solution x with initial condition $x(0^-) = x_0$ and this solution produces a zero output. $\lhd$

We conclude this section with an example which shows that each of the four subspaces in the representation of $\mathscr{M}_{0^-}^{\sigma_1}$ as in Theorem 8.18 plays a crucial role for observability, i.e., the intersection of only three of the four subspace will in general not yield the trivial subspace.

Example 8.20 ([35]) Let the switched DAE (8.1) be given by

$$\left.\begin{aligned} \begin{bmatrix} 1&0&0&0\\0&1&0&0\\0&0&0&0\\0&0&0&1 \end{bmatrix}\dot{x} &= \begin{bmatrix} 1&0&0&0\\0&1&0&0\\0&0&1&0\\0&0&0&0 \end{bmatrix}x + \begin{bmatrix}1\\0\\1\\1\end{bmatrix}u \\ y &= \begin{bmatrix}0&0&0&1\end{bmatrix}x \end{aligned}\right\} \quad t \in (-\infty, 0),$$

$$\left.\begin{aligned} \begin{bmatrix} 0&0&0&0\\0&1&0&0\\0&0&1&0\\1&0&0&0 \end{bmatrix}\dot{x} &= \begin{bmatrix} 1&0&0&0\\0&1&0&0\\0&0&1&0\\0&0&0&1 \end{bmatrix}x + \begin{bmatrix}0\\1\\1\\0\end{bmatrix}u \\ y &= \begin{bmatrix}0&1&0&1\end{bmatrix}x \end{aligned}\right\} \quad t \in [0, \infty).$$

Neither subsystem is observable in the classical sense. But it is possible to determine the exact value of the state trajectory with the switching signal (8.8). The consistency, differential, and impulse projectors as well as O_p^{diff}, O_p^{imp} for each of the two subsystems are:

$$\Pi_0 = \Pi_0^{\text{diff}} = \begin{bmatrix} 1&0&0&0\\ 0&1&0&0\\ 0&0&0&0\\ 0&0&0&1 \end{bmatrix}, \qquad \Pi_1 = \Pi_1^{\text{diff}} = \begin{bmatrix} 0&0&0&0\\ 0&1&0&0\\ 0&0&1&0\\ 0&0&0&0 \end{bmatrix},$$

$$\Pi_0^{\text{imp}} = \begin{bmatrix} 0&0&0&0\\ 0&0&0&0\\ 0&0&1&0\\ 0&0&0&0 \end{bmatrix}, \qquad \Pi_1^{\text{imp}} = \begin{bmatrix} 1&0&0&0\\ 0&0&0&0\\ 0&0&0&0\\ 0&0&0&1 \end{bmatrix},$$

$$O_0^{\text{diff}} = \begin{bmatrix} 0&0&0&1\\ 0&0&0&0\\ 0&0&0&0\\ 0&0&0&0 \end{bmatrix}, \qquad O_1^{\text{diff}} = \begin{bmatrix} 0&1&0&1\\ 0&1&0&0\\ 0&1&0&0\\ 0&1&0&0 \end{bmatrix},$$

$$O_0^{\text{imp}} = \begin{bmatrix} 0&0&0&0\\ 0&0&0&0\\ 0&0&0&0\\ 0&0&0&0 \end{bmatrix}, \qquad O_1^{\text{imp}} = \begin{bmatrix} 1&0&0&0\\ 0&0&0&0\\ 0&0&0&0\\ 0&0&0&0 \end{bmatrix},$$

and the four subspaces from Theorem 8.18 are:

$$\begin{aligned} \operatorname{im}\Pi_0 &= \operatorname{span}\{e_1, e_2,\ 0\ , e_4\}, \\ \ker O_0^{\text{diff}} &= \operatorname{span}\{e_1, e_2, e_3,\ 0\ \}, \\ \ker O_1^{\text{diff}}\Pi_1 &= \operatorname{span}\{e_1,\ 0\ , e_3, e_4\}, \\ \ker O_1^{\text{imp}} &= \operatorname{span}\{\ 0\ , e_2, e_3, e_4\}. \end{aligned}$$

where $e_i \in \mathbb{R}^4$, $i = 1, 2, 3, 4$, is the corresponding natural basis vector.

Clearly, $\operatorname{im}\Pi_0 \cap \ker O_0^{\text{diff}} \cap \ker O_1^{\text{diff}}\Pi_1 \cap \ker O_1^{\text{imp}} = \{0\}$ and the switched system is observable for σ_1 according to Theorem 8.18. Furthermore, each of the four subspaces $\mathfrak{C}_0$, $\ker O_0^{\text{diff}}$, $\ker O_1^{\text{diff}}\Pi_1$ and $\ker O_1^{\text{imp}}$ is necessary to obtain a trivial intersection. In fact, each subspace restricts exactly one state variable. In view of Remark 8.10, note that the switched system with a reversed mode sequence is not observable because

$$\operatorname{im}\Pi_1 \cap \ker O_1^{\text{diff}} \cap \ker O_0^{\text{diff}}\Pi_0 \cap \ker O_0^{\text{imp}} = \operatorname{span}\{e_3\} \neq \{0\}.$$

As an illustration of constructing state trajectories from the knowledge of the output and the input, let us consider an input,[1] $u(t) = e^{2t} + \delta_{-1} + \delta_0$, and assume that the following output is produced by the system with σ_1 specified in (8.8):

[1] Note that, for simplicity, we are misusing the notation by writing $u(t) = e^{2t} + \delta_{-1} + \delta_0$ because u is a piecewise-smooth distribution and therefore only the evaluations $u(t-)$, $u(t+)$, $u[t]$ are well defined. The correct way of writing would be to write $\hat{u}(t) = e^{2t}$ and $u = \hat{u}_{\mathbb{D}} + \delta_{-1} + \delta_0$.

$$y(t) = \begin{cases} -1, \ t \in (-\infty, -1), \\ 0, \ t \in [-1, 0), \\ e^t + e^{2t} + \delta_0, \ t \in [0, \infty). \end{cases}$$

The closed form solution for the state variables, parameterized by $a, b, c \in \mathbb{R}$, is given as follows:

$$x_1(t) = \begin{cases} -e^{t+1} + (a-1)e^t + e^{2t}, & t \in (-\infty, -1), \\ (a-1)e^t + e^{2t}, & t \in [-1, 0), \\ 0, & t \in [0, \infty), \end{cases}$$

$$x_2(t) = \begin{cases} e^t b, & t \in (-\infty, 0), \\ e^t + e^{2t} + (b-1)e^t, & t \in [0, \infty), \end{cases}$$

$$x_3(t) = \begin{cases} -e^{2t} - \delta_{-1}, & t \in (-\infty, 0), \\ -e^t + e^{2t}, & t \in [0, \infty), \end{cases}$$

$$x_4(t) = \begin{cases} \frac{1}{2}e^{2t}c - 1, & t \in (-\infty, -1), \\ \frac{1}{2}e^{2t}c, & t \in [-1, 0), \\ -a\delta_0, & t \in [0, \infty). \end{cases}$$

First note that $x_3(0^-) = -1$, which corresponds to the fact that in the homogeneous case the consistency space $\operatorname{im}\Pi_0$ restricts $x_3(0^-)$ to be zero. Since $\ker O_0^{\text{diff}}$ restricts $x_4(0^-)$, we would expect that $y(0^-)$, $\dot{y}(0^-)$, ..., determine $x_4(0^-)$. In fact, $0 = y(0^-) = x_4(0^-)$. The space $\ker O_1^{\text{diff}}\Pi_1$ restricts $x_2(0^-)$, and hence by using the values for $y^{(i)}(0^+)$, we are able to reconstruct $x_2(0^-)$: $2 = y(0^+) = x_2(0^+) + x_4(0^+) = 1 + b = 1 + x_2(0^-)$, i.e., $x_2(0^-) = 1$. Finally, $\ker O_1^{\text{imp}}$ restricts $x_1(0^-)$, therefore, the information from the impulse of y at zero can be used to determine $x_1(0^-)$: $\delta_0 = y[0] = x_2[0] + x_4[0] = -a\delta_0$, hence $-1 = a = x_1(0^-)$. Altogether, we were able to determine $x(0^-)$ which together with the knowledge of u and the regularity of the matrix pairs (E_0, A_0), (E_1, A_1) makes it possible to uniquely reconstruct the whole state x. $\triangleleft$

8.5 Observability for General Switching Signals

We now consider a general switching signal $\sigma \in \Sigma$. Let

$$\Sigma_{\mathbb{N}} := \left\{ \sigma \in \Sigma \;\middle|\; \sigma\big|_{[t_i, t_{i+1})} = i \quad \text{for } i = 0, 1, 2, \ldots \right\} \tag{8.15}$$

where $0 = t_1 < t_2 < t_3 < \ldots$ are the switching times of σ (note however Remark 8.1 when the switching signal has only finitely many switches) and $t_0 := -\infty$. When considering a fixed switching signal we can, without loss of generality, restrict our attention to switching signals of the class $\Sigma_{\mathbb{N}}$ by a suitable relabeling of the modes.

8.5.1 Observability Characterization

In this subsection, we present a characterization of observability of system (8.1) and (8.2) with a switching signal $\sigma \in \Sigma_{\mathbb{N}}$. Instead of treating both the cases separately, our goal is to present a single result that generalizes to both system classes. Towards this end, we introduce some notation that alternates its meaning depending on the system class under consideration. For switched ODEs, (8.1), we let, for $i \in \mathbb{N}$,

$$F_i := A_i, \quad J_i := G_i \tag{8.16a}$$

denote the flow matrix and the jump matrix for subsystem $i \in \mathscr{P}$, respectively. The local unobservable space at the ith switch for system (8.1) is given by:

$$\mathscr{M}^i := \ker O_{i-1} \cap \ker O_i G_i. \tag{8.16b}$$

Analogously, for switched DAEs (8.1), we let:

$$F_i := A_i^{\text{diff}} \quad J_i := \Pi_i \tag{8.17a}$$

and the corresponding local unobservable space at the ith switch is:

$$\mathscr{M}^i := \text{im}\,\Pi_{i-1} \cap \ker O_{i-1}^{\text{diff}} \cap \ker O_i^{\text{diff}} \Pi_i \cap \ker O_i^{\text{imp}}, \tag{8.17b}$$

with the notation as in Theorem 8.18.

According to Lemma 8.9 and Theorem 8.18, $\mathscr{M}^i$ is the unobservable space if there would be only a single switch from mode $i - 1$ to mode i. Note that this local unobservable space does not depend on the actual switching time t_i.

We now combine the local unobservable spaces of the first m switches as follows:

$$\boxed{\mathscr{M}_m^m := \mathscr{M}^m,} \tag{8.18a}$$

$$\boxed{\mathscr{M}_i^m := \mathscr{M}^i \cap J_i^{-1}(e^{-F_i \tau_i} \mathscr{M}_{i+1}^m), \quad m > i \geq 1,} \tag{8.18b}$$

where $\tau_i := t_{i+1} - t_i$.

The intuition behind the sequence (8.18) is as follows: Starting at the mth switch, we go backward in time and combine the local knowledge from each of the previous switches to obtain knowledge of the initial value $x(0^-)$. In fact, the local unobservable space at the mth switch is moved by the flow of the $(m - 1)$-st mode (via the negative exponential) to obtain the unobservable space at t_{m-1}^+ taking into account the knowledge of the mth switch only. With the preimage of the jump map J_{m-1} this information is translated to the information at t_{m-1}^-. Combining this with thelocal unobservable space at the $(m-1)$-st switch given by $\mathcal{M}^{m-1}$, we get a refined unobservable space at the $(m-1)$-st switch taking into account the information obtained from the switches at t_{m-1} and t_m. In particular $y_{(t_{m-2},t_{m+1})} \equiv 0$ implies $x(t_{m-1}^-) \in \mathcal{M}_{m-1}^m$. Repeating this argument we arrive at the following implication, for all $i \in \mathbb{N}$:

$$y_{(t_{i-1},t_{m+1})} \equiv 0 \quad \Rightarrow \quad x(t_i^-) \in \mathcal{M}_i^m.$$

In fact, we can formulate a stronger result using the following notation:

$$\Sigma_m := \{\sigma_m \mid \exists \sigma \in \Sigma_{\mathbb{N}} : \sigma_m := \sigma \text{ on } (-\infty, t_m) \text{ and } \sigma_m(t) = m \text{ on } [t_m, \infty)\}. \tag{8.19}$$

Theorem 8.21 (Geometric observability characterization, [34, 36]) *Consider the switched system* (8.1) *(respectively* (8.1)*) with switching signal* $\sigma \in \Sigma_{\mathbb{N}}$ *and corresponding* $\sigma_m \in \Sigma_m$ *for some* $m \in \mathbb{N}$*. Then the unobservable spaces* $\mathcal{M}_{0^-}^{\sigma}$ *and* $\mathcal{M}_{0^-}^{\sigma_m}$ *as in Definition 8.5 fulfill*

$$\boxed{\mathcal{M}_{0^-}^{\sigma} \subseteq \mathcal{M}_{0^-}^{\sigma_m} = \mathcal{M}_1^m,}$$

where $\mathcal{M}_1^m$ *is obtained using* (8.18) *together with* (8.16) *(resp.* (8.17)*). In particular* (8.1) *(resp.* (8.2) *is observable for* σ *if, and only if, there exists* $m \in \mathbb{N}$ *such that*

$$\mathcal{M}_1^m = \{0\}. \tag{8.20}$$

Example 8.22 Consider the switched system (8.1) characterized by:

$$A_0 = A_2 = \begin{bmatrix} 0 & 0 \\ 0 & 0 \end{bmatrix}, \; A_1 = \begin{bmatrix} 0 & 1 \\ -1 & 0 \end{bmatrix}$$
$$C_0 = C_2 = \begin{bmatrix} 1 & 0 \end{bmatrix}, \; C_1 = \begin{bmatrix} 0 & 0 \end{bmatrix}$$

with $G_i = I$, $H_i = 0$, $B_i = 0$, $D_i = 0$ for $i = 0, 1, 2$. It is noted that none of the pairs (A_i, C_i) is observable. Consider the switching signal σ with the mode sequence $0 \to 1 \to 2$ with switching times t_1, t_2 such that $\tau_1 := t_2 - t_1 \neq k\pi$ for any $k \in \mathbb{N}$. Clearly, $\mathcal{M}_2^2 := \operatorname{span}\{e_2\}$ and

$$\mathcal{M}_1^2 = \operatorname{span}\{e_2\} \cap e^{-A_1\tau_1}\operatorname{span}\{e_2\} = \operatorname{span}\{e_2\} \cap \operatorname{span}\left\{\left(\begin{smallmatrix} \sin\tau_1 \\ \cos\tau_1 \end{smallmatrix}\right)\right\}.$$

Thus, $\mathcal{M}_{0^-}^{\sigma} = \mathcal{M}_1^2 = \{0\}$ and system (8.1) is observable, provided $\tau_1 \neq k\pi$. Note that this switched system cannot be made observable with only a single switch. ◁

8.5.2 Removing Dependency from Switching Times

The observability condition (8.20) given in Theorem 8.21 is for a fixed switching signal σ given by (8.15), in particular, we fixed the mode sequence. It is entirely possible that for the same mode sequence the system is observable for certain switching times and unobservable for others. So it would be more useful to know whether the observability property holds for all switching signals with the same switching sequence. It can be shown that if there is a switching signal that satisfies $\mathscr{M}_{0-}^{\sigma} = \{0\}$, then the set of switching signals, with the same mode sequence, for which $\mathscr{M}_{0-}^{\sigma} \neq \{0\}$, is nowhere dense. The result is formally stated as follows:

Theorem 8.23 (Genericity of observability, [18, Theorem 2]) *If for* $m \in \mathbb{N}$ *and some* σ_m *given by* (8.19) *the switched system* (8.1) *(resp.* (8.2)*) is observable then the set*

$$\Sigma^o := \{\sigma \in \Sigma_{\mathbb{N}} \mid (8.1)(resp.\ (8.2))\ is\ observable\ for\ \sigma_m\ given\ by\ (8.19)\}$$

is open and dense in the set of all switching signals in $\Sigma_{\mathbb{N}}$*, where the topology on* $\Sigma_{\mathbb{N}}$ *is given in the Appendix 8.6.4.*

In other words, the above result indicates that the condition (8.20) is somewhat robust with respect to the switching times.

Next, we want to derive conditions for observability which are independent of switching times, while keeping the mode sequence fixed as in (8.15). As stated in Theorem 8.21, condition (8.20), which depends on the switching times, is necessary and sufficient for observability. To obtain a relaxed version of (8.20) for deriving conditions independent of switching times, one must introduce some degree of conservatism. For this reason, we only arrive at a sufficient condition and a necessary condition for a fixed mode sequence, which do not depend on switching times.

Corollary 8.24 (Sufficient condition for observability, [34, 36]) *For system* (8.1) *(resp.* (8.2)*) with* σ *given by* (8.15) *and* $m \in \mathbb{N}$ *define the following sequence of subspaces:*

$$\begin{aligned} \overline{\mathscr{M}}_m^m &:= \mathscr{M}^m, \\ \overline{\mathscr{M}}_i^m &:= \mathscr{M}^i \cap J_i^{-1}\left\langle F_i \,\middle|\, \overline{\mathscr{M}}_{i+1}^m \right\rangle, \qquad m > i \geq 1, \end{aligned}$$

where $F_i, J_i, \mathscr{M}^i$ *are defined via* (8.16) *(resp.* (8.17)*), and* $\langle F \mid \mathscr{M} \rangle$ *denotes the smallest* F*-invariant subspace which contains* $\mathscr{M}$*, for some matrix* F *and subspace* $\mathscr{M}$*. Then,*

$$\overline{\mathscr{M}}_i^m \supseteq \mathscr{M}_i^m \quad \forall i = 1, \ldots, m$$

hence, system (8.1) (resp. (8.2)*) is observable for* σ *if there exists* $m \in \mathbb{N}$ *such that* $\overline{\mathscr{M}}_1^m = \{0\}$.

It is natural to ask how much conservatism has been introduced in obtaining the sufficient condition. If the condition in Corollary 8.24 holds, that is $\overline{\mathscr{M}}_1^m = \{0\}$, then the system (8.1), or (8.1), is observable for *all* σ with mode sequence 1 through m, regardless of the switching times t_i. To address the reverse implication, note that system (8.1) is uniformly (with respect to switching times) observable for σ_m given by (8.19), if and only if,

$$\bigcup_{\sigma_m \in \Sigma_m} \mathscr{M}_{0^-}^{\sigma_m} = \bigcup_{\tau_1,\dots,\tau_{m-1}>0} \mathscr{M}_1^m = \{0\}. \tag{8.21}$$

However, in order to check the above condition in practice, a difficulty arises due to the fact that the union of two subspaces is not necessarily a subspace because the resulting union is not closed under addition in general. In case, if $\bigcup_{\tau_i,\dots,\tau_{m-1}>0} \mathscr{M}_i^m$, for each $1 \le i \le m-1$ is a subspace, then $\overline{\mathscr{M}}_i^m = \bigcup_{\tau_i,\dots,\tau_{m-1}>0} \mathscr{M}_i^m$, and in that case $\overline{\mathscr{M}}_1^m = \{0\}$ implies observability for all σ_m regardless of switching times. The following example provides an illustration of these arguments:

Example 8.25 We reconsider the switched system from Example 8.22. It was highlighted there that for some special switching signals the system is not observable; and indeed the sufficient condition of Corollary 8.24 is not satisfied because $\overline{\mathscr{M}}_1^2 = \overline{\mathscr{M}}_2^2 = \operatorname{span}\{e_2\} \neq \{0\}$. Now, in the first mode of Example 8.22, let us replace the matrix A_1 by $\tilde{A}_1 := \left[\begin{smallmatrix} 0 & 1 \\ 0 & 0 \end{smallmatrix}\right]$. For the resulting switched system, we get again $\overline{\mathscr{M}}_1^2 = \overline{\mathscr{M}}_2^2 = \operatorname{span}\{e_2\} \neq \{0\}$. Hence, the sufficient condition in Corollary 8.24 is violated; however, the resulting switched system is observable for all $\tau_1 > 0$ (which can be seen from Theorem 8.21). The source of this gap is the fact that the set

$$\bigcup_{\tau_1>0} e^{-\tilde{A}_1\tau_1}\mathscr{M}_2^2 = \bigcup_{\tau_1>0} e^{-\tilde{A}_1\tau_1}\operatorname{span}\{e_2\} = \left\{ \left(\begin{smallmatrix} -\tau_1\xi \\ \xi \end{smallmatrix} \right) \,\middle|\, \tau_1 > 0, \xi \in \mathbb{R} \right\}$$

is not a subspace and its intersection with $\mathscr{M}^1 = \operatorname{span}\{e_2\}$ is just $\{0\}$. $\lhd$

Having developed a sufficient condition using subspaces that contain $\mathscr{M}_i^m$ (for all switching times), we now obtain a necessary condition in terms of subspaces contained in $\mathscr{M}_i^m$.

Corollary 8.26 (Necessary condition for observability, [34, 36]) *For system* (8.1) (resp. (8.2)*) with $\sigma \in \Sigma_{\mathbb{N}}$ and $m \in \mathbb{N}$ define the following sequence of subspaces:*

$$\begin{aligned} \underline{\mathscr{M}}_m^m &:= \mathscr{M}^m, \\ \underline{\mathscr{M}}_i^m &:= \mathscr{M}^i \cap J_i^{-1}\langle \underline{\mathscr{M}}_{i+1}^m \mid F_i \rangle, \qquad m > i \ge 1. \end{aligned}$$

where $F_i, J_i, \mathscr{M}_i$ are defined via (8.16) (resp. (8.17))*, and $\langle \mathscr{M} \mid F \rangle$ is the largest F-invariant subspace contained within $\mathscr{M}$, for some subspace $\mathscr{M}$ and matrix F. Then,*

$$\underline{\mathscr{M}}_i^m \subseteq \mathscr{M}_i^m \quad \forall i = 1, \ldots, m,$$

hence if system (8.1) *(resp. (8.2)) is observable for* σ *then there exists* $m \in \mathbb{N}$ *such that* $\underline{\mathscr{M}}_1^m = \{0\}$.

The natural question now to ask is whether $\underline{\mathscr{M}}_1^m = \{0\}$ always guarantees the existence of a switching signal that renders the switched system observable (i.e., whether the switched system is controlled observable under the constraint that $\sigma \in \Sigma_{\mathbb{N}}$). It can be shown that

$$\underline{\mathscr{M}}_1^m = \bigcap_{\tau_1,\ldots,\tau_{m-1}>0} \mathscr{M}_1^m, \tag{8.22}$$

from where one sees that the right-hand side may be $\{0\}$ even though there exists no $\sigma \in \Sigma_{\mathbb{N}}$ with dwell times $\tau_1, \tau_2, \cdots, \tau_{m-1} > 0$ such that $\mathscr{M}_1^m = \{0\}$. However, in the next section, we present a connection between the necessary condition in Corollary 8.26 and weakly controlled observability.

Example 8.27 Consider the switched system (8.1) characterized by:

$$A_0 = \begin{bmatrix} 0 & 1 \\ -1 & 0 \end{bmatrix}, \; A_1 = \begin{bmatrix} 0 & 1 \\ 0 & 0 \end{bmatrix}, \; A_2 = \begin{bmatrix} 0 & 0 \\ 0 & 0 \end{bmatrix}$$
$$C_0 = \begin{bmatrix} 0 & 0 \end{bmatrix}, \quad C_1 = \begin{bmatrix} 0 & 0 \end{bmatrix}, \; C_2 = \begin{bmatrix} 1 & 0 \end{bmatrix}$$

with $G_i = I$, $H_i = 0$, $B_i = 0$, $D_i = 0$ for $i = 0, 1, 2$. Then, for $\sigma^m \in \Sigma_m$, with $m = 2$, we obtain that $\underline{\mathscr{M}}_2^2 = \ker O_2 \cap \ker O_1 = \mathrm{span}\{e_2\}$ and $\underline{\mathscr{M}}_1^2 = \langle \mathrm{span}\{e_2\} \mid A_1 \rangle = \{0\}$. However, it is seen that $\mathscr{M}_1^2 = \mathrm{span}\left\{\left(\begin{smallmatrix} -\tau_1 \\ 1 \end{smallmatrix}\right)\right\} \neq \{0\}$, so that (8.20) does not hold for any $\tau_1 > 0$, showing that the system is not observable for $\sigma^m \in \Sigma_m$ with $m = 2$ even though the necessary condition of Corollary 8.26 is satisfied. On the other hand, if the mode sequence is repeated at least once, then the system is observable for the new switching signal. ◁

8.5.3 Conditions for Controlled and Weakly Controlled Observability

Since (weakly) controlled observability does not consider an a priori fixed switching signal we have to consider again a general switching signal

$$\sigma : \mathbb{R} \to \{1, 2, \ldots, \mathtt{p}\}$$

instead of a switching signal given by (8.15). Here $\mathtt{p} \in \mathbb{N}$ denotes the *finite* number of different modes.

8.5.3.1 Classical Switched Systems

Observability of classical switched systems was studied in [31], where the authors call (8.1) observable if any initial value can be distinguished from zero (via a suitable switching signal). Due to linearity, this notion is equivalent to our notion of weakly controlled observability. We will present the algebraic and geometric conditions derived in [31] in the following.

We start by defining the following generalized observability matrices $\mathfrak{O}_i$, $i \in \mathbb{N}$ as follows

$$\mathfrak{O}_0 = [C_1/C_2/\cdots/C_{\mathbf{p}}] \tag{8.23a}$$

$$\mathfrak{O}_{i+1} = [\mathfrak{O}_0/\mathfrak{O}_i A_1/\mathfrak{O}_i A_2/\cdots/\mathfrak{O}_i A_{\mathbf{p}}],\ i \in \mathbb{N}. \tag{8.23b}$$

Note that $\mathfrak{O}_i$ is of size $\mathbf{py}\frac{\mathbf{p}^{i+1}-1}{\mathbf{p}-1} \times n$, where $\mathbf{y} \in \mathbb{N}$ denotes the output dimension. Let $\mathscr{K}_i = \ker \mathfrak{O}_i$. It is then easy to see that, for $i \in \mathbb{N}$,

$$\boxed{\mathscr{K}_i = \bigcap_{k=0}^{i} \bigcap_{p_0,\ldots,p_k} \ker C_{p_0} A_{p_1} \cdots A_{p_k}.}$$

Furthermore, $\mathscr{K}_{i+1} \subseteq \mathscr{K}_i$ and there exists $k^* < n$ such that $\mathscr{K}^* := \mathscr{K}_{k^*} = \mathscr{K}_k$ for all $k \geq k^*$. It follows that $\mathscr{K}^*$ is the largest subset of $\mathscr{K}_0 = \bigcap_{p\in\mathscr{P}} \ker C_p$ which is invariant for each A_p, $p \in \mathscr{P}$, i.e., it is the largest subspace which satisfies

$$\forall p \in \mathscr{P}: \quad \mathscr{K}^* \subseteq \ker C_p \text{ and } A_p\mathscr{K}^* \subseteq \mathscr{K}^*.$$

Weakly controlled observability can now be characterized as follows:

Theorem 8.28 (Weakly controlled observability for systems without jumps, [31]) *The switched system* (8.1) *without jumps is weakly controlled observable if, and only if,*

$$\mathscr{K}^* = \{0\}$$

i.e., there is no nontrivial $\{\ A_p \mid p \in \mathscr{P}\ \}$*-invariant subspace contained in the intersection of the null spaces of* C_p, $p \in \mathscr{P}$.

The intuition behind this result is as follows: An initial value $x_0 \in \mathbb{R}^n$ is indistinguishable from zero if, and only if, the corresponding output satisfies $y \equiv 0$ for all switching signals, i.e.,

$$\{0\} = \left\{ C_{p_k} e^{A_{p_k}\tau_k} e^{A_{p_{k-1}}\tau_{k-1}} \cdots e^{A_{p_1}\tau_1} x_0 \;\middle|\; k \in \mathbb{N},\, p_1,\ldots,p_k \in \mathscr{P}, \tau_1,\ldots\tau_k > 0 \right\}.$$

Due to analyticity of the map $(\tau_1, \tau_2, \ldots, \tau_k) \mapsto C_{p_k} e^{A_{p_k}\tau_k} e^{A_{p_{k-1}}\tau_{k-1}} \cdots e^{A_{p_1}\tau_1} x_0$ (in particular around zero) it follows that $x_0 \in \mathbb{R}^n$ is indistinguishable from zero if, and only if,

$$C_{p_i} A_{p_{i-1}} A_{p_{i-2}} \cdots A_{p_1} x_0 = 0 \quad \forall i \in \mathbb{N} \quad \forall p_1, p_2, \ldots, p_i \in \mathscr{P}.$$

The latter is nothing else but the condition

$$x_0 \in \ker \mathfrak{O}_i = \mathscr{K}_i, \quad \forall i \in \mathbb{N},$$

or equivalently,

$$x_0 \in \bigcap_{i \in \mathbb{N}} \mathscr{K}_i = \mathscr{K}^*.$$

Note that the condition $\mathscr{K}^* = \{0\}$ is equivalent to

$$\operatorname{rank} \mathfrak{O}_{n-1} = n,$$

which shows the similarities to the classical full rank assumption of the Kalman observability matrix for nonswitched linear ODEs. In view of Remark 8.3(iv), it is also clear that full rank of the observability matrix of any mode implies full rank of $\mathfrak{O}_{n-1}$.

The authors of [31] also study controllability of switched systems and they establish a duality between observability and controllability [31, Corollary 4.28]. For the problem of controllability it can be shown that a single switching signal and a final time $t_f > 0$ exists such that for all initial states one finds an input u that gives $x(t_f) = 0$ ([31, Remark 4.23], see also [15]). We ca use this result to establish the following important equivalence.

Theorem 8.29 (Weakly controlled and controlled observability, no jumps) *If the switched system* (8.1) without jumps *is weakly controlled observable, then there exists a single switching signal σ for which any initial value can be distinguished from zero. In particular, for classical switched systems weakly controlled observability is equivalent to controlled observability.*

Proof For the classical switched system

$$\dot{x} = A_\sigma x, \quad x(0) = x_0 \in \mathbb{R}^n, \quad y = C_\sigma x$$

consider the dual switched system

$$\dot{z} = -A_\sigma^\top z + C_\sigma^\top v, \quad z(0) = z_0 \in \mathbb{R}^n. \tag{8.24}$$

As shown in [31], controllability of (8.24) is equivalent to weakly controlled observability of the classical switched systems and implies existence of a single switching signal σ^* and $t_f > 0$ such that for all $z_0 \in \mathbb{R}^n$ there exists an input v such that the corresponding solution of (8.24) satisfies $z(t_f) = 0$. Note that for any solution x of the classical switched system and any solution z of (8.24) we have

$$\frac{\mathrm{d}}{\mathrm{d}t}(z^\top x) = (-z^\top A_{\sigma^*} + v^\top C_{\sigma^*})x + z^\top A_{\sigma^*} x = v^\top y,$$

and therefore

$$z^\top(t_f)x(t_f) - z_0^\top x_0 = \int_0^{t_f} v^\top y.$$

Hence, the assumption that there exists x_0 with vanishing output implies $z_0^\top x_0 = z^\top(t_f)x(t_f)$. Controllability of (8.24) now implies that for all $z_0 \in \mathbb{R}^n$ we can achieve $z(t_f) = 0$; hence $x_0 = 0$ must hold. This shows that for the switching signal σ^* any initial value is distinguishable from zero, i.e., the classical switched system is controlled observable. □

The above equivalence between weakly controlled observability and controlled observability is rather significant because with a simple algebraic test (independent of the switching signal), one can conclude existence of a specific switching signal which makes the switched system observable and this allows the construction of an observer (see Chap. 7 in this book or [36]).

Remark 8.30 (Generalized Kalman observability decomposition, [22]) Assume that $\dim \mathscr{K}^* = n - o$, and let $b_1, \ldots, b_n$ be a basis in $\mathbb{R}^n$ such that $b_{o+1}, \ldots, b_n$ span $\mathscr{K}^*$. Since $\mathscr{K}^*$ is an A_p-invariant subspace and it is contained in $\ker C_p$ for all $p \in \mathscr{P}$, in this new basis the matrices A_p, B_p, and C_p can be rewritten as

$$A_p = \begin{bmatrix} A_p^{\mathrm{O}}, & 0 \\ A_p', & A_p'' \end{bmatrix}, C_p = \begin{bmatrix} C_p^{\mathrm{O}}, & 0 \end{bmatrix}, B_p = \begin{bmatrix} B_p^{\mathrm{O}} \\ B_p' \end{bmatrix},$$

where $A_p^{\mathrm{O}} \in \mathbb{R}^{o \times o}$, $B_p^{\mathrm{O}} \in \mathbb{R}^{o \times \mathbf{u}}$, and $C_p^{\mathrm{O}} \in \mathbb{R}^{\mathbf{y} \times o}$. The corresponding switched system

$$\begin{aligned} \dot{z}(t) &= A_{\sigma(t)}^{\mathrm{O}} z(t) + B_{\sigma(t)}^{\mathrm{O}} u(t) \\ y(t) &= C_{\sigma(t)}^{\mathrm{O}} z(t) + D_{\sigma(t)} u(t) \end{aligned} \tag{8.25}$$

is then weakly controlled observable and it is equivalent to the original switched system without jumps in the following sense: for every solution (x, y, σ, u) of (8.1), there exists a solution of (8.25) of the form (z, y, σ, u) and vice versa, i.e., the input-output behavior of (8.1) and (8.25) coincide. ◁

Note that the above observability reduction method is in fact an algorithm and can be implemented, see [22] for more details on the numerical implementation.

8.5.3.2 Switched Systems with Jumps

Weakly controlled observability for switched ODEs with jumps was studied in [25]. The system class of so called linear hybrid systems studied therein is more general than (8.1) in two major aspects: (1) The switching signal itself is generated by a finite automaton and observability of the active mode is part of the observability definition; (2) the jump maps depend on the modes before and after the switch, this allows in particular to also study modes with different state space dimensions. For details on this framework we refer to Chap. 3 in this book. Under the assumption that the discrete output map is identity (i.e., observability of the discrete state is trivially fulfilled) and under the assumption that the jump maps only depend on the mode directly after the switch (this implies that the state space dimensions must be equal) we can obtain conditions for weakly controlled observability for (8.1).

To this end, we define for every $p \in \mathscr{P}$ and every matrix C with n columns the Kalman observability matrix of (A_p, C) by

$$O_p(C) = [C/CA_p/\dots/CA_p^{n-1}].$$

For each $i = 0, 1, 2, \dots$, and every $p \in \mathscr{P}$ define

$$\begin{aligned}\mathfrak{O}_{p,0} &= O_p(C_p),\\ \mathfrak{O}_{p,i+1} &= O_p\big([C_p \,/\, \mathfrak{O}_{1,i}G_1 \,/\, \mathfrak{O}_{2,i}G_2 \,/\, \cdots \,/\, \mathfrak{O}_{\mathbf{p},i}G_{\mathbf{p}}]\big).\end{aligned}$$

It is not difficult to see that $O_{p,i}$ consists of "rows" of the form $C_pA_p^{\alpha}$ for $p \in \mathscr{P}$ and $\alpha \in \{0, 1, 2, \dots, n-1\}$ or rows of the form

$$\boxed{C_{p_k}A_{p_k}^{\alpha_k}G_{p_k}A_{p_{k-1}}^{\alpha_{k-1}}G_{p_{k-1}}\cdots A_{p_1}^{\alpha_1}G_{p_1}A_p^{\alpha_0}}$$

where $1 \leq k \leq i$, $p_1, \dots, p_k \in \mathscr{P}$, $\alpha_0, \dots, \alpha_k \in \{0, 1, \dots, n-1\}$. With similar arguments as in Sect. 8.5.3.1 it follows that

$$x_0 \in \ker \mathfrak{O}_{p,i} \quad \forall p \in \mathbb{N}\ \forall i \in \mathbb{N}$$

is equivalent to indistinguishability of x_0 from zero. Furthermore, it can be shown (taking into account that $O_p(C)$ is the largest A_p-invariant subspace contained in the kernel of C) that $\ker \mathfrak{O}_{p,i+1} \subseteq \ker \mathfrak{O}_{p,i}$. Hence, we arrive at the following characterization of weakly controlled observability for switched systems (8.1).

Theorem 8.31 (Weakly controlled observability characterization, [25]) *The switched system* (8.1) *is weakly controlled observable if, and only if,*

$$\bigcap_{p\in\mathscr{P}}\bigcap_{i=0}^{\infty}\ker \mathfrak{O}_{p,i} = \{0\}.$$

Note that the condition of Theorem 8.31 can be checked numerically, see [22] for details.

Unfortunately, for the case with state jumps, it is not yet clear whether weak controlled observability implies controlled observability. We conjecture that if the reset maps are invertible, then weak controlled observability implies controlled observability.

Similar to Remark 8.30, an observer reduction can be carried out. However, the resulting reduced switched system will have different state space dimension and does not fit anymore into the framework considered here.

We conclude this section by presenting a possible connection between the algebraic condition for weakly controlled observability as established in Theorem 8.31 and the geometric observability conditions obtained in Sect. 8.5.2.

Conjecture 8.32 *Consider the switched system* (8.1). *For $m > 0$ and the switching signal $\sigma : \mathbb{R} \to \mathscr{P}$ let $\underline{\mathscr{M}}_1^{\sigma,m}$ denote the analog of space $\underline{\mathscr{M}}_1^m$ constructed in Corollary 8.26 with mode sequence $\sigma(t_1), \sigma(t_2), \ldots, \sigma(t_m)$ instead of $1, 2, \ldots, m$. Then* $\ker \mathfrak{O}_{p,m}$ *is the largest A_p-invariant subspace contained in $\underline{\mathscr{M}}_1^{\sigma,m}$ for all switching signals with $\sigma(0) = p$.*

8.5.4 Forward Observability

As mentioned earlier, observability deals with recovering the state trajectory of the system at all times. The weaker notion of forward observability is concerned with recovering the state on a certain interval of the form (T, ∞), and is particularly useful in designing observers for switched systems (c.f. Chap. 7 in this book). The geometric conditions for characterization of forward observability for system (8.1) (resp. (8.2)) can be obtained through parallel development. Towards that end, consider the following sequence of subspaces:

$$\boxed{\mathscr{N}^1 := J_1 \mathscr{M}^1,} \tag{8.26a}$$

$$\boxed{\mathscr{N}^{i+1} := J_{i+1}(\mathscr{M}^{i+1} \cap e^{F_i \tau_i} \mathscr{N}^i), \quad i > 0,} \tag{8.26b}$$

where F_i, J_i, $\mathscr{M}^i$ are defined in (8.16) (resp. (8.17)).

The intuition behind this sequence of subspaces is as follows: The subspace $\mathscr{N}^i$ contains all forward unobservable states at the ith switching instant where we use all the knowledge up to the ith switching instant. At the next switching instant we propagate forward the information from $\mathscr{N}^i$ and intersect it with the locally unobservable subspace $\mathscr{M}^{i+1}$. Using then the jump map J_{i+1} gives the next forward unobservable subspace $\mathscr{N}^{i+1}$. This procedure is significantly different to the subspace iteration in (8.18) as the iterations do not proceed backward in time. We can now characterize forward observability with the help of the subspace iteration (8.26).

Theorem 8.33 (Forward observability characterization, [34, 36]) *Consider the switched system* (8.1)*(respectively* (8.1)*) with switching signal* σ *given by* (8.15). *Then the forward unobservable space at the* m*th switch of* σ *is given by* $\mathscr{N}^m$, *i.e., for* σ_m *defined in* (8.19)*, it holds that*

$$\boxed{\mathscr{N}_{t_m^+}^{\sigma_m} = \mathscr{N}^m,}$$

where $\mathscr{N}^m$ *is obtained using* (8.16) *(resp.* (8.17)*) and* (8.26)*. In particular* (8.1) *(resp. (8.2)) is forward observable for* σ *if, and only if, there exists* $m \in \mathbb{N}$ *such that*

$$\mathscr{N}^m = \{0\}. \tag{8.27}$$

The conditions for forward observability independent of switching times could be developed similarly as in the previous section.

Corollary 8.34 (*Sufficient condition for forward observability*) *For system* (8.1) *(resp.* (8.2)*) with* σ *given by* (8.15) *and* $m \in \mathbb{N}$ *define the following sequence of subspaces:*

$$\begin{aligned}\overline{\mathscr{N}}^1 &:= J_1\mathscr{M}^1,\\ \overline{\mathscr{N}}^{i+1} &:= J_{i+1}\left(\mathscr{M}^{i+1} \cap \left\langle F_i \,\middle|\, \overline{\mathscr{N}}^i\right\rangle\right), \qquad i > 1,\end{aligned}$$

where $F_i, J_i, \mathscr{M}^i$ *are defined via* (8.16) *(resp.* (8.17)*). Then,*

$$\overline{\mathscr{N}}^i \supseteq \mathscr{N}^i \quad \forall i \geq 1,$$

hence, system (8.1) *(resp.* (8.2)*) is forward observable for* σ *if there exists* $m \in \mathbb{N}$ *such that* $\overline{\mathscr{N}}^m = \{0\}$.

Corollary 8.35 (*Necessary condition for forward observability*) *For system* (8.1) *(resp.* (8.2)*) with* σ *given by* (8.15) *and* $m \in \mathbb{N}$ *define the following sequence of subspaces:*

$$\begin{aligned}\underline{\mathscr{N}}^1 &:= J_1\mathscr{M}^1,\\ \underline{\mathscr{N}}^{i+1} &:= J_{i+1}\left(\mathscr{M}^{i+1} \cap \left\langle \underline{\mathscr{N}}^i \,\middle|\, F_i\right\rangle\right), \qquad i > 1,\end{aligned}$$

where $F_i, J_i, \mathscr{M}^i$ *are defined via* (8.16) (resp. (8.17)*). Then,*

$$\underline{\mathscr{N}}^i \subseteq \mathscr{N}^i \quad \forall i \geq 1,$$

hence if system (8.1)(8.1) *(resp.* (8.2)*) is forward observable for* σ *then there exists* $m \in \mathbb{N}$ *such that* $\underline{\mathscr{N}}^m = \{0\}$.

8.6 Appendix

8.6.1 Some Basic Facts Concerning Linear Algebra

Let $\mathcal{N}$, $\mathcal{M}$ be some linear subspaces of $\mathbb{R}^n$ and A, B matrices of suitable size. Then the following properties are easy to verify:

(i) $A^{-1}(\ker B) = \ker BA$
(ii) $A(A^{-1}(\mathcal{N})) = \mathcal{N} \cap \operatorname{im} A$
(iii) $A(\mathcal{N} \cap \mathcal{M}) \subseteq A\mathcal{N} \cap A\mathcal{M}$ with equality if, and only if (c.f. [47, Sect. 0.4]),

$$(\mathcal{N} + \mathcal{M}) \cap \ker A = \mathcal{N} \cap \ker A + \mathcal{M} \cap \ker A;$$

the latter holds, for example, if $\ker A \subseteq \mathcal{N}$.

8.6.2 The Wong Sequences and the QWF

Consider a regular matrix pair (E, A). The Wong sequences [1, 6, 46] are defined as

$$\begin{aligned} \mathcal{V}_0 &:= \mathbb{R}^n, & \mathcal{V}_{i+1} &:= A^{-1}(E\mathcal{V}_i), & i \in \mathbb{N}, \\ \mathcal{W}_0 &:= \{0\}, & \mathcal{W}_{i+1} &:= E^{-1}(A\mathcal{W}_i), & i \in \mathbb{N}. \end{aligned}$$

It is easily seen that the Wong sequence are nested and terminate after finitely many steps:

$$\begin{aligned} &\mathcal{V}_0 \supset \mathcal{V}_1 \supset \ldots \supset \mathcal{V}_{k^*} = \mathcal{V}_{k^*+1} = \ldots \\ &\mathcal{W}_0 \subset \mathcal{W}_1 \subset \ldots \subset \mathcal{W}_{\ell^*} = \mathcal{W}_{\ell^*+1} = \ldots \end{aligned}$$

Let $\mathcal{V}^* := \bigcap_{i \in \mathbb{N}} \mathcal{V}_i = \mathcal{V}_{k^*}$ and $\mathcal{W}^* := \bigcup_{i \in \mathbb{N}} \mathcal{W}_i = \mathcal{W}_{\ell^*}$. It can be shown [7] that (E, A) is regular if, and only if,

$$\mathcal{V}^* \oplus \mathcal{W}^* = \mathbb{R}^n = E\mathcal{V}^* \oplus A\mathcal{W}^*. \tag{8.28}$$

In that case it also holds that $k^* = \ell^* = \nu$ is the index of the matrix pair (E, A). Let V, W be full (column) rank matrices such that $\operatorname{im} V = \mathcal{V}^*$ and $\operatorname{im} W = \mathcal{W}^*$. Because of (8.28) the matrices

$$T := [V, W], \quad S := [EV, AW]^{-1} \tag{8.29}$$

are then invertible matrices.

Theorem 8.36 (QWF, *[6]*) *Consider a regular matrix pair* (E, A) *and the corresponding Wong sequences and (invertible) matrices* S, T *as in* (8.29). *Then*

$$(SET, SAT) = \left(\begin{bmatrix} I & 0 \\ 0 & N \end{bmatrix}, \begin{bmatrix} J & 0 \\ 0 & I \end{bmatrix} \right)$$

where N is nilpotent.

Remark 8.37 In the case of a singular matrix pair, the Wong sequences can also be used to obtain the *quasi-Kronecker form*, see [7, 8]. ◁

8.6.3 Distribution Theory

We recall the basic definitions and properties of classical distributions as formalized by Schwartz [29]. The space of *test functions* (i.e., smooth functions $\varphi : \mathbb{R} \to \mathbb{R}$ with compact support) is denoted by $\mathscr{C}_0^\infty$, the *space of distributions* is the dual space of the space of test functions, i.e.,

$$\mathbb{D} := \left\{ D : \mathscr{C}_0^\infty \to \mathbb{R} \;\middle|\; D \text{ is linear and continuous} \right\}.$$

Note that continuity requires a topology on the space of test function. However, in practice, the continuity is tested via sequential continuity: A linear map $D : \mathscr{C}_0^\infty \to \mathbb{R}$ is continuous if, and only if, the sequence of real numbers $D(\varphi_n)$ converges to zero as $n \to \infty$ for any sequence $(\varphi_n)_{n \in \mathbb{N}}$ of test function fulfilling the following two properties:

(i) The support of each φ_n is contained in a common compact set.
(ii) For each $i \in \mathbb{N}$ the sequence $\varphi_n^{(i)}$ converges uniformly to the zero function as $n \to \infty$.

The main two properties of distributions are (1) that they can be interpreted as generalized functions and (2) that they are arbitrarily often differentiable. To be more precise, let $\mathscr{L}_{1,\,\mathrm{loc}}$ be the space of locally integrable functions, then the mapping

$$\mathscr{L}_{1,\,\mathrm{loc}} \to \mathbb{D},\; f \mapsto f_{\mathbb{D}} := \left(\varphi \mapsto \int_{\mathbb{R}} f\varphi \right)$$

is well defined (i.e., $f_{\mathbb{D}}$ is indeed a distribution) and an injective homomorphism. The simplest distribution which is not induced by a function is the *Dirac impulse* given by $\delta(\varphi) := \varphi(0)$, or, in general for $t \in \mathbb{R}$, $\delta_t(\varphi) := \varphi(t)$ for $\varphi \in \mathscr{C}_0^\infty$. For $i \geq 1$, the ith derivative of an arbitrary distribution $D \in \mathbb{D}$ is given by

$$D^{(i)}(\varphi) := -D^{(i-1)}(\varphi'), \quad \varphi \in \mathscr{C}_0^\infty,$$

where we take $D^{(0)}(\varphi) = D(\varphi)$. Distributions can be multiplied with smooth functions:

$$(\alpha D)(\varphi) := D(\alpha\varphi), \quad \alpha \in \mathscr{C}^\infty, D \in \mathbb{D}, \varphi \in \mathscr{C}_0^\infty.$$

Let $\mathscr{C}^\infty_{\text{pw}}$ be the space of piecewise-smooth function, where $\alpha : \mathbb{R} \to \mathbb{R}$ is called piecewise-smooth when there exists a locally finite ordered set $S = \{ s_i \in \mathbb{R} \mid i \in \mathbb{Z} \}$ and smooth functions $\alpha_i \in \mathscr{C}^\infty$, $i \in \mathbb{Z}$, such that $\alpha = \sum_{i\in\mathbb{Z}} (\alpha_i)_{[s_i, s_{i+1})}$. The space of *piecewise-smooth distributions* is then given by

$$\mathbb{D}_{\text{pw}\mathscr{C}^\infty} := \left\{ f_{\mathbb{D}} + \sum_{\tau\in T} D_\tau \;\middle|\; \begin{array}{l} f \in \mathscr{C}^\infty_{\text{pw}},\, T \subseteq \mathbb{R} \text{ locally finite,} \\ \forall \tau \in T : D_\tau \in \operatorname{span}\{\delta_\tau, \delta'_\tau, \delta''_\tau, \ldots\} \end{array} \right\}.$$

The properties of $\mathbb{D}_{\text{pw}\mathscr{C}^\infty}$ and corresponding definitions are summarized in the following, where $D = f_{\mathbb{D}} + \sum_{\tau\in T} D_\tau \in \mathbb{D}_{\text{pw}\mathscr{C}^\infty}$ and $t \in \mathbb{R}$:

(i) Closed under differentiation: $D' \in \mathbb{D}_{\text{pw}\mathscr{C}^\infty}$.
(ii) Left- and right-evaluation: $D(t^+) := f(t)$, $D(t^-) := f(t^-)$.
(iii) Impulsive part: $D[t] := D_t$ if $t \in T$, $D[t] = 0$ otherwise.
(iv) Restriction to interval: $D_I := (f_I)_{\mathbb{D}} + \sum_{\tau\in T\cap I} D_\tau$, where $I \subseteq \mathbb{R}$ is some interval.
(v) Multiplication with piecewise-smooth function: $\alpha D := \sum_{i\in\mathbb{Z}} \alpha_i D_{[s_i, s_{i+1})}$, where $\alpha = \sum_{i\in\mathbb{Z}} (\alpha_i)_{[s_i, s_{i+1})}$ as above; in particular, $\alpha\delta_t = \alpha(t)\delta_t$.

For more details see [37, 38].

8.6.4 Topology on the Space of Switching Signals

Analogously as in [36] we define for $m \in \mathbb{N}$ the (pseudo-)metric

$$d_m(\sigma, \sigma') = \sum_{i=1}^{m-1} |\tau_i - \tau'_i|, \quad \sigma, \sigma' \in \Sigma_{\mathbb{N}},$$

where $\tau_i := t_{i+1} - t_i > 0$ and $\tau'_i = t'_{i+1} - t'_i > 0$ and t_i, t'_i are the switching times of σ and σ', respectively. For each $m \in \mathbb{N}$ this metric induces a topology on $\Sigma_{\mathbb{N}}$ which is isomorphic to the usual topology of $\mathbb{R}^{m-1}$ restricted to the open positive orthant.

References

1. Armentano, V.A.: The pencil *(sE-A)* and controllability-observability for generalized linear systems: a geometric approach. SIAM J. Control Optim. **24**, 616–638 (1986)
2. Babaali, M., Egerstedt, M.: Observability of switched linear systems. In: Hybrid Systems: Computation and Control, Lecture Notes in Computer Science, vol. 2993, pp. 48–63. Springer (2004)
3. Babaali, M., Pappas, G.J.: Observability of switched linear systems in continuous time. In: Hybrid Systems: Computation and Control. Lecture Notes in Computer Science, vol. 3414, pp. 103–117. Springer, Berlin (2005)

4. Baglietto, M., Battistelli, G., Scardovi, L.: Active mode observation of switching systems based on set-valued estimation of the continuous state. Int. J. Robust Nonlinear Control **19**(14), 1521–1540 (2009)
5. Balluchi, A., Benvenuti, L., Di Benedetto, M., Sangiovanni-Vincentelli, A.: Observability for hybrid systems. In: Proceedings of 42nd IEEE Conference Decision and Control, Hawaii, USA, vol. 2, pp. 1159–1164 (2003). doi:10.1109/CDC.2003.1272764
6. Berger, T., Ilchmann, A., Trenn, S.: The quasi-Weierstraß form for regular matrix pencils. Lin. Alg. Appl. **436**(10), 4052–4069 (2012). doi:10.1016/j.laa.2009.12.036
7. Berger, T., Trenn, S.: The quasi-Kronecker form for matrix pencils. SIAM J. Matrix Anal. Appl. **33**(2), 336–368 (2012). doi:10.1137/110826278
8. Berger, T., Trenn, S.: Addition to "The quasi-Kronecker form for matrix pencils". SIAM J. Matrix Anal. Appl. **34**(1), 94–101 (2013). doi:10.1137/120883244
9. Boukhobza, T., Hamelin, F.: Observability of switching structured linear systems with unknown input. A graph-theoretic approach. Automatica **47**(2), 395–402 (2011)
10. Chaib, S., Boutat, D., Benali, A., Barbot, J.P.: Observability of the discrete state for dynamical piecewise hybrid systems. Nonlin. Anal. Th. Meth. Appl. **63**(3), 423–438 (2005)
11. Collins, P., van Schuppen, J.H.: Observability of piecewise-affine hybrid systems. In: Alur, R., Pappas G.J. (eds.) Hybrid Systems: Computation and Control, Lecture Notes in Computer Science, vol. 2993, pp. 265–279. Springer-Verlag (2004)
12. De Santis, E.: On location observability notions for switching systems. Syst. Control Lett. **60**(10), 807–814 (2011). doi: http://dx.doi.org/10.1016/j.sysconle.2011.06.004
13. Fliess, M., Join, C., Perruquetti, W.: Real-time estimation for switched linear systems. In: Proceedings of 47th IEEE Conference on Decision and Control, Cancun, Mexico, pp. 941–946 (2008). doi:10.1109/CDC.2008.4739053
14. Gantmacher, F.R.: The Theory of Matrices (Vol. I & II). Chelsea, New York (1959)
15. Ji, Z., Wang, L., Guo, X.: On controllability of switched linear systems. IEEE Trans. Autom. Control **53**(3), 796–801 (2008). doi:10.1109/TAC.2008.917659
16. Lee, C., Ping, Z., Shim, H.: Real time switching signal estimation of switched linear systems with measurement noise. In: Proceedings of 12th European Control Conference 2013, Zurich, Switzerland, pp. 2180–2185 (2013)
17. Liu, B., Marquez, H.: Controllability and observability for a class of controlled switching impulsive systems. IEEE Trans. Autom. Control **53**(10), 2360–2366 (2008)
18. Medina, E.A., Lawrence, D.A.: State estimation for linear impulsive systems. In: Proceedings of American Control Conference 2009, pp. 1183–1188 (2009)
19. Medina, E.A., Lawrence, D.A.: Reachability and observability of linear impulsive systems. Automatica **44**, 1304–1309 (2008)
20. Mincarelli, D., Floquet, T.: On second-order sliding mode observers with residuals' projection for switched systems. In: Proceedings of 51st IEEE Conference on Decision and Control, Maui, USA, pp. 1929–1934 (2012). doi:10.1109/CDC.2012.6426719
21. Orani, N., Pisano, A., Franceschelli, M., Giua, A., Usai, E.: Robust reconstruction of the discrete state for a class of nonlinear uncertain switched systems. Nonlinear Anal. Hybrid Syst. **5**(2), 220–232 (2011). doi:10.1016/j.nahs.2010.10.011
22. Petreczky, M.: Realization theory of hybrid systems. Ph.D. thesis, Vrije Universiteit, Amsterdam (2006)
23. Petreczky, M.: Realization theory of linear and bilinear switched systems: A formal power series approach—part I: realization theory of linear switched systems. ESAIM Control Optim. Calc. Var., pp. 410–445 (2011)
24. Petreczky, M.: Realization theory of linear and bilinear switched systems: A formal power series approach—part II: bilinear switched systems. ESAIM Control Optim. Calc. Var., pp. 446–471 (2011)
25. Petreczky, M., van Schuppen, J.H.: Realization theory for linear hybrid systems. IEEE Trans. Autom. Control **55**(10), 2282–2297 (2010)
26. Petreczky, M., van Schuppen, J.H.: Span-reachability and observability of bilinear hybrid systems. Automatica **46**(3), 501–509 (2010). doi:10.1016/j.automatica.2010.01.008

27. Ríos, H., Davila, J., Fridman, L.: High-order sliding mode observers for nonlinear autonomous switched systems with unknown inputs. J. Franklin Inst. **349**(10), 2975–3002 (2012)
28. De Santis, E., Di Benedetto, M.D., Pola, G.: A structural approach to detectability for a class of hybrid systems. Automatica **45**(5), 1202–1206 (2009). doi:10.1016/j.automatica.2008.12.014
29. Schwartz, L.: Théorie des Distributions I, II. No. IX, X in Publications de l'institut de mathématique de l'Universite de Strasbourg. Hermann, Paris (1950, 1951)
30. Shim, H., Tanwani, A.: Hybrid-type observer design based on a sufficient condition for observability in switched nonlinear systems. Int. J. Robust Nonlinear Control (Special Issue on High Gain Observers and Nonlinear Output Feedback Control **24**(6), 1064–1089, 2014). doi:10.1002/rnc.2901
31. Sun, Z., Ge, S.S.: Switched linear systems. Communications and Control Engineering. Springer-Verlag, London (2005). doi:10.1007/1-84628-131-8
32. Tanwani, A., Liberzon, D.: Invertibility of switched nonlinear systems. Automatica 46(12), 1962–1973 (2010). doi:10.1016/j.automatica.2010.08.002
33. Tanwani, A., Trenn, S.: An observer for switched differential-algebraic equations based on geometric characterization of observability. In: Proceedings of 52nd IEEE Conference on Decision and Control, Florence, Italy, pp. 5981–5986 (2013). doi:10.1109/CDC.2013.6760833
34. Tanwani, A., Trenn, S.: Observability of switched differential-algebraic equations for general switching signals. In: Proceedings of 51st IEEE Conference on Decision and Control, Maui, USA, pp. 2648–2653 (2012). doi:10.1109/CDC.2012.6427087
35. Tanwani, A., Trenn, S.: On observability of switched differential-algebraic equations. In: Proceedings of 49th IEEE Conference on Decision and Control, Atlanta, USA, pp. 5656–5661 (2010). doi:10.1109/CDC.2010.5717685
36. Tanwani, A., Shim, H., Liberzon, D.: Observability for switched linear systems: Characterization and observer design. IEEE Trans. Autom. Control **58**(4), 891–904 (2013). doi:10.1109/TAC.2012.2224257
37. Trenn, S.: Distributional differential algebraic equations. Ph.D. thesis, Institut für Mathematik, Technische Universität Ilmenau, Universitätsverlag Ilmenau, Ilmenau, Germany (2009). URL http://www.db-thueringen.de/servlets/DocumentServlet?id=13581
38. Trenn, S.: Regularity of distributional differential algebraic equations. Math. Control Signals Syst. **21**(3), 229–264 (2009). doi:10.1007/s00498-009-0045-4
39. Trenn, S.: Switched differential algebraic equations. In: Vasca, F., Iannelli L. (eds.) Dynamics and Control of Switched Electronic Systems—Advanced Perspectives for Modeling, Simulation and Control of Power Converters, chap. 6, pp. 189–216. Springer, London (2012). doi:10.1007/978-1-4471-2885-4_6
40. Trenn, S.: Stability of switched DAEs. In: Daafouz, J., Tarbouriech, S., Sigalotti, M. (eds.) Hybrid Systems with Constraints, Automation—Control and Industrial Engineering Series, pp. 57–83. Wiley, London (2013). doi:10.1002/9781118639856.ch3
41. Trenn, S., Wirth, F.R.: Linear switched DAEs: Lyapunov exponent, converse Lyapunov theorem, and Barabanov norm. In: Proceedings of 51st IEEE Conference on Decision and Control, Maui, USA, pp. 2666–2671 (2012). doi:10.1109/CDC.2012.6426245
42. Vidal, R., Chiuso, A., Sastry, S.: Observability and identifiability of jump linear systems. In: Proceedings of 41st IEEE Conference on Decision and Control, pp. 3614–3619 (2002)
43. Vidal, R., Chiuso, A., Soatto, S., Sastry, S.: Observability of linear hybrid systems. In: Hybrid Systems: Computation and Control. Lecture Notes in Computer Science, vol. 2623, pp. 526–539. Springer, Berlin (2003)
44. Vu, L., Liberzon, D.: Invertibility of switched linear systems. Automatica **44**(4), 949–958 (2008)
45. Weierstraß, K.: Zur Theorie der bilinearen und quadratischen Formen. Berl. Monatsb. pp. 310–338 (1868)
46. Wong, K.T.: The eigenvalue problem $\lambda Tx + Sx$. J. Diff. Eqns. **16**, 270–280 (1974)
47. Wonham, W.M.: Linear Multivariable Control: A Geometric Approach, 3rd edn. Springer, New York (1985)

48. Xie, G., Wang, L.: Necessary and sufficient conditions for controllability and observability of switched impulsive control systems. IEEE Trans. Autom. Control **49**(6), 960–966 (2004). doi:10.1109/TAC.2004.829656
49. Yu, L., Zheng, G., Boutat, D., Barbot, J.P.: Observability and observer design for a class of switched systems. IET Control Theory Appl. **5**(9), 1113–1119 (2011). doi:10.1049/iet-cta.2010.0475

Chapter 9
Sampled-Data Control of Switched Affine Systems

Laurentiu Hetel and Emilia Fridman

Abstract This chapter considers the stabilization problem for switched affine systems with a sampled-data switching law. The switching law is assumed to be a function of the system state at sampling instants. Sampling interval may be subject to variations or uncertainty. Switching law design criteria, taking into account the sampled-data implementation and uncertainties, are provided. Numerical examples illustrate the approach.

9.1 Introduction

In the last decade, the design of switching controllers has represented an important problem in the hybrid system community [17, 19]. This problem is very challenging for the case of switched affine systems where, generally, the different subsystems do not share a common equilibrium point. The study is motivated by the wide range of applications to power electronics [2, 15]. Different stabilization solutions exist in the literature based on the existence of stable convex combinations [2], on optimal control methods [8, 16], or on the use of sliding modes [3, 20]. A characterization of the set of attainable equilibrium points using quadratic Lyapunov functions and conic switching laws has been provided in [2].

The next phase toward practical application of switching control is to study its sampled-data implementation. For results on sampled-data control we point to the discrete-time methods in [1, 11], the input delay-approach [5, 6, 18], the robust control approach [12, 14] and the impulsive system method [13, 21]. Recently, increasing attention has been given to the sampled-data control of switched systems [7–9]. This aspect is crucial in the switched affine system context since, due to sampling, one can

L. Hetel (✉)
Laboratoire d'Automatique Génie Informatique et Signal (CNRS LAGIS UMR 8219), Ecole Centrale de Lille, Avenue Paul Langevin, BP 48, 59651 Villeneuve d'Ascq Cedex, France
e-mail: laurentiu.hetel@ec-lille.fr

E. Fridman
Department of Electrical Engineering and Systems, Tel Aviv University, 69978 Ramat Aviv, Tel Aviv, Israel
e-mail: emilia@eng.tau.ac.il

M. Djemai and M. Defoort (eds.), *Hybrid Dynamical Systems*, Lecture Notes in Control and Information Sciences 457, DOI 10.1007/978-3-319-10795-0_9

no longer drive the state exponentially toward the equilibrium point, but only toward a limit cycle or to some attractive compact set containing the equilibrium. Moreover, in many practical applications timing imperfections due to sampling jitters or delays in switching law may affect the control performances and even induce instability. Recently, a discrete-time method has been provided in [8] based on Control Lyapunov functions and the use of a nonlinear optimization solvers for quadratically constrained quadratic program (QCQP). However, it is still a problem to choose the Control Lyapunov function so as to optimize the robustness with respect to sampling or the size of the attractive set around the desired equilibrium. Moreover, although discrete-time methods guarantee that the inter-sample behaviour remains close to the ideal equilibrium, for an accurate study, it is of interest to exactly describe the continuous-time behaviour in between sampling instants. For the case of systems with sampled-data switching control, this study is very challenging. The switching control is often described by a discrete-event system with transitions ruled by a partition of (continuous variables) state space. Then, the sampling usually induces a delay in the discrete-event system variable. This phenomenon may imply a mismatch in the system control, in the sense that one system mode may be active in other state zones than the one for which the non-sampled control has been designed. If not appropriately taken into account, the sampling may be a source of poor performance and even instability.

The goal of the chapter is to present a continuous-time switching design method that ensures the robustness with respect to sampling and to potential implementations imperfections (jitters, uncertainty etc.). Simple LMI criteria are given to optimize the choice of Lyapunov functions. First, we present the case of switched linear system; next, we treat the case of switched affine systems. We proposed preliminary results in [10].

The chapter is organized as follows: in Sect. 9.2 we provide a mathematical formulation of the problem under study. In Sect. 9.3 a solution is provided for the case of linear switched systems. The solution is generalized to the case of switched affine systems in Sect. 9.4. Numerical examples are given in Sect. 9.5.

Notations. We denote the transpose of a matrix M by M^T. By $\mathbf{I}$ (or $\mathbf{0}$) we denote the identity (or the null) matrix with the appropriate dimension. $|\cdot|$ denotes the Euclidean vector norm. For a square symmetric matrix, $M \succ 0$ ($M \prec 0$) indicates that M is positive (negative) definite. By $\text{eig}_{\min}(M)$ we denote the minimum eigenvalue of a square symmetric matrix M. For a given set $\mathscr{F}$, the symbol $co\mathscr{F}$ denotes the convex hull of the set.

9.2 Problem Formulation

Consider two positive integers, n and N, N matrices $A_1, A_2, \ldots, A_N \in \mathbb{R}^{n\times n}$ and N vectors $B_1, B_2, \ldots, B_N \in \mathbb{R}^n$. The matrices $A_i, i = 1, \ldots, N$, are not necessary Hurwitz. We are interested in the class of switched systems described by

$$\dot{x}(t) = A_\sigma x(t) + B_\sigma, \forall t \in \mathbb{R}^+, \; x(0) = x_0 \in \mathbb{R}^n, \tag{9.1}$$

where $x : \mathbb{R} \to \mathbb{R}^n$ represents the system state and $\sigma : \mathbb{R}^N \to \mathscr{I} = \{1, 2, \ldots, N\}$ a switching control.

Consider the unit simplex:

$$\Lambda = \left\{ \lambda = [\lambda_1, \lambda_2, \ldots, \lambda_N] \in \mathbb{R}^N, \ \lambda_i \geq 0, \ \sum_{i=1}^{N} \lambda_i = 1 \right\}, \tag{9.2}$$

the convex combinations of matrices

$$A(\lambda) = \sum_{i=1}^{N} \lambda_i A_i, \ B(\lambda) = \sum_{i=1}^{N} \lambda_i B_i, \lambda \in \Lambda \tag{9.3}$$

and the subset of Λ associated to Hurwitz matrices:

$$\Lambda_H = \left\{ \lambda \in \Lambda : \ \exists\, P \succ 0, \ \text{s.t. } A^T(\lambda)P + PA(\lambda) \prec 0 \right\}.$$

We recall now the existing control design procedure for switched system in the continuous time case:

Lemma 9.1 [2] *Consider system* (9.1)*, some* $\lambda \in \Lambda_H$ *and the vector*

$$x_e = -A^{-1}(\lambda)\, B(\lambda). \tag{9.4}$$

Let P *be a solution of the LMIs:*

$$P = P^T \succ 0, \ A^T(\lambda)P + PA(\lambda) \prec 0 \tag{9.5}$$

and consider the following definition of the switching law:

$$\sigma(x) \in \arg\min_{i \in \mathscr{I}} (x - x_e)^T P (A_i x + B_i). \tag{9.6}$$

Then, for any initial condition x_0*, the solution of system* (9.1), (9.6) *satisfies the relation* $\lim_{t\to\infty} x(t) = x_e$.

Assume now that the values of system state are available at sample times $\{t_k\}_{k\in\mathbb{N}}$, with

$$t_0 = 0, \ t_k < t_{k+1}, \ \forall\, k \in \mathbb{N}, \ \lim_{k\to\infty} t_k = \infty.$$

Ideally, the interval between two sampling instants is constant, i.e., $t_{k+1} - t_k = T$. However, in practice the implementation may be subject to uncertainties and we should assume that the sampling interval is time-varying, $T_k = t_{k+1} - t_k$ with $0 < T_k \leq T_{\max}$ where $T_{\max}$ is a known bound. We assume that the sequence of sampling times t_k does not admit any accumulation points, i.e., $\lim_{k\to\infty} t_k = \infty$.

Consider the notation $x_k = x(t_k)$. With a sampled-data implementation, the switched system becomes:

$$\dot{x}(t) = A_{\sigma(x_k)}x(t) + B_{\sigma(x_k)}, \forall t \in [t_k, t_{k+1}) \in \mathbb{R}^n. \tag{9.7}$$

$$\sigma(x_k) \in \arg\min_{i \in \mathscr{I}} (x_k - x_e)^T P (A_i x_k + B_i).$$

The problem under study is formulated as follows:

Problem 9.2 Consider system (9.7) and assume that $\Lambda_H \neq \{\varnothing\}$. Given $T_{\max}$, and some $\lambda \in \Lambda_H$, can we find a switching function $\sigma : \mathbb{R}^n \to \mathscr{I}$ and a constant $C > 0$ s.t. for $t \to \infty$ the trajectories of the resulting system approach to the ball $|x - x_e|^2 < CT_{\max}$ where $x_e = -A^{-1}(\lambda) B(\lambda)$?

9.3 Switched Linear Case

In this section, we provide a solution to a simplified case where $B_i = 0, \ \forall i \in \mathscr{I}$, i.e.,

$$\dot{x}(t) = A_{\sigma(x_k)}x(t), \forall t \in [t_k, t_{k+1}) \in \mathbb{R}^n. \tag{9.8}$$

We provide robust stabilization condition based on the existence of quadratic candidate Lyapunov functions $V(x) = x^T Px$ and of switching laws of the form

$$\sigma(x_k) \in \arg\min_{i \in \mathscr{I}} x_k^T P A_i x_k, \tag{9.9}$$

where $P \succ 0$ is a matrix to be designed. Note that under such a switching law, the state space is covered by conic regions

$$\mathscr{S}_{i,j} = \left\{ x \in \mathbb{R}^n : x^T P(A_j - A_i)x \geq 0 \right\}, \ i, j \in \mathscr{I}$$

and that the system evolves along the mode $\dot{x} = A_i x$ for all $t \in [t_k, t_{k+1})$ if

$$x_k \in \mathscr{Z}_i = \cap_{j \in \mathscr{I}} \mathscr{S}_{i,j}, \forall \ i \in \mathscr{I}.$$

The main technical difficulty for providing robust stabilization conditions lies in the fact that the system state $x(t)$ may exit the region $\mathscr{Z}_i$ in the sampling interval and enter another region $\mathscr{Z}_j$. Before presenting the robust design conditions for the switched linear case, we give next a useful technical lemma in which we show how the initial condition may be taken into consideration to study the decay of a quadratic function $x^T Px$ along the trajectories of a linear time invariant system with initial conditions in conic regions.

Lemma 9.3 *Consider m symmetric matrices* $Q_i \in \mathbb{R}^{n\times n}, i = 1,\ldots,m$ *the linear time-invariant system* $\dot{x}(t) = Ax(t),\ \forall\, t \in [0,h], x(0) = x_0 \neq 0$, *with* $A \in \mathbb{R}^{n\times n}$,

$$x_0^T Q_i x_0 \leq 0,\ i = 1,\ldots,m, \tag{9.10}$$

and the notations:

$$\bar{A} = \begin{bmatrix} A & \mathbf{0}_{n,n} \\ A & \mathbf{0}_{n,n} \end{bmatrix}, \bar{Q}_i = \begin{bmatrix} Q_i & -Q_i \\ -Q_i & Q_i \end{bmatrix},\ i = 1,\ldots,m. \tag{9.11}$$

Assume that there exist $P, U_1, U_3 \succ 0$ *in* $\mathbb{R}^{n\times n}$, $U_2 \in \mathbb{R}^{n\times n}$, *m scalars* $\varepsilon_i > 0, i = 1,\ldots,m$, *and a real* γ *s.t.*

$$U = \begin{bmatrix} U_1 & U_2 \\ U_2^T & U_3 \end{bmatrix} \succ \mathbf{0}, \tag{9.12}$$

$$\begin{bmatrix} A^T P + PA & -U_2 - U_1 \\ -U_2^T - U_1 & U_1 - U_3 \end{bmatrix} - \sum_{i=1}^{m} \varepsilon_i \bar{Q}_i \prec \mathbf{0}, \tag{9.13}$$

$$\begin{bmatrix} A^T P + PA - U_1(1-e^{\gamma h}) & -U_2 - U_1 e^{\gamma h} \\ -U_2^T - U_1 e^{\gamma h} & U_1 e^{\gamma h} - U_3 \end{bmatrix} - \sum_{i=1}^{m} \varepsilon_i \bar{Q}_i \prec \mathbf{0} \tag{9.14}$$

$$\bar{A}^T U + U\bar{A} \prec \gamma U. \tag{9.15}$$

Then, there exists a sufficiently small scalar $\alpha > 0$ *s.t. the function* $V(x(t)) = x^T(t)Px(t)$ *satisfies the relation* $\dot{V}(x(t)) < -\alpha V(x(t)),\ \forall t \in [0,h]$.

Proof Consider the notations $\rho(t) = x(t) - x_0$. Note that

$$\begin{bmatrix} \dot{x}(t) \\ \dot{\rho}(t) \end{bmatrix} = \begin{bmatrix} A & \mathbf{0}_{n,n} \\ A & \mathbf{0}_{n,n} \end{bmatrix} \begin{bmatrix} x(t) \\ \rho(t) \end{bmatrix}, \forall t \geq 0. \tag{9.16}$$

From Eq. (9.15), using the fact that $\rho(0) = 0$, we obtain that

$$\begin{bmatrix} x(t) \\ \rho(t) \end{bmatrix}^T \begin{bmatrix} U_1 & U_2 \\ U_2^T & U_3 \end{bmatrix} \begin{bmatrix} x(t) \\ \rho(t) \end{bmatrix} = z^T U z \leq e^{\gamma t} x_0^T U_1 x_0, \forall t \geq 0 \tag{9.17}$$

with $z(t) = \left[x^T(t)\ \rho^T(t)\right]^T$. Since $x_0 = x(t) - \rho(t)$, this leads to the following relation

$$\omega(x(t),\rho(t)) = \begin{bmatrix} x(t) \\ \rho(t) \end{bmatrix}^T \begin{bmatrix} -U_1(1-e^{\gamma t}) & -U_2 - U_1 e^{\gamma t} \\ -U_2^T - U_1 e^{\gamma t} & U_1 e^{\gamma t} - U_3 \end{bmatrix} \begin{bmatrix} x(t) \\ \rho(t) \end{bmatrix} \geq 0. \tag{9.18}$$

Similarly, (9.10) may be expressed as

$$\begin{bmatrix} x(t) \\ \rho(t) \end{bmatrix}^T \begin{bmatrix} Q_i & -Q_i \\ -Q_i & Q_i \end{bmatrix} \begin{bmatrix} x(t) \\ \rho(t) \end{bmatrix} \leq 0, \; i = 1, \ldots, m. \tag{9.19}$$

Further we prove that if conditions (9.12)–(9.15) hold, then $\dot{V}(x(t)) < -\alpha V(x(t))$, for all $0 \leq t \leq h$ and for all initial conditions satisfying (9.10). Since (9.19) $\Leftrightarrow$ (9.10), it is sufficient to show that

$$\dot{V}(x(t)) + \alpha V(x(t)) + \omega(x(t), \rho(t)) < 0, \; \forall\, 0 \leq t \leq h, \tag{9.20}$$

whenever the set of conditions (9.19) holds, with ω in (9.18). Using the S - procedure, the following sufficient condition is obtained:

$$\begin{bmatrix} A^T P + PA - U_1(1 - e^{\gamma t}) + \alpha P & -U_2 - U_1 e^{\gamma t} \\ -U_2^T - U_1 e^{\gamma t} & U_1 e^{\gamma t} - U_3 \end{bmatrix} - \sum_{i=1}^{m} \varepsilon_i \bar{Q}_i \prec \mathbf{0}, \; \forall t \in [0, h]. \tag{9.21}$$

Since the previous inequality is linear in $e^{\gamma t}$, it is sufficient to test it for $t \in \{0, h\}$, which leads to

$$\begin{bmatrix} A^T P + PA + \alpha P & -U_2 - U_1 \\ -U_2^T - U_1 & U_1 - U_3 \end{bmatrix} - \sum_{i=1}^{m} \varepsilon_i \bar{Q}_i \prec \mathbf{0},$$

$$\begin{bmatrix} A^T P + PA + \alpha P - U_1(1 - e^{\gamma h}) & -U_2 - U_1 e^{\gamma h} \\ -U_2^T - U_1 e^{\gamma h} & U_1 e^{\gamma h} - U_3 \end{bmatrix} - \sum_{i=1}^{m} \varepsilon_i \bar{Q}_i \prec \mathbf{0} \tag{9.22}$$

Note that by continuity of eigenvalues, if there exists a matrix P solution to the set of inequalities (9.13), (9.14) then there exists a sufficiently small α s.t. inequalities (9.21) hold. □

Note that the matrix A does not need to be a Hurwitz matrix. The main idea is to use an augmented vector $z(t) = \left[x^T(t) \; \rho^T(t)\right]^T$, with $\rho(t) = x(t) - x_0$, that takes into account the relation between the initial condition and the system state. In the augmented space, the system behavior is described by the equation $\dot{z} = \bar{A} z$, i.e., (9.16). Next, an estimation of the growth rate for a quadratic function of the augmented vector, $z^T U z$, is provided by the tuning parameter γ in (9.15). This makes it possible to use the constraints on initial condition (9.10) and the boundedness of system growth by reexpressing them according to the augmented state, with the matrix $\bar{Q}_i$ in (9.11) and with the positivity condition (9.18), respectively. It leads to study the decay of V under the constraints $\omega > 0$ and $z^T \bar{Q}_i z \leq 0$, $i = 1, \ldots, m$. This approach is applied to the case of switched systems with sampled switching law in the following proposition.

Proposition 9.4 *Consider the system* (9.8) *with* $t_{k+1} - t_k \in (0, T_{\max}]$ *and* N *reals* γ_i, $i \in \mathscr{I}$. *Assume that* $\Lambda_H \neq \{\varnothing\}$, *and consider some* $\lambda \in \Lambda_H$. *Assume that there exist* $P, U_{1,i}, U_{3,i} \succ 0$ *in* $\mathbb{R}^{n\times n}$ *and* $U_{2,i} \in \mathbb{R}^{n\times n}$, $i \in \mathscr{I}$, *s.t. the following set of LMIs is satisfied*

$$U_i = \begin{bmatrix} U_{1,i} & U_{2,i} \\ U_{2,i}^T & U_{3,i} \end{bmatrix} \succ \mathbf{0}, \tag{9.23}$$

$$\begin{bmatrix} A^T(\lambda)P + PA(\lambda) & -U_{2,i} - U_{1,i} - \Psi_i(\lambda) \\ -U_{2,i}^T - U_{1,i} - \Psi_i(\lambda) & U_{1,i} - U_{3,i} + \Psi_i(\lambda) \end{bmatrix} \prec \mathbf{0}, \tag{9.24}$$

$$\begin{bmatrix} H_i & -U_{2,i} - U_{1,i}e^{\gamma_i T_{\max}} - \Psi_i(\lambda) \\ -U_{2,i}^T - U_{1,i}e^{\gamma_i T_{\max}} - \Psi_i(\lambda) & U_{1,i}e^{\gamma_i h} - U_{3,i} + \Psi_i(\lambda) \end{bmatrix} \prec \mathbf{0} \tag{9.25}$$

and

$$\bar{A}_i^T U_i + U_i \bar{A}_i \prec \gamma_i U_i, \; i \in \mathscr{I} \tag{9.26}$$

with

$$H_i = A^T(\lambda)P + PA(\lambda) - U_{1,i}(1 - e^{\gamma_i T_{\max}}), \tag{9.27}$$

$$\Psi_i(\lambda) = (A(\lambda) - A_i)^T P + P(A(\lambda) - A_i), \tag{9.28}$$

$$\bar{A}_i = \begin{bmatrix} I & I \end{bmatrix}^T A_i \begin{bmatrix} \mathbf{0} & \mathbf{0} \end{bmatrix}, \; i \in \mathscr{I}. \tag{9.29}$$

Then, system (9.8) *with the switching law* (9.9) *is exponentially stable.*

Proof Consider the candidate Lyapunov function $V(x) = x^T Px$. From Eq. (9.9), $\sigma(x_k)$ can take the value $i \in \mathscr{I}$ whenever $x_k^T \left\{ (A_i - A_j)^T P + P(A_i - A_j) \right\} x_k \le 0$, $\forall j \in \mathscr{I}$. For the given $\lambda \in \Lambda$, multiply each of the inequalities by λ_j and sum. Adding

$$\lambda_i x_k^T \left\{ (A_i - A_i)^T P + P(A_i - A_i) \right\} x_k = 0 \tag{9.30}$$

to the obtained inequality and using the relation $\sum_{j\in\mathscr{I}} \lambda_j A_i = A_i$ leads to

$$x_k^T \left\{ (A_i - A(\lambda))^T P + P(A_i - A(\lambda)) \right\} x_k = -x_k^T \Psi_i(\lambda) x_k \le 0. \tag{9.31}$$

To show exponential stability it is sufficient to show that there exists a positive scalar α s.t.

$$\dot{V}(x(t)) < -\alpha V(x(t)), \; \forall t \in [t_k, t_k + T_{\max}], \tag{9.32}$$

along the solutions of $\dot{x}(t) = A_i x(t)$, whenever $-x_k^T \Psi_i(\lambda) x_k \leq 0$. Moreover

$$A_i^T P + PA_i + \Psi_i(\lambda) = A^T(\lambda)P + PA(\lambda).$$

Then, applying Lemma 9.3 with $A_i \to A$, $T_{\max} \to h$, $m = 1$, $-\Psi_i(\lambda) \to Q_1$, $1 \to \varepsilon_1, x_k \to x_0$, for all $i \in \mathcal{I}$, the set of conditions (9.24), (9.25) implies that there exists a sufficiently small $\alpha > 0$ s.t. the function $V(x) = x^T Px$ satisfies the relation $\dot{V} < -\alpha V$ when the switching law is defined by (9.9), i.e., it is a Lyapunov function for the switched linear system (9.8), (9.9) for any sampling interval $t_{k+1} - t_k \in (0, T_{\max}]$. □

The solution depends on N tuning parameters γ_i corresponding to the growth rate for each individual subsystem. Note that by imposing the initial values of growth rate γ_i in the inequalities (9.23)–(9.26), the stability conditions become LMIs. For reducing the computational complexity, the LMIs may also be tested with a common tuning parameter $\gamma_i = \gamma$, $\forall\, i \in \mathcal{I}$ (the largest growth rate). The extension to the affine case is provided in the following section.

9.4 Switched Affine Case

In this section, we present a robust control design method for switched affine systems with sampled-data switching law. Due to the sampling, the system state cannot be driven to the ideal equilibrium point, but only to a bounded set containing the equilibrium.

Without loss of generality we may consider system's (9.7) robust stabilization (to a bounded set) with respect to equilibrium point $x_e = 0$, i.e., we consider that there exists $\lambda \in \Lambda_H$ s.t. $B(\lambda) = \mathbf{0}$. Note that the robust stabilization problem with respect to an ideal equilibrium point $x_e = -A^{-1}(\lambda)B(\lambda) \neq \mathbf{0}$ (for the continuous time case), may always be reformulated as a null equilibrium point problem by considering the error state vector $e(t) = x(t) - x_{\text{eq}}$ and the model representing the error dynamics

$$\dot{e}(t) = A_{\tilde{\sigma}(e(t_k))} e(t) + \tilde{B}_{\tilde{\sigma}(e(t_k))}, \forall t \in [t_k, t_{k+1}) \tag{9.33}$$

with $\tilde{\sigma}\,(e(t_k)) = \sigma\,(x_k)$ and $\tilde{B}_i = A_i x_{\text{eq}} + B_i$, $\forall i \in \mathcal{I}$, for which $\sum_{i \in \mathcal{I}} \lambda_i \tilde{B}_i = \mathbf{0}$.

We look for switching laws of the form

$$\sigma\,(x_k) \in \arg\min_{i \in \mathcal{I}} x_k^T P\,(A_i x_k + B_i)\,, \tag{9.34}$$

where $P \succ 0$ is a matrix to be designed. Design conditions for (9.34) are provided as follows.

Theorem 9.5 *Consider system* (9.7), (9.34) *with* $t_{k+1} - t_k \in (0, T_{\max}]$. *Assume that* $\Lambda_H \neq \{\varnothing\}$ *and a* $\lambda \in \Lambda_H$ *s.t.* $B(\lambda) = 0$. *Given scalar tuning parameters* γ, $\chi > 0$, *assume that there exist matrices* $P, U_{1,i}, U_{3,i} \succ 0$ *in* $\mathbb{R}^{n\times n}$, $U_{2,i} \in \mathbb{R}^{n\times n}$ *and scalars* $\beta, \varepsilon > 0$ *s.t. the set of LMIs*

$$U_i = \begin{bmatrix} U_{1,i} & U_{2,i} \\ U_{2,i}^T & U_{3,i} \end{bmatrix} \succ \mathbf{0}, \begin{bmatrix} \bar{A}_i^T U_i + U_i \bar{A}_i - \chi U_i & U_i \bar{B}_i \\ \bar{B}_i^T U_i & -\varepsilon \end{bmatrix} \prec \mathbf{0} \tag{9.35}$$

$$\mathscr{M}_i(\lambda) + \mathscr{N}_i(0) \prec \mathbf{0}, \; \mathscr{M}_i(\lambda) + \mathscr{N}_i(T_{\max}) \prec \mathbf{0} \tag{9.36}$$

is feasible for all $i \in \mathscr{I}$, *with*

$$\Psi_i(\lambda) = (A(\lambda) - A_i)^T P + P\,(A(\lambda) - A_i)\,, \tag{9.37}$$

$$\bar{A}_i = \begin{bmatrix} A_i & \mathbf{0}_{n,n} \\ A_i & \mathbf{0}_{n,n} \end{bmatrix}, \; \bar{B}_i = \begin{bmatrix} B_i \\ B_i \end{bmatrix}, \tag{9.38}$$

$$\mathscr{M}_i(\lambda) = \begin{bmatrix} A^T(\lambda)P + PA(\lambda) + 2\gamma P & -T_{\max}\Psi_i(\lambda) & \mathbf{0} \\ * & T_{\max}^2 \Psi_i(\lambda) & T_{\max} P B_i \\ * & * & -\beta T_{\max} \end{bmatrix}, \tag{9.39}$$

$$\mathscr{N}_i(\tau) = \begin{bmatrix} U_{1,i}\,(e^{\chi\tau} - 1) & -T_{\max}\left(U_{1,i}e^{\chi\tau} + U_{2,i}\right) & \mathbf{0} \\ * & T_{\max}^2\left(U_{1,i} - U_{3,i}\right)e^{\chi\tau} & \mathbf{0} \\ * & * & \frac{\varepsilon}{\chi}\,(e^{\chi\tau} - 1) \end{bmatrix}, \; \forall\, i \in \mathscr{I}. \tag{9.40}$$

Then $\lim_{t\to\infty} x(t) \in \mathscr{E}_\lambda^{T_{\max}}$ *with*

$$\mathscr{E}_\lambda^{T_{\max}} = \left\{ x \in \mathbb{R}^n : x^T P x \le T_{\max}\frac{\beta}{2\gamma} \right\}. \tag{9.41}$$

Moreover, the set $\mathscr{E}_\lambda^{T_{\max}}$ *is positively invariant in the sense that if for some* $t^a > 0$ *we have* $x(t^a) \in \mathscr{E}^{T_{\max}}$ *then* $x(t) \in \mathscr{E}^{T_{\max}}$ *for all* $t \ge t^a$.

Proof Step 1: Evaluation of the system growth rate. We consider the evolution of an augmented system including the error $\rho(t) := x(t) - x_k, \forall\, t \in [t_k, t_{k+1})$. Denote $\eta(t) = \left(x^T(t)\; \rho^T(t)\right)^T$. The augmented system has the form:

$$\dot{\eta}(t) = \bar{A}_{\sigma(x_k)}\eta(t) + \bar{B}_{\sigma(x_k)}, \; \forall\, t \in [t_k, t_{k+1}). \tag{9.42}$$

The set of LMIs (9.35) are used to provide an upper estimate for the growth of the norm $\eta^T(t)U_i\eta(t)$ for a given sub-system $i \in \mathscr{I}$. Assume that the LMIs (9.35) are satisfied. Then

$$\begin{bmatrix} \eta(t) \\ 1 \end{bmatrix}^T \begin{bmatrix} \bar{A}_i^T U_i + U_i\bar{A}_i - \chi U_i & U_i\bar{B}_i \\ \bar{B}_i^T U_i & -\varepsilon \end{bmatrix} \begin{bmatrix} \eta(t) \\ 1 \end{bmatrix} < 0$$

for all $i \in \mathscr{I}$. Therefore, when $\sigma(x_k) = i$, this leads to

$$\frac{\mathrm{d}}{\mathrm{d}t}\left(\eta^T(t)U_i\eta(t)\right) < \chi\eta(t)^T U_i\eta(t) + \varepsilon, \tag{9.43}$$

for all $i \in \mathscr{I}$. Using standard manipulations and the notation $\tau(t) := t - t_k$ we obtain

$$\eta^T(t)U_i\eta(t) \le \eta^T(t_k)U_i\eta(t_k)e^{\chi\tau} + \frac{\varepsilon}{\chi}\left(e^{\chi\tau} - 1\right) \tag{9.44}$$

for all $t \in [t_k, t_{k+1})$. Since $\rho(t_k) = 0$ and $x(t_k) = x(t) - \rho(t)$

$$\xi^T(t)\mathscr{N}_i(\tau(t))\xi(t) \ge 0,\ t \in [t_k, t_{k+1}), \tag{9.45}$$

with $\xi(t) = \left[x^T(t)\ \ \rho^T(t)T_{\max}^{-1}\ \ 1\right]^T$. The inequality (9.45) takes into account the growth of the sampling error $\rho(t)$ with respect to the system state $x(t)$.

Step 2: Taking into account the relation between the active subsystem and system state. Note that $\sigma = i$ for all $t \in [t_k, t_{k+1})$ if

$$2x_k^T P\left(A_j x_k + B_j\right) - 2x_k^T P\left(A_i x_k + B_i\right) \ge 0, \forall j \in \mathscr{I}. \tag{9.46}$$

Multiplying these inequalities by $\lambda_j, j \in \mathscr{I}$ and summing we have

$$2x_k^T P\left(A(\lambda) - A_i\right)x_k - 2x_k^T PB_i \ge 0,\ \forall j \in \mathscr{I}. \tag{9.47}$$

Replacing $x_k = x(t) - \rho(t)$ we obtain the following relation

$$\xi^T(t)\begin{bmatrix} \Psi_i(\lambda) & -T_{\max}\Psi_i(\lambda) & -PB_i \\ * & T_{\max}^2\Psi_i(\lambda) & T_{\max}PB_i \\ * & * & \mathbf{0} \end{bmatrix}\xi(t) \ge 0. \tag{9.48}$$

Step 3: Decay of Lyapunov function outside $\mathscr{E}_\lambda^{T_{\max}}$. Consider the function $V(x) = x^T Px$. To guarantee the exponential converges to the set $\mathscr{E}_\lambda^{T_{\max}}$ in (9.41) and to prove the positive invariance of $\mathscr{E}_\lambda^{T_{\max}}$ we show that the set of LMIs (9.35), (9.36) guarantee

$$\dot{V}(x(t)) + 2\gamma V(x(t)) - \beta T_{\max} < 0 \tag{9.49}$$

for all $t \in [t_k, t_{k+1})$ and all $k \in \mathbb{N}$. Adding (9.45) and (9.48), using the fact that

$$\Psi_i(\lambda) + A_i^T P + PA_i = A^T(\lambda)P + PA(\lambda)$$

and expressing according to the vector ξ leads to the sufficient condition

$$\xi^T\left(\mathscr{M}_i(\lambda) + \mathscr{N}_i(\tau)\right)\xi < 0, \tag{9.50}$$

$\forall \tau \in [0, T_{\max}]$ and all $\xi \neq 0$. Since $e^{\chi\tau} \in co\left\{1, e^{\chi T_{\max}}\right\}$ for all $\tau \in [0, T_{\max}]$, it suffice to test the conditions only in the extreme points, as in (9.36). □

The proposed design solution is based on the existence of a function $V(x) = x^T P x$, which is strictly decreasing for all x outside the set $\mathscr{E}_\lambda^{T_{\max}}$. Note that several relaxations are possible for the stabilization criteria. Improvement may be obtained by using more complex Lyapunov functions or by imposing the decay of the function $V(x)$ only with respect to its value at sampling times, i.e., $V(x(t)) \leq V(x(t_k))$ for all $t \in [t_k, t_{k+1})$ instead of a monotonous decay. We show how to implement such conditions in the following theorem.

Theorem 9.6 *Consider system* (9.7), (9.34) *with* $t_{k+1} - t_k \in (0, T_{\max}]$ *and a given scalar tuning parameter* $\gamma > 0$. *Assume that* $\Lambda_H \neq \{\varnothing\}$, *that there exists a* $\lambda \in \Lambda_H$ *s.t.* $B(\lambda) = 0$. *Let there exist matrices* $P, U \succ 0$ *in* $\mathbb{R}^{n\times n}$, *a scalar* $\beta > 0$ *s.t. the LMIs*

$$\begin{bmatrix} A^T(\lambda)P + PA(\lambda) + 2\gamma P + T_{\max}A_i^T U A_i & T_{\max}A_i^T U B_i \\ * & T_{\max}\left(B_i^T U B_i - \beta \boldsymbol{I}\right) \end{bmatrix} \prec \mathbf{0}, \tag{9.51}$$

$$\begin{bmatrix} A^T(\lambda)P + PA(\lambda) + 2\gamma P & \mathbf{0} & -T_{\max}\Psi_i(\lambda) \\ * & -T_{\max}\beta\mathbf{I} & T_{\max}B_i^T P \\ * & * & -T_{\max}Ue^{-2\gamma T_{\max}} + T_{\max}^2\Psi_i(\lambda) \end{bmatrix} \prec \mathbf{0}, \tag{9.52}$$

$\forall i \in \mathscr{I}$, *with*

$$\Psi_i(\lambda) = (A(\lambda) - A_i)^T P + P(A(\lambda) - A_i), \; i \in \mathscr{I}.$$

Then $\lim_{t\to\infty} x(t) \in \mathscr{E}_\lambda^{T_{\max}}$. *Moreover, for the case of* $B_i = 0, \forall i = 1, \ldots, N$, *if there exist matrices* $P, U \succ 0$ *in* $\mathbb{R}^{n\times n}$ *s.t. the LMIs*

$$A^T(\lambda)P + PA(\lambda) + 2\gamma P + T_{\max}A_i^T U A_i \prec \mathbf{0}, \tag{9.53}$$

$$\begin{bmatrix} A^T(\lambda)P + PA(\lambda) + 2\gamma P & -T_{\max}\Psi_i(\lambda) \\ * & -T_{\max}Ue^{-2\gamma T_{\max}} + T_{\max}^2\Psi_i(\lambda) \end{bmatrix} \prec \mathbf{0}, \; i \in \mathscr{I} \tag{9.54}$$

are feasible then system (9.8), (9.9) *is exponentially stable at* $x = 0$.

Proof Step 1: convergence towards the set $\mathscr{E}_\lambda^{T_{\max}}$. The proof uses the function $V(x) = x^T P x$ and a continuous function $w : \mathbb{R}^+ \to \mathbb{R}$, $k \in \mathbb{N}$, differentiable over $[t_k, t_{k+1})$, with $w(t_k) = 0$ and $w(t) \geq 0, \forall t \in (t_k, t_{k+1}), \forall\, k \in \mathbb{N}$, satisfying the following condition:

$$\frac{\mathrm{dV}}{\mathrm{d}t} + \frac{\mathrm{d}w}{\mathrm{d}t} + 2\gamma\left(V(x(t)) + w(t)\right) < \beta T_{\max}, \forall t \in [t_k, t_{k+1}). \tag{9.55}$$

We show further that (9.55) implies that $\lim_{t\to\infty} x(t) \in \mathscr{E}_\lambda^{T_{\max}}$. Assume that the condition (9.55) holds and consider the function

$$z(t) = e^{2\gamma t}\left(V(x(t)) + w(t) - \frac{\beta}{2\gamma}T_{\max}\right).$$

Then

$$\dot{z}(t) = e^{2\gamma t}\left(\frac{\mathrm{d}V}{\mathrm{d}t} + \frac{\mathrm{d}w}{\mathrm{d}t} + 2\gamma\left(V(x(t)) + w(t)\right) - \beta T_{\max}\right) < 0 \tag{9.56}$$

$\forall t \in [t_k, t_{k+1}), \forall\, k \in \mathbb{N}$, which implies that $z(t) < z(0), \forall t > 0$, that is

$$V(x(t)) < e^{-2\gamma t}\left(V(x(0)) + w(0) - \frac{\beta}{2\gamma}T_{\max}\right) + \frac{\beta}{2\gamma}T_{\max} - w(t), \forall\, t > 0. \tag{9.57}$$

Since $\gamma > 0$, $w(t) \geq 0$ and $w(0) = w(t_0) = 0$ this leads to

$$V(x(t)) < e^{-2\gamma t}V(x(0)) + \frac{\beta}{2\gamma}T_{\max}, \forall\, t > 0, \tag{9.58}$$

i.e., the system solution converges exponentially to $\mathscr{E}_\lambda^{T_{\max}}$.

Step 2: derivation of LMI conditions. We show next that the sufficient matrix inequalities conditions (9.51), (9.52) imply (9.55). For $U \succ 0$, $\gamma > 0$ and $k \in \mathbb{N}$ consider the notations $T_k := t_{k+1} - t_k$, $\tau(t) := t - t_k$, $\forall t \in [t_k, t_{k+1})$, and a function $w(t)$ of the form

$$w(t) = (T_k - \tau(t))\int_{t_k}^{t} e^{2\gamma(s-t)}\dot{x}^T(s)U\dot{x}(s)ds \geq 0, \forall t \in [t_k, t_{k+1}). \tag{9.59}$$

Consider system (9.7) and assume that $\sigma(x_k) = i$. Denote $\theta(t) = (x(t) - x_k)\,\tau^{-1}(t)$, for all $t \in [t_k, t_{k+1})$. Using the Jensen inequality it is seen that

$$\begin{aligned}\frac{\mathrm{d}w}{\mathrm{d}t} + 2\gamma w(t) &\leq (T_k - \tau)\dot{x}^T U\dot{x} - \tau\theta^T U\theta e^{-2\gamma T_{\max}}\\ &\leq (T_{\max} - \tau)\left(A_i x + B_i\right)^T U\left(A_i x + B_i\right) - \tau\theta^T U\theta e^{-2\gamma T_{\max}}.\end{aligned}$$

Furthermore, with $\sigma = i$ for $t \in [t_k, t_{k+1})$,

$$\dot{V} = 2x^T P\left(A_i x + B_i\right),\ \forall t \in [t_k, t_{k+1}).$$

Therefore, a sufficient condition for (9.55) to hold with $0 < t_{k+1} - t_k \le T_{\max}$ and $\sigma = i$ is

$$\begin{bmatrix} x \\ 1 \\ \theta \end{bmatrix}^T \begin{bmatrix} \Theta_i(\tau) & PB_i + (T_{\max} - \tau)A_i^T UB_i & \mathbf{0} \\ * & (T_{\max} - \tau)B_i^T UB_i - \beta T_{\max}\mathbf{I} & \mathbf{0} \\ * & * & -\tau U e^{-2\gamma T_{\max}} \end{bmatrix} \begin{bmatrix} x \\ 1 \\ \theta \end{bmatrix} < 0 \tag{9.60}$$

for all $\tau \in [0, T_{\max}]$, where

$$\Theta_i(\tau) = A_i^T P + PA_i + 2\gamma P + (T_{\max} - \tau)A_i^T UA_i.$$

Note that $\sigma = i$ for all $t \in [t_k, t_{k+1})$ if x_k satisfies the relation

$$2x_k^T P\left(A_j x_k + B_j\right) - 2x_k^T P\left(A_i x_k + B_i\right) \ge 0, \ \forall j \in \mathscr{I}. \tag{9.61}$$

Multiplying by $\lambda_j, j \in \mathscr{I}$ and summing we obtain

$$2x_k^T P\left(A(\lambda) - A_i\right) x_k - 2x_k^T PB_i \ge 0, \ \forall j \in \mathscr{I}. \tag{9.62}$$

Replacing $x_k = x(t) - \tau(t)\theta(t)$ leads to

$$\begin{bmatrix} x(t) \\ 1 \\ \theta(t) \end{bmatrix}^T \begin{bmatrix} \Psi_i(\lambda) & -PB_i & -\tau(t)\Psi_i(\lambda) \\ * & \mathbf{0} & \tau(t)B_i^T P \\ * & * & \tau^2(t)\Psi_i(\lambda) \end{bmatrix} \begin{bmatrix} x(t) \\ 1 \\ \theta(t) \end{bmatrix} \ge 0. \tag{9.63}$$

Since $\sigma = i$ whenever (9.63) holds, using the S - procedure by adding the left side of (9.63) to the left side of (9.60), the following sufficient condition for stabilization is obtained

$$z^T \Omega_i(\tau, \tau^2) z < 0, \ \forall \tau \in [0, T_{\max}] \tag{9.64}$$

where $z(t) = \left[x^T \ 1 \ \theta^T\right]^T$ and

$$\Omega_i(\tau, \tau^2) = \begin{bmatrix} \Theta_i(\tau) & (T_{\max} - \tau)A_i^T UB_i & -\tau\Psi_i(\lambda) \\ * & (T_{\max} - \tau)B_i^T UB_i - \beta T_{\max}\mathbf{I} & \tau B_i^T P \\ * & * & -\tau U e^{-2\gamma T_{\max}} + \tau^2\Psi_i(\lambda) \end{bmatrix}.$$

Moreover,

$$\Omega_i(\tau, \tau^2) \in co\left\{\Omega(0,0), \Omega(T_{\max}, 0), \Omega(T_{\max}, T_{\max}^2)\right\}, \forall \tau \in [0, T_{\max}],$$

therefore the set of inequalities

$$z^T \Omega_i(0,0)z < 0,\ z^T \Omega_i(T_{\max}, 0)z < 0,\ z^T \Omega_i(T_{\max}, T_{\max}^2)z < 0,\ i \in \mathscr{I} \quad (9.65)$$

are sufficient to guarantee that the function $V(x)$ satisfies the condition (9.55) along the solutions of system (9.7) with the switching law (9.34). Note that the inequality $\Omega_i(T_{\max}, T_{\max}^2) \prec 0$ implies that $\Omega_i(T_{\max}, 0) \prec 0$. Therefore the LMIs (9.51) and (9.52) are sufficient to guarantee that the inequalities (9.65) hold.

The case of $B_i = 0$, $\forall i \in \mathscr{I}$, can be derived using the same steps and taking $\beta = 0$. □

The provided conditions are robust to uncertainties and variations of the sampling interval. If the set of LMIs (9.51), (9.52) are satisfied, then the system is stable with any unknown time varying sampling interval in $(0, T_{\max}]$. The previous theorem is based on the condition (9.55). This condition may be interpreted as $\dot{V}(x(t)) + \dot{w}(t) < 0, \forall t \in [t_k, t_{k+1})$ whenever $x(t) \in \mathbb{R}^n \setminus \mathscr{E}_\lambda^{T_{\max}}$. Integrating this relation from t_k to t_{k+1} guarantees that $V(x_{k+1}) < V(x_k)$ whenever $x(t) \in \mathbb{R}^n \setminus \mathscr{E}_\lambda^{T_{\max}}, \forall t \in [t_k, t_{k+1})$. Then, the function $V(x) = x^T P x$ may be related to a Lyapunov function for the discrete-time system representation whereas the function $\bar{V} = V + w$ may be related to a Lyapunov-Krasovskii functional as in [5].

Remark 9.7 Theorems 9.5 and 9.6 provide a solution to Problem 9.2 with $C = \beta\,(2eig_{min}(P)\gamma)^{-1}$. The size of the obtained attractive ellipsoid (9.41) is linear in $T_{\max}$. Furthermore, when $T_{\max}$ approaches 0 the obtained stabilization conditions (the set of LMIs (9.51), (9.52)) are reduced to the conditions for the ideal continuous-time case in [2], with $x(t) \to 0$ as $t \to \infty$. The main difference between Theorem 9.5 and Theorem 9.6 is the fact that in Theorem 9.6 we only guarantee the exponential convergence to the attractive ellipsoid (9.41) and not the positive invariance.

The LMIs (9.51), (9.52) presented in this section are not affine in the systems parameters. This makes difficult their use for time-varying parametric uncertainties. However, the conditions may be modified to cope with this important case. The main idea is to use the descriptor method [4] to obtain LMIs which are affine in the system matrices. Consider the system

$$\dot{x} = \tilde{A}_{\sigma(x_k)}x + \tilde{B}_{\sigma(x_k)},\ \forall\, t \in [t_k, t_{k+1}) \quad (9.66)$$

where $\forall i \in \mathscr{I}$ the system matrices have the form

$$\tilde{A}_i = A_i + \sum_{j=1}^{m} \alpha_i^j(t)\bar{A}_i^j,\ \tilde{B}_i = B_i + \sum_{j=1}^{m} \alpha_i^j(t)\bar{B}_i^j$$

with $\alpha_i^j(t) \geq 0$, $\sum_{j=1}^{m} \alpha_i^j(t) = 1$ and $m \in \mathbb{N}^+$. The matrices A_i, B_i represent the nominal system parameters, while $\bar{A}_i^j, \bar{B}_i^j$ represent the vertex of perturbations with respect to the nominal value. We assume that there exists a stable convex combination of nominal matrices $A(\lambda) = \sum_{i=1}^{N} \lambda_i A_i$ s.t. $\sum_{i=1}^{N} \lambda_i B_i = 0$. Robust stabilization conditions are given as follows.

Corollary 9.8 *Consider system* (9.66), (9.34) *with* $t_{k+1} - t_k \in [0, T_{\max}]$ *a given scalar tuning parameter* $\gamma > 0$. *Assume that* $\Lambda_H \neq \{\varnothing\}$ *and that there exists a* $\lambda \in \Lambda_H$ *s.t.* $B(\lambda) = 0$. *Let there exist matrices* $P, U \succ 0$, *matrices* $P_2, P_3 \in \mathbb{R}^{n\times n}$ *and a scalar* $\beta > 0$ *s.t. the LMIs*

$$\begin{bmatrix} T_{\max}U - P_3 - P_3^T & P_3^T\tilde{A}_i^j - P_2 & P_3^T\tilde{B}_i^j \\ * & K_i^j(\lambda) & P_2^T\tilde{B}_i^j + P\bar{B}_i^j \\ * & * & -\beta T_{\max} \end{bmatrix} \prec \mathbf{0} \tag{9.67}$$

$$\begin{bmatrix} -P_3 - P_3^T & P_3^T\tilde{A}_i^j - P_2 & P_3^T\tilde{B}_i^j & \mathbf{0} \\ * & K_i^j(\lambda) & P_2^T\tilde{B}_i^j + P\bar{B}_i^j & -T_{\max}\Psi_i(\lambda) \\ * & * & -\beta T_{\max} & T_{\max}B_i^T P \\ * & * & * & -T_{\max}Ue^{-2\gamma T_{\max}} + T_{\max}^2\Psi_i(\lambda) \end{bmatrix} \prec \mathbf{0} \tag{9.68}$$

are feasible for all $i \in \mathscr{I}$, $j = 1, \ldots, m$, *with the notations*

$$\tilde{A}_i^j = A_i + \bar{A}_i^j,\ \tilde{B}_i^j = B_i + \bar{B}_i^j,$$

$$K_i^j(\lambda) = \left(\tilde{A}_i^j\right)^T P_2 + P_2^T\tilde{A}_i^j + \left(A(\lambda) + \bar{A}_i^j\right)^T P + P\left(A(\lambda) + \bar{A}_i^j\right) + 2\gamma P$$

$\forall i \in \mathscr{I}$, $j = 1, \ldots, m$ *and* $\Psi_i(\lambda)$ *as in Theorem* 9.6. *Then* $\lim_{t\to\infty} x(t) \in \mathscr{E}_\lambda^{T_{\max}}$, *where* $\mathscr{E}_\lambda^{T_{\max}}$ *is defined in* (9.41).

Proof Following the arguments in the proof of Theorem 9.6, with $V(x) = x^T Px$ and w in (9.59) a sufficient condition for (9.55) is

$$2x^T P\left(\tilde{A}_i x + \tilde{B}_i\right) + 2\gamma x^T Px + (T_{\max} - \tau)\dot{x}^T U\dot{x} - \tau\theta^T U\theta e^{-2\gamma T_{\max}} - \beta T_{\max} < 0$$

whenever $\sigma = i$. This inequality may be reexpressed as:

$$z^T \begin{bmatrix} (T_{\max} - \tau)U & \mathbf{0} & \mathbf{0} & \mathbf{0} \\ * & \tilde{A}_i^T P + P\tilde{A}_i + 2\gamma P & P\tilde{B}_i & \mathbf{0} \\ * & * & -\beta T_{\max} & \mathbf{0} \\ * & * & * & -\tau Ue^{-2\gamma T_{\max}} \end{bmatrix} z < 0 \tag{9.69}$$

with $z^T = \left[\dot{x}^T \; x^T \; 1 \; \theta^T\right]$. Using the descriptor form,

$$0 = 2\left(x^T P_2^T + \dot{x}^T P_3^T\right)\left(\tilde{A}_i x + \tilde{B}_i - \dot{x}\right),$$

which is the same as

$$z^T \begin{bmatrix} -P_3^T - P_3 & P_3^T\tilde{A}_i - P_2 & P_3^T\tilde{B}_i & \mathbf{0} \\ * & \tilde{A}_i^T P_2 + P_2^T\tilde{A}_i & P_2^T\tilde{B}_i & \mathbf{0} \\ * & * & \mathbf{0} & \mathbf{0} \\ * & * & * & \mathbf{0} \end{bmatrix} z = 0. \tag{9.70}$$

Summing (9.69), (9.70), (9.63) and using convexity arguments, the LMIs (9.67), (9.68) are obtained. □

9.5 Numerical Examples

Example 9.9 Consider a switched system with unstable modes described by

$$A_1 = \begin{bmatrix} 0.1 & 0.3 \\ 0.6 & -0.2 \end{bmatrix},\; A_2 = \begin{bmatrix} -0.13 & -0.16 \\ -0.33 & 0.03 \end{bmatrix},\; B_1 = \begin{bmatrix} 0 \\ \frac{2}{31} \end{bmatrix},\; B_2 = \begin{bmatrix} 0 \\ -\frac{2}{69} \end{bmatrix}. \tag{9.71}$$

For $\lambda_1 = 0.31$ and $\lambda_2 = 0.69$, $x_e = 0$. Using Proposition 9.4 and Theorem 9.5 with $\gamma_1 = \gamma_2 = 0.9$ for the linear and affine case, respectively, stabilization may be achieved up to $T = 0.04$. Using Theorem 9.6, the LMIs are feasible for sampling intervals up to $T_{\max} = 0.08$. The numerical simulation confirm the obtained theoretical results. An illustration for which the attractive ellipsoid is small relatively to the sampling interval is shown in Fig. 9.1 with $T_{\max} = 10^{-2}$.

Example 9.10 Consider the DC-DC power converter in [8],

$$\frac{\mathrm{d}x}{\mathbf{dt}} = \mathscr{A}_{\sigma(x_k)} x(\mathbf{t}) + \mathscr{B}_{\sigma(x_k)} \tag{9.72}$$

where

$$\mathscr{A}_1 = \begin{bmatrix} 0 & 1/L \\ -1/C & -1/(RC) \end{bmatrix},\; \mathscr{A}_2 = \begin{bmatrix} 0 & 0 \\ 0 & -1/(RC) \end{bmatrix},\; \mathscr{B}_1 = \begin{bmatrix} 0 \\ 0 \end{bmatrix},\; \mathscr{B}_2 = \begin{bmatrix} E/L \\ 0 \end{bmatrix}, \tag{9.73}$$

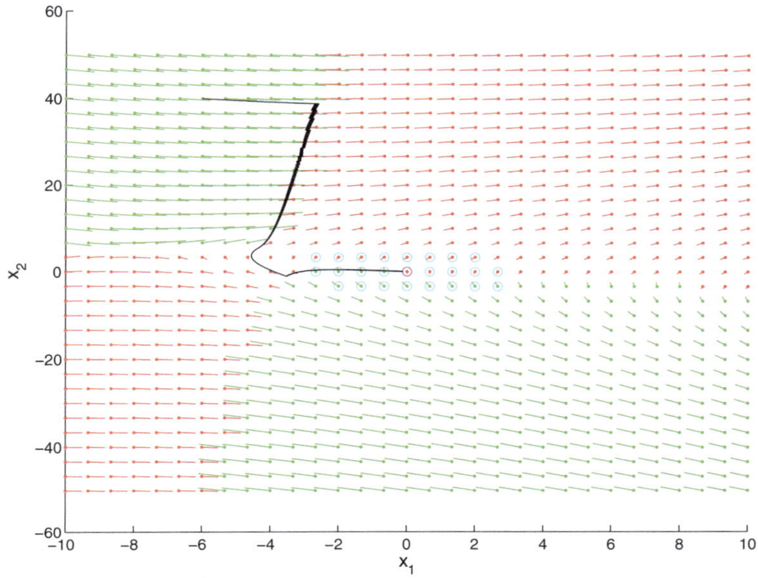

Fig. 9.1 Phase space and state partitions for the switched affine system in Example 9.9 (*green*—$\sigma = 1$, *red*—$\sigma = 2$). The ellipsoid $\mathscr{E}_{\lambda}^{T_{\max}}$ is marked in *cyan*. An example of system evolution from an arbitrary initial condition is represented in *black*

with $E = 6\,V$, $R = 50 \pm 10\,\Omega$, $L = 20\,\text{mH}$ and $C_0 = 220\,\mu\text{F}$ and a maximum sampling interval $\mathbf{T}_{\max} = 2.5 \times 10^{-5}s$. Considering that the nominal value of R is 50Ω,

$$x_e = -A(\lambda)^{-1}B(\lambda) = [0.24 \quad -6]^T$$

with $\lambda_1 = \lambda_2 = 0.5$. To cope with large numerical values, a time scale change $t = \varepsilon\mathbf{t}$ with $\varepsilon = 10^4$ is used. A system (9.66) is obtained with $\tilde{A}_i = \varepsilon^{-1}\mathscr{A}_i$, $\tilde{B}_i = \varepsilon^{-1}\mathscr{B}_i$ and $T_{\max} = \varepsilon\mathbf{T}_{\max}$. Note that the trajectories are invariant with respect to time scaling. Furthermore, the switching laws are equivalent, since

$$\arg\min(x - x_e)^T P\,(\varepsilon A_i x + \varepsilon B_i) = \arg\min(x - x_e)^T P\,(A_i x + B_i)\,.$$

Using Corollary 9.8 a stabilizing control may be designed. The ellipsoid $\mathscr{E}_{\lambda}^{T_{\max}}$ is described by

$$P = \begin{bmatrix} 9.4 & 0.15 \\ 0.15 & 0.10 \end{bmatrix} \tag{9.74}$$

and $c = \dfrac{\beta T_{\max}}{2\gamma} = 0.7$. An illustration of system evolution is shown in Fig. 9.2.

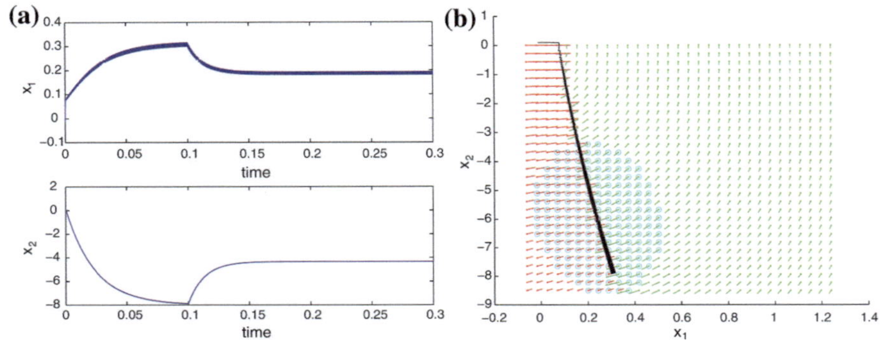

Fig. 9.2 Example 2: **a** evolution of systems states with a control law obtained based on Theorem 9.5 and a change in the resistor value from 40Ω to 60Ω; **b** phase space and state partitions (*green*—$\sigma = 1$, *red*—$\sigma = 2$). The ellipsoid $\mathscr{E}_{\lambda}^{T_{\max}}$ is marked in *cyan*. Note that due to sampling and to parametric uncertainties the system state in *black* does not converges to the equilibrium point the *center* of the ellipsoid, but only to a bounded region in $\mathscr{E}_{\lambda}^{T_{\max}}$. This region is not far from the border of $\mathscr{E}_{\lambda}^{T_{\max}}$, which shows the efficacy of the results

9.6 Conclusion

This chapter presented a continuous-time switching design method for switched affine system. The robustness with respect to sampling and to potential implementations imperfections such as sampling jitters or parametric uncertainties is taken into account. The method is based on an LMI optimization of Lyapunov functions, and is illustrated by numerical examples.

Acknowledgments The research leading to these results has received funding from the European Community's 7th Framework Programme (grant agreement No 257462) HYCON2 Network of Excellence and Israel Science Foundation (grant No 754/10).

References

1. Åström, K.J., Wittenmark, B.: Computer-Controlled Systems. Prentice Hall, New Jersy (1997)
2. Deaecto, G.S., Geromel, J.C., Garcia, F.S., Pomilio, J.A.: Switched affine systems control design with application to dc-dc converters. Control Theory Appl. IET **4**(7), 1201–1210 (2010). July
3. Defoort, M., Djemai, M., Floquet, T., Perruquetti, W.: On finite time observer design for multicellular converter. In: 11th IVSS, pp. 56–61 (2010)
4. Fridman, E.: New Lyapunov-Krasovskii functionals for stability of linear retarded and neutral type systems. Syst. Control. Lett. **43**, 309–319 (2001)
5. Fridman, E.: A refined input delay approach to sampled-data control. Automatica **46**(2), 421–427 (2010)
6. Fridman, E., Seuret, A., Richard, J.-P.: Robust sampled-data stabilization of linear systems. Automatica **40**, 1441–1446 (2004)

7. Fujioka, H., Kao, C.-Y., Almer, S., Jonsson, U.: Robust tracking with performance for pwm systems. Automatica **45**(8), 1808–1818 (2009)
8. Hauroigne, P., Riedinger, P., Iung, C.: Switched affine systems using sampled-data controllers: robust and guaranteed stabilization. IEEE Trans. Autom. Control. **56**(12), 2929–2935 (2011)
9. Hetel, L., Daafouz, J., Iung, C.: Stabilization of arbitrary switched linear systems with unknown time varying delays. IEEE Trans Autom Control **51**(10), 1668–1674 (2006)
10. Hetel, L., Fridman, E.: Sampled-data control of switched affine systems: a continuous-time approach. In: Proceedings of the 7th IFAC Symposium on Robust Control Design, Aalborg, Denmark (2012)
11. Hetel, L., Kruszewski, A., Perruquetti, W., Richard, J.: Discrete and intersample analysis of systems with aperiodic sampling. IEEE Trans. Autom. Control. **56**(7), 1696–1701 (2011)
12. Mirkin, L.: Some remarks on the use of time-varying delay to model sample-and-hold circuits. IEEE Trans. Autom. Control. **52**(6), 1109–1112 (2007)
13. Naghshtabrizi, P., Teel, A., Hespanha, J.P.: Exponential stability of impulsive systems with application to uncertain sampled-data systems. Syst. Control. Lett. **57**(5), 378–385 (2008)
14. Oishi, Y., Fujioka, H.: Stability and stabilization of aperiodic sampled-data control systems using robust linear matrix inequalities. Automatica **46**(8), 1327–1333 (2010)
15. Patino, D., Riedinger, P., Iung, C.: Predictive control approach for multicellular converters. In: 34th Annual Conference of IEEE, Industrial Electronics, IECON 2008, pp. 3309–3314, november 2008
16. Patino, D., Riedinger, P., Iung, C.: Practical optimal state feedback control law for continuous-time switched affine systems with cyclic steady state. Int. J. Control. **82**(7), 1357–1376 (2009)
17. Pettersson, S., Lennartson, B.: Stability and robustness for hybrid systems. In: Proceedings of the 35th Conference on Decision and Control, pp. 1202–1207, December 1996
18. Seuret, A.: A novel stability analysis of linear systems under asynchronous samplings. Automatica **48**(1), 177–182 (2012)
19. Shorten, R., Wirth, F., Mason, O., Wulff, K., King, C.: Stability criteria for switched and hybrid systems. Invited paper for SIAM Review **49**(4), 545–592 (2007)
20. Sira-Ramirez, H., Rios-Bolivar, M.: Sliding mode control of dc-to-dc power converters via extended linearization. IEEE Trans. Circuits Syst. I: Fundam. Theor. Appl. **41**(10), 652–661 (1994)
21. Sivashankar, N., Khargonekar, P.P.: Robust stability and performance analysis of sampled-data systems. IEEE Trans. Autom. Control. **38**(1), 58–69 (1993)

Chapter 10
Continuous–Discrete-Time Observers for a Class of Uniformly Observable Systems

Mondher Farza, Mohammed M'Saad and Krishna Busawon

Abstract This chapter addresses the observer design problem for a class of continuous-time dynamical systems with nonuniformly sampled measurements. More specifically, an observer is proposed that runs in continuous-time with an output error correction term that is updated in a mixed continuous-discrete fashion. The proposed observer is actually an impulsive system as it is described by a set of differential equations with instantaneous state impulses corresponding to the measured samples and their estimates. In addition, it is shown that such an impulsive system can be put under the form of a hybrid system composed of a continuous-time high gain observer coupled with an inter-sample output predictor. The proposed observer present two design features that are worth noting: First, the observer calibration is achieved through the tuning of a scalar design parameter. Second, the exponential convergence to zero of the observation error is established under a well-defined condition on the maximum value of the sampling partition diameter. Simulations results dealing with a flexible joint robot arm are given in order to highlight the performance of the proposed observer.

10.1 Introduction

The observer design problem for nonlinear dynamical systems has received considerable attention over the last couple of decades. However, most of the existing contributions deal mainly with dynamical systems with continuous-time measurements [1–6]. There are, in fact, very few works that tackle the observer design problem for continuous-time dynamical systems with nonuniformly sampled measurements in a

M. Farza · M. M'Saad
GREYC Laboratory (UMR 6072 CNRS), Université de Caen and ENSICAEN,
6 Bd Maréchal Juin, 14050 Caen Cedex, France
e-mail: mondher.farza@unicaen.fr

K. Busawon (✉)
Engineering and Environment, Northumbria University,
Newcastle upon Tyne NE1 8ST, UK
e-mail: krishna.busawon@unn.ac.uk

M. Djemai and M. Defoort (eds.), *Hybrid Dynamical Systems*, Lecture Notes in Control and Information Sciences 457, DOI 10.1007/978-3-319-10795-0_10

consistent and systematic manner. In fact, in many practical control applications, we are faced with the issue of controlling a continuous-time system with measurements available at discrete instants of time. Such is the case for chemical and biotechniological processes in general, whereby the output measurements are available in discrete time. In this context, an early contribution, in this context, was made using a high gain observer for a class of nonlinear systems that are observable for any input [7]. The design was first carried out by assuming continuous-time output measurements before being appropriately modified to the case where these measurements are only available at sampling instants. Based on the aforementioned contribution, many other observers have been proposed for special classes of continuous-time systems with discrete-time output measurements [8–10]. In all these contributions, the strategy employed to design the continuous-discrete observer is as follows: a dynamical system, that is similar to the original system, is used to provide a state prediction for the time interval between two consecutive sampling instants. At the sampling instant, the output measurements are employed to update the state prediction provided by the dynamical system.

On the other hand, an output predictor approach to cope with the nonavailability of the output measurements between the sampling instants has been proposed in [11]. The main idea of such an approach consists in using a continuous time-observer that uses a suitable output predictor. Specifically, the output prediction is provided by the solution of a scalar ordinary differential equation (ODE) between two successive sampling instants with the value of the measured output sample as initial condition. The underlying observer is a hybrid system that is able to recover the continuous time observer properties for relatively fast sampling. Another approach has been adopted in [12] to deal with the observer design for a particular class of nonlinear systems with sampled measurements using an appropriate impulsive dynamical system. The underlying correction term is constructed as the product of a constant gain by the difference between the estimated and measured values of the last output sample. The convergence analysis as well as the determination of the observer gain are obtained using suitable LMI tools similar to those described in [13].

The aim of this chapter is to address the continuous–discrete-time observer design problem for a class of multi-output nonlinear systems with sampled measurements. The considered systems are observable for any input and the sampling process is simply characterized by an upper bound on the corresponding sampling intervals. More specifically, a continuous–discrete-time observer, which is basically a redesigned version of the well-known high gain continuous-time observer for uniformly obserable systems, is proposed. This observer shares the impulsive nature of the observer given in [12] under some specific modifications of the observer gain matrix. However, in contrast to the observer given in [12], the proposed observer gain is time-varying and depends on the sampling periods. Furthermore, the convergence analysis is performed under some well-defined assumptions, whereas in [12] the adopted hypotheses are hidden by the LMI feasibility requirements. More specifically, it has been shown that the proposed impulsive observer can be written under the form of a hybrid system similar to the continuous–discrete-time observer proposed in [11]. There are two special features of the proposed continuous–discrete-time observer that must be highlighted with respect to the existing ones. The first one is concerned with the ease with which the observer gain is updated at sampling instants together with the simplicity of its implementation, due to the structure of the underlying continuous time observer. The second feature is related to the approach adopted for the convergence analysis of the observer, which is simple and provides precise expressions of the upper bound of the sampling partition diameter and of the exponential convergence rate.

The chapter is organized as follows. In the next section, the class of systems considered and the notations used throughout this chapter is introduced together with the concise convergence results on the continuous-time high gain observer on which the proposed continuous–discrete-time observer is based upon. A technical lemma is given, which is subsequently used to establish the fundamental result of the chapter. Section 3 is devoted the observer design of the continuous–discrete-time observer. It is also shown how one can recover a hybrid form of the proposed impulsive continuous–discrete-time observer, emphasizing thereby the close relationship with its underlying continuous-time high gain observer. In Sect. 4, the effectiveness of the proposed observer is highlighted via simulation results involving a robotic arm with flexible joint. Finally, some concluding remarks are given in Sect. 5.

Notation: Throughout the chapter, I_p and 0_p will denote the p-dimensional identity and zero matrices respectively and $\|\cdot\|$ denotes the euclidian norm.

10.2 Problem Statement and Preliminaries

Consider the class of multi-outputs nonlinear systems that can be transformed, via a change of coordinates, into the following triangular form

$$\begin{cases} \dot{x}(t) = Ax(t) + \varphi(u(t), x(t)) \\ y(t_k) = Cx(t_k) = x^1(t_k) \end{cases} \tag{10.1}$$

with

$$x = \begin{pmatrix} x^1 \\ \vdots \\ x^{q-1} \\ x^q \end{pmatrix} \in \mathbb{R}^n, \quad \varphi(u, x) = \begin{pmatrix} \varphi^1(u, x^1) \\ \varphi^2(u, x^1, x^2) \\ \vdots \\ \varphi^{q-1}(u, x^1, \ldots, x^{q-1}) \\ \varphi^q(u, x) \end{pmatrix}$$

$$A = \begin{pmatrix} 0_p & I_p & & 0_p \\ \vdots & & \ddots & \ddots \\ \vdots & & & \ddots & I_p \\ 0_p & 0_p & \ldots & 0_p \end{pmatrix}, \quad C = \begin{pmatrix} I_p & 0_p & \ldots & 0_p \end{pmatrix}$$

where $x^i \in \mathbb{R}^p$ for $i \in [1,\ q]$ are the state variables blocks, $u(t) \in U$ a compact subset of $\mathbb{R}^s$ denotes the system input and $y \in \mathbb{R}^p$ denotes the system output that is available at the sampling times that satisfy $0 \leq t_0 < \cdots < t_k < t_{k+1} < \cdots$ with time-varying intervals $\tau_k = t_{k+1} - t_k$ and $\lim_{k\to\infty} t_k = +\infty$.
The function φ assumes a triangular structure with respect to the state decomposition and satisfies the following assumption:
A1. The functions φ^i for $i \in [1,\ q]$ are globally Lipschitz with respect to x uniformly in u, i.e., there exists $L > 0$ such that the following inequality hold for $i \in [1,\ q]$ and x and $\bar{x} \in \mathbb{R}^n$

$$\|\varphi^i(u, x) - \varphi^i(u, \bar{x})\| \leq L\|x - \bar{x}\|$$

We assume that the time intervals τ_k's are bounded away from zero and are upperly bounded by the upper bound of the sampling partition diameter, τ_M, i.e.,

$$0 < \tau_k = t_{k+1} - t_k \leq \tau_M, \quad \forall k \geq 0 \tag{10.2}$$

As it is mentioned in the introduction, the main objective is to design an exponential continuous–discrete-time observer that can provide a consistent continuous-time estimation of the entire state of system (10.1) by using the output measurements that are available at the sampling instants only. Additionally, an expression for the upper bound of the sampling partition diameter under which the exponential convergence of the proposed observer is guaranteed will be provided. Such a property is of fundamental importance from the parameter design specification point of view.

At this stage, there are two points that are worth noting on the considered class of systems. First, the class of systems described by (10.1) may seem very restrictive since it assumes a non prime dimension ($n = pq$) and in all the q blocks, the sub-blocks x^k have the same dimension p. This is not the case as it is shown in [14]. In fact, it is shown in [14] that system (10.1) is a normal form that characterizes a class of uniformly observable nonlinear systems that can be put under the form (10.1) via an injective map (see e.g., [14, 15] for more details). The second point is related to Assumption A1 that may also seem very restrictive since the Lipschitz conditions are, in general, satisfied locally. However, these conditions can be omitted in the case where the trajectory $x(t)$ of system (10.1) lies in a bounded set Ω. In such a case, one can extend the nonlinearities $\varphi(u, x)$ such that the extension becomes globally Lipschitz on the entire state $\mathbb{R}^n$. One can refer to [4, 16] and references therein for more details on how to carry globally Lipschitz prolongations in the context of the state observation.

10.2.1 Recall on Continuous-Time High-Gain Observer Design

Before proposing the continuous–discrete-time impulsive observer, we shall recall the main design steps of the continuous-time high gain observer for system (10.1) in the case where the output measurements are available in continuous time; that is

$$\dot{x}(t) = Ax(t) + \varphi(u(t), x(t)) \quad \text{with} \quad y(t) = Cx(t) = x^1(t) \tag{10.3}$$

A continuous-time observer proposed in [7, 17] for system (10.3) is given by

$$\dot{\hat{x}}(t) = A\hat{x}(t) + \varphi(u(t), \hat{x}(t)) - \theta\Delta_\theta^{-1}K(C\hat{x}(t) - y(t)) \tag{10.4}$$

where $\hat{x} \in \mathrm{I\!R}^n$ denotes the state estimate which is decomposed in q blocks $\hat{x}^i \in \mathrm{I\!R}^p$ for $i \in [1, q]$. The gain matrix K is chosen such that $\bar{A} \triangleq A - KC$ is Hurwitz. The matrix Δ_θ is block-diagonal and is defined as

$$\Delta_\theta = diag\left(I_p, \frac{1}{\theta}I_p, \ldots, \frac{1}{\theta^{q-1}}I_p\right) \tag{10.5}$$

with $\theta \geq 1$ being a positive scalar. A detailed proof of the exponential convergence of the observation error can be found in [17]. In what follows, an outline of the convergence proof is given in order to point out the relationship between the design of this observer and that of the continuous–discrete-time one given in the next section. To this end, let $\tilde{x} \triangleq \hat{x} - x$ be the observation error. Then, one has $\dot{\tilde{x}} = (A - \theta\Delta_\theta^{-1}KC)\tilde{x} + \Phi(u, \hat{x}, x)$, where $\Phi(u, \hat{x}, x) = \varphi(u, \hat{x}) - \varphi(u, x)$. Setting $\bar{x} \triangleq \Delta_\theta\tilde{x}$, while taking into account the following identities

$$\Delta_\theta A\Delta_\theta^{-1} = \theta A \quad \text{and} \quad C\Delta_\theta^{-1} = C \tag{10.6}$$

one obtains $\dot{\bar{x}} = \theta\bar{A}\bar{x} + \Delta_\theta\Phi(u, \hat{x}, x)$. Since $\bar{A}$ is Hurwitz, there exist a symmetric positive definite (SPD) matrix P and a positive real μ such that

$$P\bar{A} + \bar{A}^T P \leq -2\mu I_n \tag{10.7}$$

This property permits to suggest the following candidate Lyapunov function $V(\bar{x}) = \bar{x}^T P\bar{x}$. Indeed, one has

$$\begin{aligned} \dot{V}(\bar{x}) &= 2\theta\bar{x}^T P\bar{A}\bar{x} + 2\bar{x}^T P\Delta_\theta\Phi(u, \hat{x}, x) \\ &\leq -2\mu\theta\|\bar{x}\|^2 + 2\bar{x}^T P\Delta_\theta\Phi(u, \hat{x}, x) \end{aligned}$$

According to the Lipschitz Assumption A1 and the triangular structure of φ, one can show that [14, 17]

$$2\bar{x}^T P\Delta_\theta\Phi(u, \hat{x}, x) \leq 2L\sqrt{n}\lambda_M\|\bar{x}\|^2$$

where L is the Lipschitz constant of φ and λ_M is the largest eigenvalue of P. This leads to the following inequality: $\dot{V}(\bar{x}) \leq -(2\mu\theta - 2L\sqrt{n}\lambda_M)\|\bar{x}\|^2$. Choosing θ

such that $2\mu\theta - 2L\sqrt{n}\lambda_M > \mu\theta$; that is

$$\theta > \theta_0 \triangleq \frac{2L\sqrt{n}\lambda_M}{\mu} \tag{10.8}$$

one obtains $\dot{V}(\bar{x}) \leq -\mu\theta\|\bar{x}\|^2 \leq -\frac{\mu\theta}{\lambda_M}V(\bar{x}(t))$. As a result, $V(\bar{x}(t)) \leq e^{-\frac{\mu\theta}{\lambda_M}(t-t_0)}$ $V(\bar{x}(t_0))$.

10.2.2 A Technical Result

In what follows, a technical lemma together with its corresponding proof is given. This lemma will be used to establish the main result of the chapter in the next section.

Lemma 10.1 *Consider a differentiable function $v : t \in R^+ \mapsto v(t) \in R^+$ satisfying the following inequality:*

$$\dot{v}(t) \leq -av(t) + b\int_{t_k}^{t} v(s)\,\mathrm{d}s \quad \forall t \in [t_k, t_{k+1}[\text{ with } k \in I\!N \tag{10.9}$$

where $0 < t_{k+1} - t_k \leq \tau_M < +\infty$ and a and b are positive reals satisfying

$$\frac{b\tau_M}{a} < 1. \tag{10.10}$$

Then,

$$v(t) \leq e^{-\eta(t-t_0)}v(t_0) \quad \text{with} \quad 0 < \eta = (a - b\tau_M)\,e^{-a\tau_M}. \tag{10.11}$$

Proof We shall first prove that $v(t) \leq v(t_k), \quad \forall t \geq t_k$ and use this property to establish the key inequality (10.11). By integrating (10.9) from t_k to t, we obtain

$$\begin{aligned} v(t) &\leq v(t_k) - a\int_{t_k}^{t} v(s)\,\mathrm{d}s + b\int_{t_k}^{t}\int_{t_k}^{s} v(\nu)\,\mathrm{d}\nu\mathrm{d}s \\ &\leq v(t_k) - a\int_{t_k}^{t} v(s)\,\mathrm{d}s + b\tau_M\int_{t_k}^{t} v(\nu)\mathrm{d}\nu \end{aligned} \tag{10.12}$$

By using the condition (10.10), i.e., $b\tau_M - a < 0$, the last inequality (10.12) becomes

$$v(t) \le v(t_k) + (b\tau_M - a) \int_{t_k}^{t} v(s)\,\mathrm{d}s \le v(t_k) \tag{10.13}$$

Now, we shall establish the inequality (10.11) using the property (10.13). Indeed, inequality (10.9) becomes

$$\dot{v}(t) \le -av(t) + b\tau_M v(t_k) \tag{10.14}$$

Solving the above inequality from t_k to t, we get

$$\begin{aligned} v(t) &\le e^{-a(t-t_k)} v(t_k) + b\tau_M \int_{t_k}^{t} e^{-a(t-s)}\,\mathrm{d}s\; v(t_k) \\ &= e^{-a(t-t_k)} v(t_k) + \frac{b\tau_M}{a}\left(1 - e^{-a(t-t_k)}\right) v(t_k) = g(t)v(t_k) \end{aligned}$$

where $g(t) \triangleq e^{-a(t-t_k)}\left(1 - \frac{b\tau_M}{a}\right) + \frac{b\tau_M}{a}$. Let us now prove that

$$v(t) \le e^{-\eta(t-t_k)} v(t_k) \tag{10.15}$$

where η is given as in (10.11). To this end, it suffices to show that $\delta(t) = \left(g(t) - e^{-\eta(t-t_k)}\right) \le 0$ for all $t \ge t_k$. This can be done by simply showing that $\dot{\delta}(t) \le 0$ since one can easily check that $\delta(t_k) = 0$. In fact, this is actually the case since

$$\begin{aligned} \dot{\delta}(t) &= \dot{g}(t) + \eta e^{-\eta(t-t_k)} \\ &= -ae^{-a(t-t_k)}\left(1 - \frac{b\tau_M}{a}\right) + \eta e^{-\eta(t-t_k)} \\ &\le -e^{-a\tau_M}\,(a - b\tau_M) + \eta = 0 \end{aligned} \tag{10.16}$$

The last equality results from the expression of η given by (10.11). Finally, since the function v is continuous on R^+, one can easily deduce (10.11) from (10.15) by successively iterating the latter. This ends the proof of the lemma.

10.3 Continuous–Discrete-Time Observer Design

An impulsive continuous–discrete-time observer for system (10.1) can derived as follows:

$$\begin{aligned}\dot{\hat{x}}(t) &= A\hat{x}(t) + \varphi(u(t), \hat{x}(t)) \\ &\quad -\theta\Delta_\theta^{-1}Ke^{-\theta K^1(t-t_k)}(C\hat{x}(t_k) - y(t_k))\end{aligned} \tag{10.17}$$

where $\hat{x} = \left(\hat{x}^{1T}, \ldots, \hat{x}^{qT}\right)^T$ is the state estimate, $K = \left(K^{1T}, \ldots, K^{qT}\right)^T$ is a gain matrix with K^i being $p \times p$ square matrices that are designed such that the matrix $\bar{A} = A - KC$ is Hurwitz and Δ_θ is the block-diagonal matrix defined in (10.5) with $\theta \geq 1$.
We can now state the following theorem.:

Theorem 10.2 *Consider the system* (10.1), (10.17) *subject to Assumption* A1 *and let* τ_M *and* θ_0 *be the parameters respectively given by* (10.2) *and* (10.8). *Then, for every bounded input, there exists a constant* $\chi(\theta) > 0$ *such that for every* $\theta > \theta_0$, *if the upper bound of the sampling partition diameter is chosen such that* $\tau_M < \chi(\theta)$, *then the observation error* $\tilde{x}(t) = \hat{x}(t) - x(t)$ *converges exponentially to zero.*

10.3.1 Proof of Theorem 10.1

First, it can be easily checked that the observation error dynamics is given by

$$\dot{\tilde{x}}(t) = A\tilde{x} + \Phi(u, \hat{x}, x) - \theta\Delta_\theta^{-1}Ke^{-\theta K^1(t-t_k)}C\tilde{x}(t_k)$$

where $\Phi(u, \hat{x}, x) = \varphi(u, \hat{x}) - \varphi(u, x)$. Setting $\bar{x} = \Delta_\theta\tilde{x}$ and using the identities (10.6) permit to obtain:

$$\begin{aligned}\dot{\bar{x}} &= \Delta_\theta A\Delta_\theta^{-1}\bar{x} + \Delta_\theta\Phi(u, \hat{x}, x) - \theta Ke^{-\theta K^1(t-t_k)}C\Delta_\theta^{-1}\bar{x}(t_k) \\ &= \theta A\bar{x} + \Delta_\theta\Phi(u, \hat{x}, x) - \theta Ke^{-\theta K^1(t-t_k)}C\bar{x}(t_k)\end{aligned} \tag{10.18}$$

Moreover, by adding and subtracting the term $\theta KC\bar{x}$ in the last equation yields

$$\begin{aligned}\dot{\bar{x}} &= \theta(A - KC)\bar{x} + \Delta_\theta\Phi(u, \hat{x}, x) \\ &\quad + \theta K\left(C\bar{x} - e^{-\theta K^1(t-t_k)}C\bar{x}(t_k)\right) \\ &= \theta\bar{A}\bar{x} + \Delta_\theta\Phi(u, \hat{x}, x) + \theta Kz\end{aligned}$$

where $z(t) = C\bar{x}(t) - e^{-\theta K^1(t-t_k)}C\bar{x}(t_k) = \bar{x}^1(t) - e^{-\theta K^1(t-t_k)}\bar{x}^1(t_k)$. Notice that $z(t_k) = 0$ and according to (10.18), the time derivative of z can be written as follows

$$\begin{aligned}\dot{z}(t) &= \dot{\bar{x}}^1(t) + K^1\theta e^{-\theta K^1(t-t_k)}\bar{x}^1(t_k) \\ &= \theta\bar{x}^2 + \Phi^1(u, \hat{x}^1, x^1)\end{aligned} \tag{10.19}$$

where $\Phi^1(u, \hat{x}^1, x^1) = \varphi^1(u, \hat{x}^1) - \varphi^1(u, x^1)$.

Now, let us consider the following candidate quadratic Lyapunov function: $V(\bar{x}) = \bar{x}^T P \bar{x}$ where P is defined as in (10.7). Proceeding as in the continuous case, one can show that

$$\begin{aligned}\dot{V}(\bar{x}) &= 2\bar{x}^T(t)P\left(\theta\bar{A}\bar{x} + \Delta_\theta\Phi(u,\hat{x},x) + \theta Kz\right) \\ &\leq -(2\mu\theta - 2L\sqrt{n}\lambda_M)\|\bar{x}\|^2 + 2\theta\bar{x}^T PKz\end{aligned} \tag{10.20}$$

Choosing θ as in (10.8), the last inequality (10.20) becomes

$$\begin{aligned}\dot{V}(\bar{x}) &\leq -\frac{\mu\theta}{\lambda_M}V(\bar{x}) + 2\theta\|P\bar{x}\|\|K\|\|z(t)\| \\ &\leq -\frac{\mu\theta}{\lambda_M}V(\bar{x}) + 2\theta\sqrt{\lambda_M}\sqrt{V(\bar{x})}\|K\|\|z(t)\|\end{aligned}$$

Furthermore, integrating Eq. (10.19) from t_k to t while using the fact that $z(t_k) = 0$ yields

$$z(t) = \int_{t_k}^{t}\left(\theta\bar{x}^2(s) + \Phi^1(u(s),\hat{x}^1(s),x^1(s))\right)\mathrm{d}s$$

Bearing in mind that $\Phi^1(u,\hat{x}^1,x^1) = \varphi^1(u,\hat{x}^1) - \varphi^1(u,x^1)$, one gets

$$\begin{aligned}\|z(t)\| &\leq \int_{t_k}^{t}\left(\theta\|\bar{x}^2(s)\| + L\|\bar{x}^1(s)\|\right)\mathrm{d}s \\ &\leq (\theta + L)\int_{t_k}^{t}\|\bar{x}(s)\|\,\mathrm{d}s \leq \frac{\theta + L}{\sqrt{\lambda_m}}\int_{t_k}^{t}\sqrt{V(\bar{x}(s))}\,\mathrm{d}s\end{aligned}$$

where λ_m is the smallest eigenvalue of P. Combining the above inequalities, one obtains

$$\begin{aligned}\dot{V}(\bar{x}(t)) \leq &-\frac{\mu\theta}{\lambda_M}V(\bar{x}(t)) \\ &+2\theta\sqrt{\frac{\lambda_M}{\lambda_m}}\|K\|(\theta + L)\sqrt{V(\bar{x}(t))}\int_{t_k}^{t}\sqrt{V(\bar{x}(s))}\,\mathrm{d}s\end{aligned}$$

or equivalently

$$\frac{d}{dt}\sqrt{V(\bar{x}(t))} \le -\frac{\mu\theta}{2\lambda_M}\sqrt{V(\bar{x}(t))} + \theta\sqrt{\frac{\lambda_M}{\lambda_m}}\|K\|(L+\theta)\int_{t_k}^{t}\sqrt{V(\bar{x}(s))}\,ds \quad (10.21)$$

Finally, let $a_\theta = \frac{\mu\theta}{2\lambda_M}$ and $b_\theta = \theta\sqrt{\frac{\lambda_M}{\lambda_m}}\|K\|(L+\theta)$ and assume that the upper diameter of the sampling partition τ_M satisfies the following condition $\tau_M < \frac{a_\theta}{b_\theta}$, i.e., $\frac{b_\theta \tau_M}{a_\theta} < 1$ or more explicitly

$$2\tau_M(L+\theta)\frac{\|K\|\lambda_M\sqrt{\lambda_M}}{\mu\sqrt{\lambda_m}} < 1 \quad (10.22)$$

Then, according to Lemma 10.1, one has $\sqrt{V(\bar{x}(t))} \le e^{-\eta t}\sqrt{V(\bar{x}(0))}$ where $0 < \eta = (a_\theta - b_\theta\tau_M)\,e^{-a_\theta\tau_M}$. The constant $\chi(\theta)$ involved in the theorem is $\chi(\theta) = \frac{a_\theta}{b_\theta} = \frac{\mu\sqrt{\lambda_m}}{2(L+\theta)\|K\|\lambda_M\sqrt{\lambda_M}}$. This completes the proof of theorem.

It is worth noting that the calibration of the continuous–discrete-time observer (10.17) is achieved through the tuning of the observer design parameter θ as in the continuous time case. The specification of this design parameter is commonly carried out by considering a suitable compromise between the convergence speed and the insensitivity with respect to noise measurements as pointed out in the earlier contributions on high gain observers [18, 19]. This compromise is generally obtained from a trial and error approach since conditions (10.8) and (10.22) are too conservative to provide reasonable specification of the involved design parameters. The admissible value of the design parameter θ can be used to determine the maximum allowed value for the sampling partition diameter according to inequation (10.22). Nonetheless, it is possible to find a higher value than the theoretical bound using an appropriate trial and error approach.

A thorough investigation addressing the challenging performance improvement issue of high gain observers noise sensitivity has been carried out in [20–22].

10.3.2 The Output Prediction Form of the Observer

Although Eq. (10.17) puts forward the impulsive form of the proposed observer, it will be shown that this observer can be rewritten under an output prediction form given by the following two dynamical systems:

$$\dot{\hat{x}}(t) = Ax(t) + \varphi(u(t), \hat{x}(t)) - \theta \Delta_\theta^{-1} K(C\hat{x}(t) - \omega(t)) \tag{10.23}$$

with

$$\begin{cases} \dot{\omega}(t) = \hat{x}^2(t) + \varphi^1(u(t), \hat{x}^1(t)) \text{ for } t \in [t_k, t_{k+1}[\\ \omega(t_k) = Cx(t_k) = y(t_k) \end{cases} \tag{10.24}$$

where $\hat{x}$, Δ_θ and K are respectively the state estimate, the block diagonal matrix and the gain matrix used in the continuous–discrete-time observer (10.17). Notice that the structure of the dynamical subsystem (10.23) is quite similar to that of the continuous time observer (10.4) with the difference that the term corresponding to the observation error in (10.4), i.e., $\left(C\hat{x}(t) - y(t)\right)$, is replaced in the dynamical system (10.23) by the term $\left(C\hat{x}(t) - \omega(t)\right)$ where $\omega(t)$ is a new variable provided by the dynamical system (10.24). The latter can be viewed as an appropriate predictor of the system output as its dynamics is very similar to that of the output estimate $\hat{x}^1(t)$ that can be written as follows

$$\dot{\hat{x}}^1(t) = \hat{x}^2(t) + \varphi^1(u(t), \hat{x}^1(t)) - \theta K^1 \left(C\hat{x}(t) - \omega(t)\right) \tag{10.25}$$

It is worth mentioning that the output predictor (10.24) has to be re-initialized at each sampling instant t_k and then resolved for $t \in [t_k, t_{k+1}[$, while Eq. (10.25) or more generally the system (10.23) is initialized only once at the time instant t_0 and then resolved for all $t \geq t_0$. Now, set $\xi(t) \triangleq C\hat{x}(t) - \omega(t) = \hat{x}^1(t) - \omega(t)$ and let us focus on the dynamics of the variable $\xi(t)$. According to Eqs. (10.24) and (10.25), one has: $\dot{\xi}(t) = -\theta K^1 \xi(t)$. Integrating this equation from t_k to $t < t_{k+1}$ yields: $\xi(t) = e^{-\theta K^1(t-t_k)} \xi(t_k)$. On the other hand, taking into account the fact that $\omega(t_k) = x^1(t_k)$, one gets $\xi(t_k) = \hat{x}^1(t_k) - \omega(t_k) = C\hat{x}(t_k) - y(t_k)$. This leads to : $\xi(t) = e^{-\theta K^1(t-t_k)} \left(C\hat{x}(t_k) - y(t_k)\right)$ and substituting $C\hat{x}(t) - \omega(t) \triangleq \xi(t)$ in Eq. (10.23) by this expression yields the impulsive observer (10.17).

10.4 Application to a Flexible Joint Robotic Arm

The performance of the proposed observer is illustrated through the same example considered in [12, 23]. Such an example involves a flexible joint robotic arm given by

$$\dot{x} = Fx + g(u, x) \text{ with } y_k = x^1(t_k) = \begin{pmatrix} x_1(t_k) \\ x_2(t_k) \end{pmatrix} \tag{10.26}$$

with $x = \begin{pmatrix} x^1 \\ x^2 \\ x^3 \end{pmatrix} \in \mathbb{R}^4$, $x^1 = \begin{pmatrix} x_1 \\ x_2 \end{pmatrix}$, $x^2 = x_3$, $x^3 = x_4$, $x_i \in \mathbb{R}$, $F = \begin{pmatrix} 0_2 & F^1 & 0_{2\times1} \\ 0_{1\times2} & 0 & 1 \\ 0_{1\times2} & 0 & 0 \end{pmatrix}$ and $g(x) = \begin{pmatrix} g^1(u, x_1, x_2) \\ 0 \\ g^3(x_2, x_3) \end{pmatrix}$, $F^1 = \begin{pmatrix} 48.6 \\ 0 \end{pmatrix}$, $g^1(u, x_1, x_2) = \begin{pmatrix} -1.25x_1 - 48.6x_2 + 21.6u \\ x_1 \end{pmatrix}$, $g^3(x_2, x_3) = 19.5x_2 - 19.5x_3 - 3.3\sin(x_3)$ and $u = \sin(t) - \sin(7t)$. As the system (10.26) is not under the form (10.1), a state transformation that puts it under this form [14, 17] is presented. Indeed, consider the following injective map $\Phi : x \in \mathbb{R}^4 \to z = \Phi(x) = \Lambda x \in \mathbb{R}^6$ where $z = \left(z^{1T}\ z^{2T}\ z^{3T}\right)^T$, with $z^1 = (z_1\ z_2)^T$, $z^2 = (z_3\ z_4)^T$, $z^3 = (z_5\ z_6)^T$ and $\Lambda = \begin{pmatrix} I_2 & 0_{2\times1} & 0_{2\times1} \\ 0_2 & F^1 & 0_{2\times1} \\ 0_2 & 0_{2\times1} & F^1 \end{pmatrix}$. Moreover, one can check the following identity: $\Lambda F = A\Lambda$ where $A = \begin{pmatrix} 0_{4\times2} & I_4 \\ 0_2 & 0_{2\times4} \end{pmatrix}$ is the anti-shift block matrix and the dynamics of z can then be written as follows:

$$\begin{cases} \dot{z}(t) = \Lambda F x(t) + \Lambda g(u(t), x(t)) = Az(t) + \Lambda g(u(t), \Lambda^+ z(t)) \\ y_k = Cz(t_k) = z^1(t_k) \end{cases} \tag{10.27}$$

where $C = [I_2\ 0_2\ 0_2]$ and $\Lambda^+ = \left(\Lambda^T\Lambda\right)^{-1}\Lambda^T$ is the left inverse of Λ. As a result, one can design an observer of the form (10.17) to estimate the unknown trajectories of the system. The equations of this observer in the original coordinates are given by

$$\dot{\hat{x}}(t) = F\hat{x} + g(u, \hat{x}) - \Lambda^+\theta\Delta_\theta^{-1}Ke^{-\theta K^1(t-t_k)}\left(C\hat{x}(t_k) - y(t_k)\right)$$

where Δ_θ and K are defined as in (10.17) with $q = 3$ and $p = 2$ since the observer was designed in the new coordinates "z". The simulation experiments have been performed for a uniform sampling partition, i.e., $t_k = k\tau = k\tau_M$. The gain matrix K has been particularly determined to assign all the modes of the matrix $\bar{A}$ at -1 and the initial conditions of the system have been set to $x(0) = [3\ 2\ 3\ \ -2]^T$ while the initial conditions of the observer were arbitrarily set to zero. In a first step, several simulation experiments have been carried out assuming continuous-time output measurements. This allows to obtain a lower bound of the design parameter θ for which the underlying continuous time observer performs well, namely $\theta = 4$. Then, another set of simulation experiments has been performed with a fixed design parameter θ and different values for the sampling period. From an experiment to the other, the value of the sampling period has been increased until reaching the maximum allowable value τ_M for witch the performances of the continuous–discrete-time observer are quite satisfactory, e.g., $\tau_M = 0.55$ s for $\theta = 4$. In particular, the simulation results obtained with $\theta = 5$ and a sampling period $\tau = 0.4$ s will be presented. Notice that the maximum allowable value for the sampling partition with

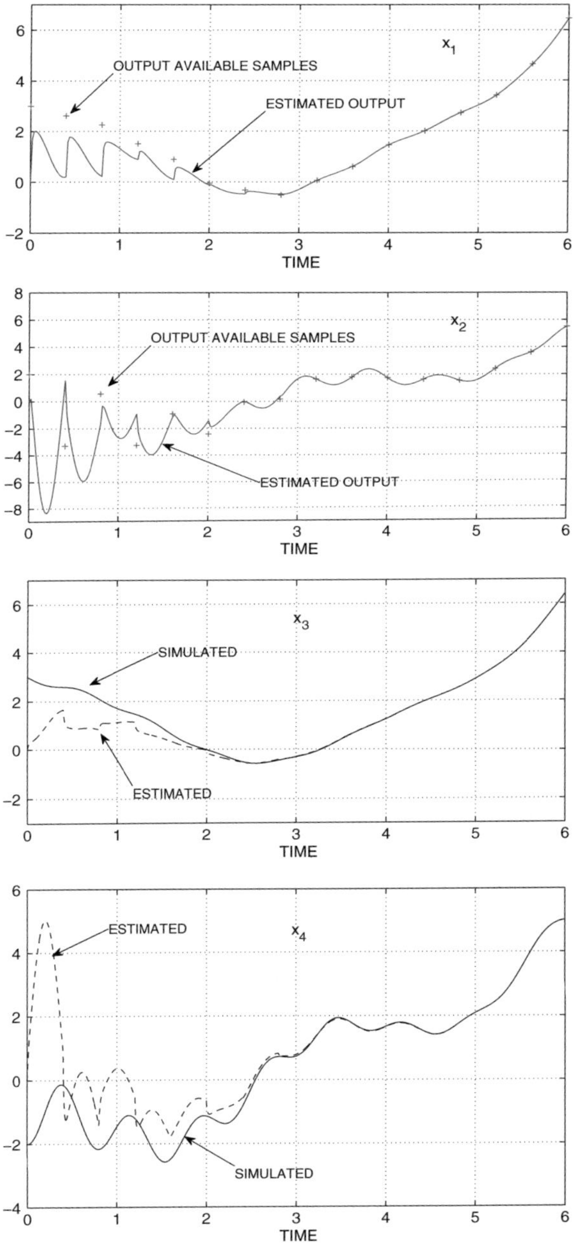

Fig. 10.1 Comparison of estimated and simulated states

$\theta = 5$ is about 0.5 s. The available measurements corresponding to both outputs x_1 and x_2 are given in Fig. 10.1 together with their continuous estimates provided by the proposed continuous–discrete-time observer. This figure also shows the unknown trajectories of x_3 and x_4 together with their estimates. Notice that the state estimates quickly converge to the unknown states confirming thereby the fundamental result derived in this chapter.

10.5 Conclusion

The main motivation of this chapter has been to provide a continuous–discrete-time observer for a class of systems that are observable for any input. The proposed observer shares the same structure as Luenberger observers up to an appropriate updating rule of the observer gain at every sampling instant. In addition, it shares the same simplicity of implementation as high gain observers. In particular, the exponential convergence to zero of the observation error has been established under a well defined condition on the upper bound of the sampling partition diameter. Even though the proposed observer has been firstly presented under an impulsive form, it has been shown that it can also be thought as a hybrid system, composed of a continuous-time observer with an inter-sample output predictor. Simulations results involving a flexible joint robotic arm have been given to demonstrate the effectiveness of the proposed continuous–discrete-time observer. It is worth noticing that the proposed design can be extended to a more general class of observable systems including system (10.1) and be properly adapted to several observer design problems; namely adaptive observers to jointly estimate the state and some unknown parameters [15]. Furthermore, the proposed continuous–discrete-time design framework can be enhanced to handle the case of long output delays. Such a problem has been particularly investigated in the context of stabilization (see e.g., [24] and references therein) but to the authors' best knowledge, there is no significant results in the observation context and the (few) available ones rather deal with continuous-time measurements [25–27]. All these issues are challenging and deserve to be investigated.

References

1. Gauthier, J., Kupka, I.: Deterministic Observation Theory and Applications. Cambridge University Press, Cambridge (2001)
2. Kazantzis, N., Kravaris, C.: Nonlinear observer design using lyapunov's auxiliary theorem. Syst. Contr. Lett. **34**, 241–247 (1998)
3. Farza, M., M'Saad, M., Triki, M., Maatoug, T.: High gain observer for a class of non-triangular systems. Syst. Contr. Lett. **60**(1), 27–35 (2011)
4. Andrieu, V., Praly, L.: On the existence of Kazantzis-Kravaris/Luenberger observers. SIAM J. Contr. Optim. **45**, 432–456 (2006)
5. Rajamani, R.: Observers for Lipschitz Nonlinear Systems. IEEE Trans. Autom. Contr. **43**(3), 397–401 (1998)

6. Krener, A.J., Isidori, A.: Linearization by output injection and nonlinear observers. Syst. Contr. Lett. **3**, 47–52 (1983)
7. Deza, F., Busvelle, E., Gauthier, J., Rakotopara, D.: High gain estimation for nonlinear systems. Syst. Contr. Lett. **18**, 295–299 (1992)
8. Nadri, M., Hammouri, H., Zaragoza, C.A.: Observer design for continuous-discrete time state affine systems up to output injection. Eur. J. Contr. **10**(3), 252–263 (2004)
9. Andrieu, V., Nadri, M.: Observer design for lipschitz systems with discrete-time measurements. In: Proceedings of 49th IEEE Conference on Decision and Control, Atlanta, Georgia, USA, 2010
10. Nadri, M., Hammouri, H., Grajales, R.: Observer design for uniformly observable systems with sampled measurements. IEEE Trans. Autom. Contr. **58**, 757–762 (2013)
11. Karafyllis, I., Kravaris, C.: From continuous-time design to sampled-data design of observers. IEEE Trans. Autom. Contr. **54**(9), 2169–2174 (2009)
12. Raff, T., Kögel, M., Allgöwer, F.: Observer with sample-and-hold updating for Lipschitz nonlinear systems with nonuniformly sampled measurements. In: Proceedings of the American Control Conference, Washington, USA, 2008
13. Naghshtabrizi, P., Hespanha, J., Teel, A.: Exponential stability of impulsive systems with application to uncertain sampled-data systems. Syst. Contr. Lett. **57**, 378–385 (2008)
14. Hammouri, H., Farza, M.: Nonlinear observers for locally uniformly observable systems. ESAIM J. Contr. Optim. Calc. Var. **9**, 353–370 (2003)
15. Farza, M., M'Saad, M., Maatoug, T., Kamoun, M.: Adaptive observers for nonlinearly parameterized class of nonlinear systems. Automatica **45**(10), 2292–2299 (2008)
16. Shim, H., Son, Y.I., Seo, J.H.: Semi-global observer for multi-output nonlinear systems. Syst. Contr. Lett. **42**, 233–244 (2001)
17. Farza, M., M'Saad, M., Rossignol, L.: Observer design for a class of MIMO nonlinear systems. Automatica **40**(1), 135–143 (2004)
18. Gauthier, J., Hammouri, H., Othman, S.: A simple observer for nonlinear systems - application to bioreactors. IEEE Trans. Auto. Contr. **37**(6), 875–880 (1992)
19. Farza, M., Busawon, K., Hammouri, H.: Simple nonlinear observers for on-line estimation of kinetic rates in bioreactors. Automatica **34**(3), 301–318 (1998)
20. Andrieu, V., Praly, L., Astolfi, A.: High gain observers with updated gain and homogeneous correction terms. Automatica **45**, 422–428 (2009)
21. Boizot, N., Busvelle, E., Gauthier, J.: An adaptive high-gain observer for nonlinear systems. Automatica **46**, 1483–1488 (2010)
22. Oueder, M., Farza, M., Abdennour, R.B., M'Saad, M.: A high gain observer with updated gain for a class of MIMO non-triangular systems. Syst. Contr. Lett. **61**, 298–308 (2012)
23. Howell, A., Hedrick, J.K.: Nonlinear observer design via convex optimization. In: Proceedings of the American Control Conference, Washington, USA, 2008
24. Karafyllis, I., Krstic, M.: Nonlinear Stabilization Under Sampled and Delayed Measurements, and with Inputs Subject to Delay and Zero-Order Hold. Private Correspondance, 2012.
25. Germani, A., Manes, C., Pepe, P.: A new approach to state observation of nonlinear systems with delayed output. IEEE Trans. Autom. Contr. **47**(1), 96–101 (2002)
26. Kazantzis, N., Wright, R.A.: Nonlinear observer design in the presence of delayed output measurements. Syst. Contr. Lett. **54**(9), 877–886 (2005)
27. Cacace, F., Germani, A., Manes, C.: An observer for a class of nonlinear systems with time varying observation delay. Syst. Contr. Lett. **59**(5), 305–312 (2010)

Chapter 11
Self-Triggered Robust Control of Nonlinear Stochastic Systems

Woihida Aggoune, Bernardino Castillo Toledo and Stefano Di Gennaro

Abstract In this chapter, results on self-triggering control for nonlinear systems are presented. Conditions guaranteeing the existence of a self-triggered control strategy stabilizing the closed-loop system are presented both for deterministic and stochastic nonlinear systems. The problems addressed are self-triggered stabilization and safety. In the stochastic case, the state equations are described by an Itô differential equation driven by a Wiener noise, where the input enters either in the deterministic dynamics or in the dynamics affected by the noise. This kind of model embraces a quite large class of systems, of particular interest since in practice. In fact, various disturbances can be modeled in this way.

11.1 Introduction

The implementation of controllers with digital devices presents many advantages, but at the same time poses some issues. In particular, one important aspect is the design of the digital controller so that the control system recovers, at least in first approximation, the same behavior of a system controlled with a continuous time controller. Many authors propose nonlinear digital controllers reproducing the performances of a certain continuous controller, viz. emulating the behavior of the continuous controller [19, 20, 26]. This very popular technique relies on the simple consideration

W. Aggoune
École Nationale Supérieure de l'Electronique et de ses Applications,
6 Avenue du Ponceau, 95014 Cedex Cergy–Pontoise, France
e-mail: aggoune@ensea.fr

B. Castillo Toledo
Centro de Investigación Y de Estudios Avanzados – CINVESTAV Del IPN,
Unidad Guadalajara, Av. Cientif íca, Col. El Baj ío, Zapopan, 45010 Jalisco, Mexico
e-mail: toledo@gdl.cinvestav.mx

S. Di Gennaro (✉)
Department of Information Engineering, Computer Science and Mathematics,
Center of Excellence DEWS, University of L'Aquila, Via G. Gronchi 18,
67100 L'Aquila, Italy
e-mail: stefano.digennaro@univaq.it

M. Djemai and M. Defoort (eds.), *Hybrid Dynamical Systems*, Lecture Notes in Control and Information Sciences 457, DOI 10.1007/978-3-319-10795-0_11

that when the (fixed) sampling period is short enough, one regains the continuous behavior. Other authors aim at designing the digital controller directly in the digital setting, although nonlinear systems cannot be discretized exactly in closed form, in general. In both cases, a relevant problem arises: the determination of the sampling period. From a theoretical point of view, the sampling period is usually considered constant, namely the new control value is computed periodically at each sampling time. This helps, from a mathematical point of view, the analysis of the sampled nonlinear system, and gives some mathematical tools to solve the design problem. However, in practice, the control law is calculated when the hardware, on which the digital algorithm is implemented, is not busy with processes with higher priority. In other cases, such as those in which a wireless sensor network is used, it would be better to calculate the control value only "when necessary". This clearly complicates the problem from an analytical point of view. But there are also practical aspects that push to deal with variable sampling. A first aspect is that a constant sampling is quite inefficient (in terms of hardware usage, communication bandwidth, energy, etc.), since it has to be considered the worst-case scenario. In fact, since the system dynamics are nonlinear, one has to ensure good performances for all the operative points. Incidentally, this reveals the need of criteria to fix the sampling time value, that otherwise has to be fixed using (possibly extensive) simulations, or considering empirical rules (20 times the system bandwidth [7]). A further aspect is that, usually, digital microcontrollers perform various tasks at the same time. This is quite typical, especially in the case of embedded systems. An example is the electronic central unit in an automobile, which has to manage different tasks, with different priority. Their scheduling is clearly of critical importance to prevent negative coupling effects of lower priority tasks when computing high level tasks, such as attitude control laws. Another important example is given by networked systems, where not only the processor time is a resource to be optimized, but also the available communication bandwidth is limited. In wireless sensor networks, already mentioned, an important issue is the minimization of the power consumption, in order to augment the life span of the network. In all these applications, the energy consumption is related to the frequency of measurements and transmission over the network. Clearly, in these cases measurement/computation/actuation data transmission should be minimized and should occur only "when necessary".

Among the various techniques proposed to face this problem, the event triggered technique seems to be promising [6, 10, 21, 22]. This technique formalize the statement "when necessary": the measurement/computation/actuation data transmission event occurs when the state of the system assumes certain values. Clearly, this technique requires the continuous measurement of the state. To circumvent this drawback, self-triggered techniques have been proposed. In this case the controller determines its next execution time, and does not require continuous measurements of the state. In particular, when the stabilization of the system origin is considered, this event is triggered only when the asymptotic stability property, as formalized by the Lyapunov approach, can be lost [3–5, 8, 9, 17, 22–25]. This approach can be also generalized to a weaker property such as safety [8, 9].

In this paper we generalize the self-triggered stabilization problem for the class of stochastic systems, where the input generically enters both in the deterministic dynamics and in those affected by noise. We also generalize the safety problem introduced in [8, 9] to the stochastic case. This class of system is of particular interest since in practice various disturbances, not measurable, may affect the dynamics. In this work we assume that the state equations are described by an Itô differential equation driven by a Wiener noise. While the stability in probability of the equilibria of stochastic differential equations has been extensively studied [12, 16, 18], as well as the self-triggered control of deterministic systems, to the best of the authors' knowledge, the self-triggered version represents a new contribution to this control strategy.

The paper is organized as follows. In Sect. 11.2, the self-triggered control problem for deterministic systems is recalled. In Sect. 11.3, we illustrate the mathematical setting and the problem formulation. In Sect. 11.4, we derive results for the asymptotic stability in probability, while in Sect. 11.5 the safety problem is studied. In Sect. 11.6, we apply the obtained results to a simple example. Some observations conclude the paper.

11.2 Recalls of Known Facts in the Deterministic Self-Triggered Control

In this section we recall some aspects of the self-triggered control in the deterministic case [8, 9]. The class of systems under study is quite generic

$$\dot{x} = f(x, u, \mu, d) \tag{11.1}$$

where $x \in \mathscr{D}_x \subset \mathbb{R}^n$ is the state vector, $u \in \mathscr{D}_u \subset \mathbb{R}^p$ is the input vector, $\mu \in \mathscr{D}_\mu \subset \mathbb{R}^r$ is a vector of uncertain parameters, $d \in \mathscr{D}_d \subset \mathbb{R}^s$ is an external bounded disturbance vector, where $\mathscr{D}_x$, $\mathscr{D}_u$ are a domains containing the origin, $\mathscr{D}_\mu$, $\mathscr{D}_d$ are compact sets containing $\mu = 0$ and $d = 0$, respectively. System (11.1) can be regarded as a system perturbed by $d \neq 0$ and $\mu \neq 0$, and the following nominal system, associated to the perturbed system (11.1), can be introduced

$$\dot{x} = f_0(x, u) \doteq f(x, u, 0, 0). \tag{11.2}$$

An obvious hypothesis, that will be assumed true in this work, is that there exists at least a state feedback control law $\kappa \colon \mathscr{D}_x \to \mathscr{D}_u$ such that closed-loop nominal system

$$\dot{x} = f_0(x, \kappa(x)) = F_0(x) \tag{11.3}$$

is (at least) locally asymptotically stable in the origin. If $F_0(x) \in C^\ell(\mathscr{D}_x)$, $\ell > 1$ an integer, this implies the existence of a Lyapunov function $V(x)$ of class

$C^1(\mathscr{D}_x)$ such that

$$\begin{aligned} \alpha_1(\|x\|) \le V(x) \le \alpha_2(\|x\|) \\ \frac{\partial V(x)}{\partial x} f_0(x, \kappa(x)) \le -\alpha_3(\|x\|) \\ \left\| \frac{\partial V(x)}{\partial x} \right\| \le \alpha_4(\|x\|) \end{aligned} \tag{11.4}$$

with $\alpha_1, \ldots, \alpha_4$ some functions of class $\mathscr{K}$ [11, 27].

Obviously, to implement the feedback control $u(t) = \kappa(x(t))$ the continuous measurement of the system state is required. If, on the contrary, the state measurement is available only in discrete time instants t_k, $k \ge 0$, the applied control is $u(t) = \kappa(x(t_k))$ for $t \in [t_k, t_{k+1})$. In general, such a control can not ensure neither the asymptotic stability at the origin, nor the simple stability. To ensure such properties, it is necessary to determine the time instants t_k in a proper way, imposing an appropriate stability condition. The set of sampling instants so determined will be denoted $\mathscr{I} = \{t_k\}_{k \ge 0}$, and the feedback control updated at the sampling instants of $\mathscr{I}$ will be denoted $u(t)|_{t_k \in \mathscr{I}}$, or simply $u_{\mathscr{I}}(t)$. In [8, 9] it is shown how to design such time instants to preserve the required stability condition also in the case of actuator delays. Since here the aim is to make a parallel with the stochastic case, for the sake of simplicity the case of systems without actuator delays.

Making use of an appropriate cost function, one could also consider the maximization problem of the inter-sampling times $t_{k+1} - t_k$, $k \ge 0$, here not addressed, which in turn corresponds to the minimization of the sensor transmission power, and of the CPU effort due to the computation.

The most stringent problem is the stability problem, consisting of determining the sequence $\mathscr{I}$ so that the feedback system is asymptotically stable in the origin.

Problem 11.1 *(Stability problem)* Given the nominal system (11.2), and a stabilizing state feedback control law κ, determine a function $\tau_s : \mathscr{D}_x \to [\tau_{\min}, \infty)$, with $\tau_{\min} > 0$ a minimum sampling time, such that if the sequence of sampling instants $\mathscr{I}$ is inductively defined by

$$t_{k+1} = t_k + \tau_s(x(t_k)) \tag{11.5}$$

then the origin of the closed-loop system (11.3) with control input signal $u_{\mathscr{I}}(t)$ is asymptotically stable.

A milder problem is the safety problem, where it is required that the state trajectory starting from a "safe set", containing the origin, remains in this safe set.

Definition 11.2 Given a state feedback control law κ, the system

$$\dot{x} = f(x, \kappa(x), \mu, d), \tag{11.6}$$

is safe with respect to the set $\mathscr{B} \subseteq \mathscr{D}_x$ for the time interval $\mathscr{T} \subseteq \mathbb{R}^+$, if $x(t) \in \mathscr{B}$, $\forall t \in \mathscr{T}$.

Problem 11.3 *(Safety problem)* Given the perturbed system (11.1), a stabilizing state feedback control law κ, and an arbitrary safe set $\mathscr{B}_\delta = \{x \in \mathbb{R}^n \mid \|x\| < \delta\} \subset \mathscr{D}_x$, determine a function $\tau_s : \mathscr{D}_x \to [\tau_{\min}, \infty)$, with $\tau_{\min} > 0$ a minimum sampling time, such that if $\mathscr{I}$ is inductively defined by (11.5), then the closed-loop system (11.6) with control input signal $u_{\mathscr{I}}(t)$ is safe with respect to $\mathscr{B}_\delta$, for the time interval $[t_0, \infty)$.

It is clear from the Definitions 11.2 and 11.11 that the existence of the minimum sampling time $\tau_{\min} > 0$ avoids undesired Zeno behaviors. The solution of the Stability Problem (11.1) is based on the following assumptions.

Assumption 11.4 *Assume that*

1. $f_0 \in C^\ell(\mathscr{D}_x \times \mathscr{D}_u)$, *with ℓ a positive integer sufficiently large;*
2. *There exists a nonempty set $\mathscr{U}$ of state feedback laws $\kappa\colon \mathscr{D}_x \to \mathscr{D}_u$, such that $\kappa \in C^\ell(\mathscr{D}_x)$ and the origin of* (11.3) *is asymptotically stable, with region of attraction a certain compact $\Omega \subset \mathscr{D}_x$.*

Assumption 11.5 *The functions $\alpha_3, \alpha_4 \in \mathscr{K}$ in* (11.4) *are such that α_3^{-1}, α_4 are Lipschitz.*

The assumption of non-emptiness of the set $\mathscr{U}$ is not restrictive since, if the nominal system can not be stabilized using a continuous feedback, then it can not be stabilized using a digital control with zero-order holders. The main limitation of Assumption 11.5, which is the same of [22], i.e., the Lipschitz condition on $\alpha_3^{-1}(\cdot)$ and $\alpha_4(\cdot)$, can be relaxed considering the Safety Problem 11.3. However, note that the conditions of Assumption 11.4 are weaker than those used in [4], i.e., the homogeneity of the closed-loop dynamics.

Theorem 11.6 *Let us consider the nominal system* (11.2). *Under Assumptions* 11.4 *and* 11.5, *there exist a state feedback control law κ, a minimum sampling time $\tau_{\min} > 0$, a function $\tau_s : \mathscr{D}_x \to [\tau_{\min}, \infty)$, such that if $\mathscr{I}$ is inductively defined by* (11.5), *then the origin of the closed-loop system* (11.3) *with control $u_{\mathscr{I}}(t)$ is asymptotically stable.*

The proof is constructive, and is given in Appendix. The main limitation of this result is that, if $\alpha_3^{-1}(\cdot)$ is not Lipschitz, the next sampling time $\tau_s(x_k)$ goes to zero as x_k approaches the equilibrium point, and this might generate Zeno behaviors.

Theorem 11.7 *Given the nominal system* (11.2) *and a safe set $\mathscr{B}_\delta$, $\delta > 0$, under Assumption* 11.4 *there exist a state feedback control law κ, a minimum sampling time $\tau_{\min} > 0$, a function $\tau_s : \mathscr{D}_x \to [\tau_{\min}, \infty)$, such that if $\mathscr{I}$ is inductively defined by* (11.5), *if $x(t_0) \in \mathscr{B}_\delta$ at the initial time t_0, then the closed-loop system* (11.3) *with control input signal $u_{\mathscr{I}}(t)$ is safe with respect to $\mathscr{B}_\delta$, for the time interval $[t_0, \infty)$.*

This theorem, which can be proved using the same reasoning of Theorem 11.6, states that if a system is asymptotically stabilizable using a continuous time state feedback control law, then it is always possible to keep the state arbitrarily close to the equilibrium point by applying a digital self-triggering strategy.

11.3 The Self-Triggered Control Problem for Stochastic Systems

We consider nonlinear stochastic systems of the form

$$\mathrm{d}x(t) = f_0(x, u)\mathrm{d}t + \sum_{j=1}^{m} g_{0j}(x, u)\mathrm{d}\xi_j(t) \tag{11.7}$$

where $x \in \mathscr{D}_x \subset \mathbb{R}^n$, $\mathscr{D}_x$ is a domain containing the origin, $u \in \mathscr{D}_u \subset \mathbb{R}^p$, $f_0,\ g_{0j} : \mathscr{D}_x \times \mathscr{D}u \to \mathbb{R}^n$, $j = 1, \ldots, m$ are sufficiently smooth vector fields, such that $f_0(0,0) = 0$, $g_{0j}(0,0) = 0$, $j = 1, \ldots, m$. Moreover, $\{\xi(t) = (\xi_1(t), \ldots, \xi_m(t))^T, t \geq 0\}$ is a standard R^m-valued Wiener process, defined on the usual complete probability space $(\Omega, \mathscr{F}, (\mathscr{F}_t)_{t\geq 0}, P)$, with $(\mathscr{F}_t)_{t\geq 0}$ the complete right-continuous filtration generated by ξ and $\mathscr{F}_0$ contains all P-null sets. It is worth stressing that in (11.7) the control appears either in the deterministic or in the stochastic terms [1, 2]. When $g_{0j} = 0$, one recovers the deterministic case.

Given a continuous state feedback control law $\kappa : \mathscr{D}_x \to \mathscr{D}_u$, the closed-loop system is

$$\mathrm{d}x(t) = f_0(x, \kappa(x))\mathrm{d}t + \sum_{j=1}^{m} g_{0j}(x, \kappa(x))\mathrm{d}\xi_j(t) \tag{11.8}$$

and we will denote by $x(t)$, $t \geq t_0$, the solution of the closed-loop system (11.8), with initial condition $x_0 = x(t_0)$. It is well-known that if the origin of system (11.8) is locally asymptotical stable in probability for a certain feedback κ, and if $f_0(x, \kappa(x)) \in C^{\ell}(\mathscr{D}_x)$, $\ell > 1$ integer, then there exists a Lyapunov function $V(x)$ of class $C^2(\mathscr{D}_x)$ such that [13, 14]

$$\begin{aligned} \alpha_1(\|x\|) \leq V(x) &\leq \alpha_2(\|x\|) \\ \mathscr{L}V(x) &\leq -\alpha_3(\|x\|) \\ \left\| \frac{\partial V(x)}{\partial x} \right\| &\leq \alpha_4(\|x\|) \\ \left\| \frac{\partial^2 V(x)}{\partial x^2} \right\| &\leq \alpha_5(\|x\|) \end{aligned} \tag{11.9}$$

with $\alpha_i \in \mathscr{K}$, $i = 1, \ldots, 5$. The infinitesimal generator associated to (11.8), obtain by differentiating V in the sense of Itô, is given by

$$\mathscr{L}V(x) = \frac{\partial V(x)}{\partial x} f_0(x, \kappa(x)) + \frac{1}{2} \sum_{j=1}^{m} \mathrm{Tr}\left(g_{0j}^T(x, \kappa(x)) \frac{\partial^2 V}{\partial x^2} g_{0j}(x, \kappa(x)) \right). \tag{11.10}$$

Here, the matrix $\partial^2 V(t,x)/\partial x^2$ is the Hessian matrix of the second order partial derivatives, and $\mathrm{Tr}(\cdot)$ denotes the trace of a matrix.

Clearly, the feedback control signal $u(t) = \kappa(x(t))$ requires continuous measurements of the state of the system. In view of an implementation of $\kappa(x(t))$ by means of digital devices, with variable sampling intervals δ_k, in the following we consider its digital version

$$u(t) = \kappa(x_k), \qquad \forall t \in [t_k, t_{k+1} = t_k + \delta_k),\ k \geq 0 \tag{11.11}$$

one needs to determine these sampling instants t_k so that the stability property of the origin is preserved in probability. Following the approach developed in [3, 4, 8, 9, 15, 17, 22], the aim is hence to determine on-line a sequence of strictly positive sampling intervals $\delta_k > 0$, i.e., a sequence $\{t_k\}_{k\geq 0}$ of sampling times, such that the origin of

$$\mathrm{d}x(t) = f_0(x, \kappa(x_k))\mathrm{d}t + \sum_{j=1}^{m} g_{0j}(x, \kappa(x_k))\mathrm{d}\xi_j(t) \tag{11.12}$$

is asymptotically stable in probability.

It is worth noting that to require $\delta_k > 0$ means that there exists a minimum sampling time $0 < \delta_{\min} \leq \delta_k$, $\forall\, k \geq 0$, which in turns will ensure that no Zeno behavior can occur. Hence, the time interval between two sampling instants is lower bounded by $\tau_{\min} > 0$.

The philosophy behind the self-triggered control is obvious: the control is performed only when necessary for guaranteeing the control objectives. This clearly reduces the transmission power of the sensing and actuation data transmissions, as well as the control effort of the digital device computing the control.

11.4 Self-Triggered Stabilizing Control

The result developed in this section is based on the following assumption, analogous to the assumptions used in [8, 9, 22] in the case of a deterministic systems.

Assumption 11.8 *Assume that*

1. $f_0, g_0 \in C^\ell(\mathscr{D}_x \times \mathscr{D}_u)$, *with* $\ell > 1$ *an integer sufficiently large;*
2. *There exists a nonempty set* $\mathscr{U}$ *of state feedback laws* $\kappa : \mathscr{D}_x \to \mathscr{D}_u$, *such that* $\kappa \in C^\ell(\mathscr{D}_x)$, $\ell > 1$, *and the origin of* (11.8) *is asymptotically stable in probability, with region of attraction a certain compact* $\Omega \subset \mathscr{D}_x$;
3. *The functions* $\alpha_3, \alpha_4, \alpha_5 \in \mathscr{K}$ *in* (11.9) *are such that* $\alpha_3^{-1}, \alpha_4, \alpha_5$ *are Lipschitz.*

The assumption of sufficient regularity of the functions f_0, g_0 is required in order to ensure the determination of the next sampling time, making use of a Taylor expansion, analogous to that used in [8, 9]. The assumption of existence of a stabilizing control is not restrictive, since if the nominal system cannot be stabilized in probability using continuous time measurements and actuations, then it is clear that the nominal system cannot be stabilized using a digital control with zero-order holders. Finally, the Lipschitz assumption on $\alpha_3^{-1}, \alpha_4, \alpha_5$ is required to write a simple stability condition, as used in [22], and represents the main limitation of this approach.

Using Assumption 11.8, one can state the following result.

Theorem 11.9 *Let us consider the nonlinear stochastic system* (11.7). *Under Assumption* 11.8, *there exist a piece-wise constant state feedback control law* (11.11), *and a sequence of strictly positive sampling intervals* $\delta_k > 0$, *such that the origin of the closed-loop system* (11.12) *is asymptotically stable in probability.*

Proof One can consider a control law $\kappa \in \mathscr{U}$, where $\mathscr{U}$ is not empty due to Assumption 11.8. Moreover, since $f_0(x, \kappa(x))$ is at least of class $C^2(\mathscr{D}_x)$, the converse theorem ensures the existence of a Lyapunov candidate satisfying (11.9). Let us fix $r > 0$ such that $B_r = \{x \in \mathscr{D}_x \mid \|x\| \leq r\} \subset \Omega$, where Ω is the domain of attraction due to the control κ. For $x_k \in B_r$, the infinitesimal generator associated to (11.12) is given by

$$\mathscr{L}V(x) = \frac{\partial V}{\partial x} f_0(x, \kappa(x_k)) + \frac{1}{2}\sum_{j=1}^{m} \mathrm{Tr}\left(g_{0j}^T(x, \kappa(x_k)) \frac{\partial^2 V}{\partial x^2} g_{0j}(x, \kappa(x_k))\right)$$

which can be rewritten considering the differences between the functions calculated in x and those calculated in x_k

$$\begin{aligned}
\mathscr{L}V(x) = {} & \frac{\partial V}{\partial x} f_0(x, \kappa(x)) + \frac{\partial V}{\partial x}\Big(f_0(x, \kappa(x_k)) - f_0(x, \kappa(x))\Big) \\
& + \frac{1}{2}\sum_{j=1}^{m} \mathrm{Tr}\left(g_{0j}^T(x, \kappa(x)) \frac{\partial^2 V}{\partial x^2} g_{0j}(x, \kappa(x))\right) \\
& + \frac{1}{2}\sum_{j=1}^{m} \mathrm{Tr}\Bigg(\Big[g_{0j}(x, \kappa(x_k)) - g_{0j}(x, \kappa(x))\Big]^T \\
& \qquad \times \frac{\partial^2 V}{\partial x^2} \Big[g_{0j}(x, \kappa(x_k)) - g_{0j}(x, \kappa(x))\Big]\Bigg) \\
& + \sum_{j=1}^{m} \mathrm{Tr}\left(\Big[g_{0j}(x, \kappa(x_k)) - g_{0j}(x, \kappa(x))\Big]^T \frac{\partial^2 V}{\partial x^2} g_{0j}(x, \kappa(x))\Big]\right).
\end{aligned} \tag{11.13}$$

The following upper bound can be considered for the quadratic terms in $g_{0j}(x, \kappa(x_k)) - g_{0j}(x, \kappa(x))$ in (11.13)

$$
\begin{aligned}
&\frac{1}{2}\mathrm{Tr}\Big(\big[g_{0j}(x,\kappa(x_k)) - g_{0j}(x,\kappa(x))\big]^T \frac{\partial^2 V}{\partial x^2}\big[g_{0j}(x,\kappa(x_k)) - g_{0j}(x,\kappa(x))\big]\Big) \\
&\quad \le \frac{n}{2}\left\|\big[g_{0j}(x,\kappa(x_k)) - g_{0j}(x,\kappa(x))\big]^T \frac{\partial^2 V}{\partial x^2}\big[g_{0j}(x,\kappa(x_k)) - g_{0j}(x,\kappa(x))\big]\right\|_\infty \\
&\quad \le \frac{n\sqrt{n}}{2}\left\|\big[g_{0j}(x,\kappa(x_k)) - g_{0j}(x,\kappa(x))\big]^T \frac{\partial^2 V}{\partial x^2}\big[g_{0j}(x,\kappa(x_k)) - g_{0j}(x,\kappa(x))\big]\right\| \\
&\quad \le \frac{n\sqrt{n}}{2}\left\|g_{0j}(x,\kappa(x_k)) - g_{0j}(x,\kappa(x))\right\|^2 \left\|\frac{\partial^2 V}{\partial x^2}\right\|
\end{aligned}
$$

for all $j = 1, \ldots, m$. Similarly, the last term of (11.13) can be bounded as follows

$$
\begin{aligned}
&\mathrm{Tr}\Big(\big[g_{0j}(x,\kappa(x_k)) - g_{0j}(x,\kappa(x))\big]^T \frac{\partial^2 V}{\partial x^2} g_{0j}(x,\kappa(x))\Big) \\
&\quad \le n\left\|\big[g_{0j}(x,\kappa(x_k)) - g_{0j}(x,\kappa(x))\big]^T \frac{\partial^2 V}{\partial x^2} g_{0j}(x,\kappa(x))\right\|_\infty \\
&\quad \le n\sqrt{n}\left\|\big[g_{0j}(x,\kappa(x_k)) - g_{0j}(x,\kappa(x))\big]^T \frac{\partial^2 V}{\partial x^2} g_{0j}(x,\kappa(x))\right\| \\
&\quad \le n\sqrt{n}\left\|g_{0j}(x,\kappa(x_k)) - g_{0j}(x,\kappa(x))\right\| \left\|g_{0j}(x,\kappa(x))\right\| \left\|\frac{\partial^2 V}{\partial x^2}\right\|.
\end{aligned}
$$

Using these bounds, from (11.13) and (11.9) one obtains

$$
\begin{aligned}
\mathscr{L}V(x) \le {} & -\alpha_3(\|x\|) + \alpha_4(\|x\|)\|d_{h,f}\| + \frac{n\sqrt{n}}{2}\alpha_5(\|x\|)\sum_{j=1}^{m}\|d_{h,g_j}\|^2 \\
& + n\sqrt{n}\alpha_5(\|x\|)\sum_{j=1}^{m}\|d_{h,g_j}\|\|g_{0j}(x,\kappa(x))\|
\end{aligned}
\tag{11.14}
$$

where

$$
\begin{aligned}
d_{h,f} &= f_0(x(t),\kappa(x(t_k))) - f_0(x(t),\kappa(x(t))) \\
d_{h,g_j} &= g_{0j}(x(t),\kappa(x(t_k))) - g_{0j}(x(t),\kappa(x(t)))
\end{aligned}
$$

$j = 1, \ldots, m$, are terms that can be regarded as perturbations due to the holding, as done in [8, 9] for the deterministic case. Also the following developments can be done in the same line. In fact, under Assumption 11.8, there exists a time interval $[t_k, t_k+\varepsilon_k]$ such that (11.12) has a unique solution $x(t)$. Hence, it is possible to expand in Taylor series the ith components $d_{h,f,i}$, $d_{h,g_j,i}$ of $d_{h,f}$, d_{h,g_j}, $j = 1, \ldots, m$, with respect to $t \in [t_k, t_k + \varepsilon_k]$, on the right of t_k, with the Lagrange remainder

$$\begin{aligned} d_{h,f,i} &= \varphi_{1,i}(x_k)(t - t_k) + \varphi_{2,i}(\bar{x}_i, x_k)(t - t_k)^2 \\ d_{h,g_j,i} &= \varphi_{3j,i}(x_k)(t - t_k) + \varphi_{4j,i}(\bar{x}_i, x_k)(t - t_k)^2 \end{aligned} \tag{11.15}$$

for $j = 1, \ldots, m$, where $\mathrm{d}_+^n(\cdot)/\mathrm{d}t^n$ denotes the n-th right derivative, and where the following quantities have been defined

$$\begin{aligned} \varphi_{1,i}(x_k) &= \left.\frac{\mathrm{d}_+ d_{h,f,i}}{\mathrm{d}t}\right|_{x(t)=x_k}, \quad \varphi_{2,i}(\bar{x}_i, x_k) = \frac{1}{2}\left.\frac{\mathrm{d}_+^2 d_{h,f,i}}{\mathrm{d}t^2}\right|_{x(t)=\bar{x}_i} \\ \varphi_{3j,i}(x_k) &= \left.\frac{\mathrm{d}_+ d_{h,g_j,i}}{\mathrm{d}t}\right|_{x(t)=x_k}, \quad \varphi_{4j,i}(\bar{x}_i, x_k) = \frac{1}{2}\left.\frac{\mathrm{d}_+^2 d_{h,g_j,i}}{\mathrm{d}t^2}\right|_{x(t)=\bar{x}_i} . \end{aligned}$$

The Taylor theorem with the Lagrange remainder ensures the existence of time instants $\bar{t}_i \in [t_k, t]$, such that the equalities (11.15) hold. In these expressions, $\bar{x}_i = x(\bar{t}_i), i = 1, \ldots, n$, is the value of the state at time $\bar{t}_i$. Denoting $\bar{x} = (\bar{x}_1, \ldots, \bar{x}_n)$, the following bounds can be considered

$$\begin{aligned} \|d_{h,f}\| &\leq \|\varphi_1(x_k)\|(t - t_k) + \|\varphi_2(\bar{x}, x_k)\|(t - t_k)^2 \\ \|d_{h,g_j}\| &\leq \|\varphi_{3j}(x_k)\|(t - t_k) + \|\varphi_{4j}(\bar{x}, x_k)\|(t - t_k)^2 \end{aligned} \tag{11.16}$$

where

$$\begin{aligned} \varphi_p(x_k) &= \Big(\varphi_{p,1}(x_k), \ldots, \varphi_{p,n}(x_k)\Big)^T, \qquad p = 1, 3j, \quad j = 1, \ldots, m \\ \varphi_q(\bar{x}, x_k) &= \Big(\varphi_{q,1}(\bar{x}_1, x_k), \ldots, \varphi_{q,n}(\bar{x}_n, x_k)\Big)^T, \quad q = 2, 4j. \end{aligned}$$

Let us now consider the level set

$$\Omega_{V(x_k)} = \big\{x \in \mathscr{D}_x \mid V(x) \leq V(x_k)\big\}$$

and let us define the quantities

$$\begin{aligned} M_p(x_k) &= \|\varphi_p(x_k)\|, \qquad p = 1, 3j, \quad j = 1, \ldots, m \\ M_q(x_k) &= \max_{\bar{x} \in \Omega_{V(x_k)}} \|\varphi_q(\bar{x}, x_k)\|, \quad q = 2, 4j. \end{aligned}$$

Since $f_0, g_{0j}, \kappa \in C^\ell$, with ℓ sufficiently large, and $\Omega_{V(x_k)}$ is a compact set, then $M_p(x_k)$ is finite for any $x_k \in \Omega_{V(x_k)}$, and $M_q(x_k) \in \mathbb{R}^+$ exists and is finite for any

$x_k \in \Omega_{V(x_k)}$. Finally, let us consider the terms

$$C_{g_{0j}} = \max_{x\in\Omega_{V(x_k)}} \left\| g_{0j}(x, \kappa(x)) \right\|, \quad j = 1, \ldots, m$$

which exist and are finite on the compact set $\Omega_{V(x_k)}$. With these positions, the infinitesimal generator (11.14) associated to (11.12) can be upper bounded by the following expression

$$\begin{aligned} \mathscr{L}V(x) \;\le\; & -\alpha_3(\|x\|) + \alpha_4(\|x\|)\|d_{h,f}\| \\ & + \frac{n\sqrt{n}}{2}\alpha_5(\|x\|) \sum_{j=1}^{m} \left(\|d_{h,g_j}\| + 2C_{g_{0j}}\|d_{h,g_j}\|\right) \\ = & -\alpha_3(\|x\|) + \alpha_4(\|x\|)\|d_{h,f}\| \\ & + \frac{n\sqrt{n}}{2}\alpha_5(\|x\|) \sum_{j=1}^{m} \left(\left(\|d_{h,g_j}\| + C_{g_{0j}}\right)^2 - C_{g_{0j}}^2\right). \end{aligned} \tag{11.17}$$

The undetermined terms in (11.17) can be dominated by negative terms, at least for small (but nonzero) time intervals. To show this, it is imposed that

$$\begin{aligned} & \alpha_4(\|x\|)\|d_{h,f}\| \le \vartheta_0\alpha_3(\|x\|) \\ & \frac{n\sqrt{n}}{2}\alpha_5(\|x\|)\left(\left(\|d_{h,g_j}\| + C_{g_{0j}}\right)^2 - C_{g_{0j}}^2\right) \le \vartheta_j\alpha_3(\|x\|) \\ & j = 1, \ldots, m \end{aligned} \tag{11.18}$$

with

$$\vartheta = \sum_{j=0}^{m} \vartheta_j < 1, \qquad \vartheta_j \in (0, 1),\; j = 0, \ldots, m.$$

Conditions (11.18) will determine a time instant $t_{k+1} = t_k + \delta_k < t_k + \varepsilon_k$, and hence a positive time interval δ_k in which the infinitesimal generator is negative definite. In fact, using (11.16), equations (11.18) are satisfied if

$$\alpha_3^{-1}\left(\frac{1}{\vartheta_0}\alpha_4(\|x\|)\left(M_1(x_k)(t - t_k) + M_2(x_k)(t - t_k)^2\right)\right) \le \|x\|$$

$$\alpha_3^{-1}\left(\frac{n\sqrt{n}}{2\vartheta_j}\alpha_5(\|x\|)\left[\left(M_{3j}(x_k)(t - t_k) + M_{4j}(x_k)(t - t_k)^2 + C_{g_{0j}}\right)^2 - C_{g_{0j}}^2\right]\right) \le \|x\|$$

for $j = 1, \ldots, m$. Since α_3^{-1}, α_4 and α_5 are Lipschitz, then equations (11.18) are satisfied if

$$\frac{1}{\vartheta_0} L_{\alpha_3^{-1}} L_{\alpha_4} \|x\| \left(M_1(x_k)(t - t_k) + M_2(x_k)(t - t_k)^2 \right) \leq \|x\|$$

$$\frac{n\sqrt{n}}{2\vartheta_j} L_{\alpha_3^{-1}} L_{\alpha_5} \|x\| \left[\left(M_{3j}(x_k)(t - t_k) + M_{4j}(x_k)(t - t_k)^2 + C_{g_{0j}} \right)^2 - C_{g_{0j}}^2 \right] \leq \|x\|$$

for all $j = 1, \ldots, m$, where $L_{\alpha_3^{-1}}, L_{\alpha_4}, L_{\alpha_5} > 0$ are the Lipschitz constants of $\alpha_3^{-1}, \alpha_4, \alpha_5$, respectively. These equations imply that (11.18) are satisfied under the sufficient conditions

$$M_1(x_k)(t - t_k) + M_2(x_k)(t - t_k)^2 \leq \frac{\vartheta_0}{L_{\alpha_3^{-1}} L_{\alpha_4}} \tag{11.19}$$

$$M_{3j}(x_k)(t - t_k) + M_{4j}(x_k)(t - t_k)^2 \leq \left(\sqrt{1 + \frac{2\vartheta_j}{n\sqrt{n}} \frac{1}{C_{g_{0j}}^2 L_{\alpha_3^{-1}} L_{\alpha_5}}} - 1 \right) C_{g_{0j}} \tag{11.20}$$

for $j = 1, \ldots, m$. Defining

$$\delta_k = \min \max \left\{ t - t_k \mid (11.19), (11.20) \text{ are satisfied}, j = 1, \ldots, m \right\}$$

and choosing $t_{k+1} = t_k + \delta_k$, then

$$\mathscr{L}V(x) \leq -(1 - \vartheta)\alpha_3(\|x\|)$$

for all $t \in [t_k, t_{k+1}]$ and for all $k \geq 0$. Therefore, the origin is asymptotically stable in probability. Equations (11.19), (11.20) are second degree inequalities in the form $a(x_k)y^2 + b(x_k)y \leq c$, where $a(x_k), b(x_k)$ are non-negative and upper bounded for each $x_k \in \mathscr{D}_x$, and c is strictly positive and upper bounded. Hence, this trivially implies that δ_k, $\forall\, k \geq 0$, are strictly positive for each $x_k \in \Omega_{V(x_k)}$, and thus a minimum dwell time does exists, so ensuring that δ_k does not go to zero as $k \to \infty$.

Remark 11.10 From the proof of the previous result, it is clear that the main difference between the deterministic and the stochastic case consists of the fact that the sampling period has to satisfy extra conditions [8, 9]. In fact, while in the deterministic case one can determine a sampling sequence $\{\delta_k\}$ solving only condition (11.19), in the stochastic case one needs to satisfy m additional conditions given by (11.20). Therefore, in the stochastic case the self-triggered control strategy will determine, in general, more restrictive (shorter) sampling times.

11.5 Self-Triggered Safety Control

The main limitation of the results developed in Sect. 11.4 is the Lipschitz continuity assumption of $\alpha_3^{-1}(\cdot)$. In fact, if $\alpha_3^{-1}(\cdot)$ is not Lipschitz, the next sampling time $t_k + \delta_k$ goes to zero as x_k approaches the equilibrium point, and this might generate Zeno behaviors. Hence, in the spirit of the self-triggered safety control addressed in [8, 9], in this section we will show that it is possible to keep the state arbitrarily close to the equilibrium point by applying a self-triggering strategy. The solution of this problem will not require the Lipschitz assumption on α_3^{-1}.

In the following definition, an invariant property is used to define that a system is almost surely (a.s.) safe with respect to a given subset of the state space.

Definition 11.11 Given a state feedback control law κ, system (11.8) is a.s. safe with respect to the set $\mathscr{B} \subseteq \mathscr{D}_x$ for the time interval $\mathscr{T} \subseteq \mathbb{R}^+$, if $x(t) \in \mathscr{B}$, $\forall t \in \mathscr{T}$ a.s.

Given the system (11.7), a stabilizing state feedback control law κ, and an arbitrary safe set $\mathscr{B}_\delta = \{x \in \mathbb{R}^n \mid \|x\| < \delta\} \subset \mathscr{D}_x$, the objective is to determine a sequence of strictly positive sampling intervals $\delta_k > 0$ and a piece-wise constant state feedback control law, as in the previous section, such that the closed-loop system (11.8) is a. s. safe with respect to $\mathscr{B}_\delta$, for the time interval $[t_0, \infty)$.

The results developed in this section are based on the following.

Assumption 11.12 *Assume that* $f_0, g_{0j} \in C^\ell(\mathscr{D}_x \times \mathscr{D}_u)$, $j = 1, \ldots, m$, *with* ℓ *a positive integer sufficiently large. Assume that there exists a nonempty set* $\mathscr{U}$ *of state feedback laws* $\kappa : \mathscr{D}_x \to \mathscr{D}_u$, *such that* $\kappa \in C^\ell(\mathscr{D}_x)$ *and the origin of the system* (11.8) *is asymptotically stable in probability.*

The following theorem states that if a system is almost surely asymptotically stabilizable, using a continuous time state feedback control law, then it is always possible to keep the state arbitrarily close to the equilibrium point by applying a digital self-triggering strategy. Note that, in order to guarantee that the state is arbitrarily close to the equilibrium point, we still need the stabilizability assumption.

Theorem 11.13 *Given the system* (11.7) *and a safe set* $\mathscr{B}_\delta$, $\delta > 0$, *under Assumption* 11.12 *there exist piece-wise constant state feedback control law* (11.11) *and a sequence of strictly positive sampling intervals* $\delta_k > 0$ *such that the closed-loop system* (11.8), (11.11) *is almost surely safe with respect to* $\mathscr{B}_\delta$, *for the time interval* $[t_0, \infty)$.

Proof The proof of Theorem 11.13 follows the same arguments of Theorem 11.9.

11.6 A Simple Illustrative Example

In order to illustrate the propose approach, let us consider the following system

$$\begin{aligned} dx &= \big(Ax + Bu + f(x, u)\big)dt + Cxdw \\ &= f_0(x, u)dt + g_0(x, u)dw \end{aligned}$$

with

$$f_0(x,u) = \begin{pmatrix} -x_1 + x_2 + x_1^2 \\ (1+x_1)u \end{pmatrix}, \quad C = \begin{pmatrix} 0 & -1 \\ -1 & 0 \end{pmatrix}.$$

Let us consider the continuous control $u = \kappa(x) = -x_2 \in \mathcal{U}$, and the Lyapunov candidate $V(x) = x^T Px$, with P solution of the Lyapunov equation

$$PA_c + A_c^T P + C^T PC = -R$$

with R a symmetric positive definite matrix, and

$$A_c = \begin{pmatrix} -1 & 1 \\ 0 & -1 \end{pmatrix}.$$

If

$$R = \begin{pmatrix} -2 & 0 \\ 0 & -3 \end{pmatrix}$$

the matrix

$$P = \begin{pmatrix} 3 & 1 \\ 1 & 4 \end{pmatrix}$$

is a solution of the Lyapunov equation, with $\lambda_{\min}^P \cong 2.382$, $\lambda_{\max}^P \cong 4.618$ the minimum and the maximum eigenvalue of P, respectively.

The infinitesimal generator associated to the previous system, for $\|x\| \leq 1/3$ satisfies

$$\begin{aligned}\mathscr{L}V &\leq -2x_1^2 - 3x_2^2 + 6|x_1|^3 + 8|x_1|x_2^2 \\ &\leq -2x_1^2 - 3x_2^2 + 6(1/3)x_1^2 + 8(1/3)x_2^2 \leq -\frac{1}{3}\|x\|^2.\end{aligned}$$

Thus, the origin of the system is almost surely exponentially stable in probability, with

$$\begin{aligned}&\alpha_1 = \lambda_{\min}^P \|x\|^2, \quad \alpha_2 = \lambda_{\max}^P \|x\|^2, \quad \alpha_3 = \|x\|^2/3 \\ &\alpha_4 = \lambda_{\max}^P \|x\|, \quad \alpha_5 = \|2P\|.\end{aligned}$$

It is clear that Assumption 11.8 is not satisfied, since α_3^{-1} is not Lipschitz. For this reason, we can not imply the existence of a stabilizing self-triggered strategy. However, Theorem 11.13 implies the existence of a self-triggered strategy that guarantees almost surely safety for an arbitrary small neighborhood of the equilibrium point. Since the origin is locally stabilizable in probability for $\|x\| \leq 1/3$, we can define the safe set as the ball $\mathscr{B}_\delta$ with $\delta = 10^{-4} < 1/3$.

Remark 11.14 Note that a better choice of the Lyapunov function may lead to a differente function α_3, with α_3^{-1} Lipschitz. In this case one may assess the asymptotical stability in probability of the origin, using Theorem 11.9.

11.7 Conclusions

In this work the self-triggered stabilization and safety problems have been generalized for the class of stochastic systems, where the dynamics are described by an Itô differential equation driven by a Wiener noise. In these equations, the input enters in both the deterministic dynamics and those affected by noise. The main difference with the deterministic case consists of the fact that the sampling period has to satisfy extra conditions, so determining a more restrictive (shorter) sampling time. In the next future, the self-triggered stabilization problem of nonlinear stochastic systems affected by parameter perturbations and subject to delays will be considered.

References

1. Aggoune, W., Contribution to the stabilization of stochastic nonlinear systems with time delays. In: Proceedings of the 18th IFAC world congress Milano, Italy, August 28-September 2, pp. 3885–3890 (2011)
2. Aggoune, W., On feedback stabilization of stochastic nonlinear systems with discrete and distributed delays. In: Proceedings of the 50th IEEE Conference on Decision and Control, and European Control Conference (CDC-ECC), Orlando, FL, USA, December 12–15, pp. 6296–6301 (2011)
3. Anta, A., Tabuada, P., Self-triggered stabilization of homogeneous control systems. In: Proceedings of the 2008 American Control Conference - ACC 2008, pp. 4129–4134 (2008)
4. Anta, A., Tabuada, P.: To sample or not to sample: self-triggered control for nonlinear systems. IEEE Trans. Autom. Control **55**(9), 2030–2042 (2010)
5. Anta, A., Tabuada, P.: Exploiting isochrony in self-triggered control. IEEE Trans. Autom. Control **57**(4), 950–962 (2010)
6. Årzén, K.E., A simple event based pid controller. In: Proceedings of the 14th IFAC World Congress, vol. 18, pp. 423–428 (1999)
7. Åström, K.J., Wittenmark, B.: Computer Controlled Systems. Prentice Hall, New Jersey (1990)
8. Di Benedetto, M.D., Di Gennaro, S., D'Innocenzo, A.: Digital self triggered robust control of nonlinear systems. Int. J. Control **86**(9), 1664–1672 (2013)
9. Di Benedetto, M.D., Di Gennaro, S., D'Innocenzo, A., Digital self triggered robust control of nonlinear systems. In: Proceedings of the 50th IEEE Conference on Decision and Control, and European Control Conference (CDC-ECC), Orlando, FL, USA, pp. 1674–1679 (2011)
10. Heemels, W.P., Sandee, J.H., Bosch, P.P.: Analysis of event-driven controllers for linear systems. Int. J. Control **81**(4), 571–590 (2008)
11. Khalil, H.K.: Nonlinear Systems, 3rd edn. Prentice Hall, New Jersey (2002)
12. Kolmanovskii, V.B., Myshkis, A.: Applied Theory of Functional Differential Equations. Kluwer Academic Publishers, Dordrecht (1992)
13. Kushner, H.J.: Stochastic Stability and Control. Academic Press, New York (1967)
14. Kushner, H.J.: Converse theorem for stochastic Lyapunov functions. SIAM J. Control Optim. **5**, 228–233 (1967)

15. Lemmon, M., Chantem, T., Hu, X., Zyskowski, M.: On Self-Triggered Full Information H-infinity Controllers, in Hybrid Systems: Computation and Control. Springer, Berlin (2007)
16. Mao, X.: Stochastic Differential Equations and Applications. Springer, Horwood (1997)
17. Mazo, M., Tabuada, P., On event-triggered and self-triggered control over sensor/actuator networks. In: Proceedings of the 47th Conference on Decision and Control, Cancun, Mexico, pp. 435–440 (2008)
18. Mohammed, S.-E.A.: Stochastic Functional Differential Equations. Longman, UK (1986)
19. Nesic, D., Gruene, L.: Lyapunov based continuous-time controller redesign for sampled-data implementation. Automatica **41**(7), 1143–1156 (2005)
20. Nesic, D., Teel, A.R.: Stabilization of sampled-data nonlinear systems via backstepping on their euler approximate model. Automatica **42**(10), 1801–1808 (2006)
21. Otanez, P.G., Moyne, J.R., Tilbury, D.M.: Using deadbands to reduce communication in networked control systems. In: Proceedings of the American Control Conference **4**, 3015–3020 (2002)
22. Tabuada, P.: Event-triggered real-time scheduling of stabilizing control tasks. IEEE Trans. Autom. Control **52**(9), 1680–1685 (2007)
23. Velasco, M., Fuertes, J., Marti, P., The self triggered task model for real-time control systems. In: Proceedings of the Real-Time Systems Symposyum - RTSS03, Work Progress Track, pp. 67–70 (2003)
24. Wang, X., Lemmon, M., State based self-triggered feedback control systems with l_2 stability. In: Proceedings of the 17th IFAC world congress, pp. 15238–15243 (2008)
25. Wang, X., Lemmon, M.: Self-triggered feedback control systems with finite-gain L_2 stability. IEEE Trans. Autom. Control **45**(3), 452–467 (2009)
26. Yamamoto, Y., Anderson, B.D.O., Nagahara, M.: Approximating sampled-data systems with applications to digital redesign. In: Proceedings of the IEEE Conference on Decision and Control, Las Vegas, NV **4**, 3724–3729 (2002)
27. Yoshizawa, T.: On the stability of solutions of a system of differential equations. Ser. A Math. **29**, 27–33 (1955)

Chapter 12
Multicellular Converter: A Benchmark for Control and Observation for Hybrid Dynamical Systems

Michael Defoort, Jeremy Van Gorp and Mohamed Djemai

Abstract Due to their cost and their reliability, multicellular converters are interesting devices to supply electrical systems in many applicative fields. The first part of this chapter deals with the controller design for switching power converters, which are a particular class of switched systems. Two controllers are introduced. The first one is based on the average model and exhibits good performances in the transient. The second one is based on a Lyapunov function to jointly control the load current and the capacitor voltages across each commutation cell while improving the steady-state behavior. The second part of this chapter deals with the observer design to solve the capacitor voltages estimation while taking into account the hybrid behavior of the converter. A hybrid observer, based on gathering partial information from individual modes of the switched system, is designed. Some simulations highlight the efficiency of the proposed control and observer schemes for the three-cells converter.

12.1 Introduction

Power electronics have been evolving through the last decades due to the developments of semiconductor power components and new systems of energy conversion [1]. Since the 1950s, power converters are used in traction system, power supplies, or numerical amplifiers. They ensure an adaptation of the electrical energy between the source and the load. Among these systems, multicellular converters, which appeared at the beginning of the 1990s [2], are based on the association in series

M. Defoort (✉) · M. Djemai
LAMIH, CNRS UMR 8201, University of Valenciennes and Hainaut-Cambresis, 59313 Valenciennes, France
e-mail: michael.defoort@univ-valenciennes.fr

M. Djemai
e-mail: mohamed.djemai@univ-valenciennes.fr

J. Van Gorp
CReSTIC, UFR SEN, Universit? de Reims Champagne Ardenne, Moulin de la Housse, BP 1039, 51687 Reims, France
e-mail: jeremy.van-gorp@univ-reims.fr

M. Djemai and M. Defoort (eds.), *Hybrid Dynamical Systems*, Lecture Notes in Control and Information Sciences 457, DOI 10.1007/978-3-319-10795-0_12

of the elementary cells of commutation with passive storage elements controlled by switches (transistors, diodes). During this last decade, these systems became more and more attractive to industrial applications, especially in high-power applications [3]. Indeed, the harmonic contents of the output signal are improved compared to the classical two levels converter technology using the same switching frequency. Furthermore, this structure enables the reduction of the losses due to commutations of power semiconductors while allowing low cost components [4, 5].

To benefit as well as possible from the large potential of the multicellular structure, an appropriate distribution of the voltages crossing each cell is needed. These voltages are generated when a suitable control of switches is applied. The control of switches allows to cancel the harmonics at the cutting frequency and to reduce the switching losses and the electromagnetic interference [6–8].

Multicellular converters have received the attention of many researchers in the area of control and observation. For instance, using the average model, a modulation strategy, i.e., pulse-width-modulation (PWM), is applied to translate the continuous control into switching orders [9]. Other strategies based on predictive control [10] and Lyapunov techniques [12]. Nevertheless, the use of an average model cannot represent the natural balancing of the terminal voltages of the capacities. Hence, it should be also interesting to directly use the exact or instantaneous model, which takes into account the evolution at every time instant of all the variables including the state of the switch [11].

The first part of this chapter deals with the controller design to ensure the regulation of the load current and an appropriate distribution of the capacitor voltages across each cell. Two schemes are introduced. The first one is based on the average model and exhibits good performances in the transient. The second one is based on an appropriate Lyapunov function to directly control the converter switches without the need of modulation technique.

Anyway, the knowledge of the capacitor voltages is always needed. It is important to note that the use of physical extra sensors in order to measure such voltages increases the cost and the complexity of the system. That is why, its estimation by means of an observer becomes an attractive and economical solution. Furthermore, occurrence of faults can be extremely detrimental, not only to the equipment and surroundings, but also to the human operator if they are not detected and isolated in time. Hence, fault detection and isolation (FDI) have been recently investigated for the multicellular converter [13]. Observer-based FDI techniques rely on the estimation of outputs from measurements with the observer in order to detect the fault.

Different observer-based methods have been developed for switched systems such as, for instance, adaptive observer [14], algebraic observer [15], first order sliding mode observer [16, 17] or super-twisting observer [18, 19]. Some of these [15, 17, 19] assume that each mode is observable and admits a state observer. Furthermore, most of the existing literature incur the need of a common Lyapunov function or a fixed amount of dwell-time between switching instants or do not give an explicit design of the observer. To remove these drawbacks, a new concept of observability, based on geometric conditions, is introduced in [20]. This concept is applied for linear switched systems [21] and nonlinear systems [22].

The second part of this chapter deals with the observer design to estimate the capacitor voltages by taking into account the hybrid behavior of the converter where only the current in the load is measured. Note that, for this system, the state components are only partially observable for every fixed configuration of the switches. The proposed scheme is based on gathering partial information from individual modes.

Some simulations highlight the efficiency of the proposed control and observer schemes for the three-cells converter.

The following notational conventions are used within this chapter. Let ν be an Euclidean subspace of $\mathbb{R}^n$ and A a matrix $n \times n$. $\nu^{\perp}$ is the orthogonal subspace of ν. A^T is the transpose of A. The pre-image of ν under A is $A^{-1}\nu = \{x|Ax \in \nu\}$. $ker(A)$ is the null space of the transformation associated to A, i.e., $ker(A) = A^{-1}\{0\}$. $Im(A)$ is the range of the transformation associated to A. $\prod_{i=j}^{k} A_i = A_j \dots A_k$.

12.2 Multicellular Converter Modeling

The multicellular converter is based on the combination of p elementary cells of commutation (Fig. 12.1). The current flows from the source E toward the output through the different switches. The converter shows, by its structure, a hybrid behavior due to the discrete variables, i.e., switches. Note that because of the presence of the floating capacitors, there are also continuous variables, i.e., currents and voltages.

It is important to highlight that in order to standardize the industrial production, the electrical switches constraints should be similar. This requirement implies a unique voltage switch constraint of $\frac{E}{p}$. Thus, it is necessary to ensure a balanced distribution of the capacitor voltages. Under these conditions, the reference voltage of the jth capacitor $j = \{1, \dots, p-1\}$ is given by:

$$V_{c_j,\text{ref}} = j\frac{E}{p} \tag{12.1}$$

The dynamics of the converter, with a load consisting in a resistance R and an inductance L, is given by the following differential equations:

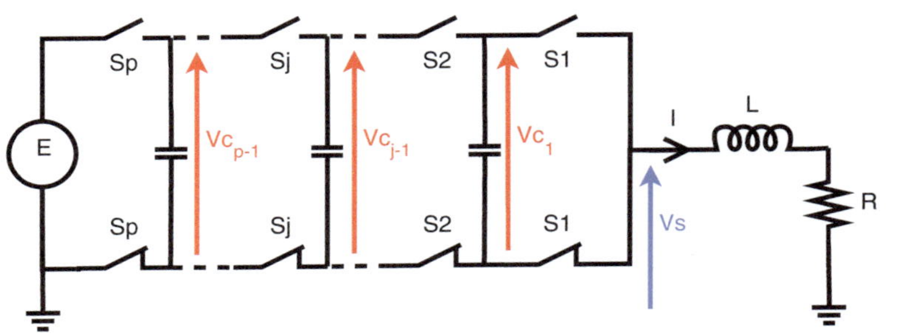

Fig. 12.1 Multicellular converter associated to an inductive load

$$\dot{I} = -\frac{R}{L}I + \frac{E}{L}S_p - \sum_{j=1}^{p-1} \frac{V_{c_j}}{L}(S_{j+1} - S_j) \tag{12.2a}$$

$$\dot{V}_{c_j} = \frac{I}{c_j}(S_{j+1} - S_j), \quad j = 1, \ldots, p-1 \tag{12.2b}$$

where I is the load current, c_j is the capacitance, V_{c_j} is the voltage in the jth capacitor and E is the voltage of the source. Each commutation cell is controlled by the binary signal $S_j \in \{0, 1\}$. Signal $S_j = 1$ means that the upper switch of the jth cell is "on" and the lower switch is "off" whereas $S_j = 0$ means that the upper switch is "off" and the lower switch is "on".

Remark 12.1 According to Fig. 12.1, one can note that the load current I is physically bounded between two known values, i.e.,

$$0 < I < \frac{E}{R} \tag{12.3}$$

The discrete control input $q = \left[q_1, \ldots, q_p\right]^T$ of the switches can be described as follows:

$$\begin{cases} q_j = S_{j+1} - S_j, & j = 1, \ldots, p-1 \\ q_p = S_p \end{cases} \tag{12.4}$$

Let us define the diffeomorphism T between the switches input q and the mode of the converter as:

$$\begin{array}{ll} \mathbb{R}^p \mapsto & \{1, \ldots, 2^p\} \\ q \;\rightarrow T(q) = 1 + \sum\limits_{j=0}^{p-1} 2^j S_{j+1} \end{array} \tag{12.5}$$

Assuming that only the load current I can be measured (i.e., $y = I$), the system can be represented as follows:

$$\dot{I} = -\frac{R}{L}I + \frac{E}{L}q_p - \sum_{j=1}^{p-1} \frac{V_{c_j}}{L}q_j \tag{12.6a}$$

$$\dot{V}_{c_j} = \frac{I}{c_j}q_j, \quad j = 1, \ldots, p-1 \tag{12.6b}$$

$$y = I \tag{12.6c}$$

Using the hybrid system formalism, one can describe system (12.6a–12.6c) as:

$$\dot{x}(t) = A_{\sigma(t)}x(t) + B_{\sigma(t)}u(t), \quad t \neq \{t_k\} \tag{12.7a}$$

$$x(t_k) = F_{\sigma(t_k^-)}x(t_k^-) \tag{12.7b}$$

$$y(t) = C_{\sigma(t)}x(t) \tag{12.7c}$$

where $x = [V_{c_1}, V_{c_2}, \ldots, V_{c_{p-1}}, I]^T \in \mathbb{R}^p$ is the state, $u(t) = E$ is the source. The switching signal $\sigma : \mathbb{R} \mapsto \{1, \ldots, 2^p\}$ is a piecewise constant and right-continuous function that changes its values at switching times t_k, $k \in \mathbb{N}$. At each instant t, it takes the mode of the converter, i.e., $\sigma(t) = T(q)$, $\forall t \in [t_k, t_{k+1})$.

Matrices $A_{\sigma(t)}$, $B_{\sigma(t)}$, $F_{\sigma(t_k^-)}$ and $C_{\sigma(t)}$ are defined as follows:

$$
\begin{aligned}
A_{T(q)} &= \begin{bmatrix} 0 & \ldots & 0 & \frac{q_1}{c_1} \\ \vdots & & \vdots & \vdots \\ 0 & \ldots & 0 & \frac{q_{p-1}}{c_{p-1}} \\ -\frac{q_1}{L} & \ldots & -\frac{q_{p-1}}{L} & -\frac{R}{L} \end{bmatrix} \\
B_{T(q)} &= \begin{bmatrix} 0 \cdots 0 \; \frac{q_p}{L} \end{bmatrix}^T \\
F_{T(q)} &= I_p \\
C_{T(q)} &= \begin{bmatrix} 0 \cdots 0 \; 1 \end{bmatrix}
\end{aligned}
$$

Note that the system state evolves without jump and is continuous.

Remark 12.2 It is assumed that there are a finite number of switching times in any finite time interval (this exclude Zeno phenomenon).

12.3 Controller Design

The control objective is to find the evolution of the discrete control input q_j, $j = \{1, \ldots, p\}$ such that the multicellular converter ensures:

- The regulation of the load current I as close as possible to the constant desired load current I_{ref}. That is to say, the error $I - I_{\text{ref}}$ is stabilized toward the origin.
- The balanced distribution of the capacitor voltages across each cell, i.e., $\forall j = \{1, \ldots, p-1\}$, $V_{c_j} - V_{c_j,\text{ref}}$ is stabilized toward the origin.

Therefore, the proposed controller will be designed in order to obtain the appropriate level of voltages at the converter output and at the terminals of the floating capacitors to regulate the output current.

Two controllers are introduced hereafter. The first one is based on a pulse-width modulation strategy and exhibits good performances in the transient. The second one is based on an appropriate Lyapunov function to directly control the converter switches without the need of modulation technique.

12.3.1 Pulse-Width-Modulation Strategy

The objective of this strategy is to jointly control the load current and the capacitor voltages across each commutation cell.

The PWM strategy is based on the average model. The idea is to generate p binary signals with phase shift of $\frac{2\pi}{p}$. p duty cycles are computed and allow to stabilize the load current and the voltages of the floating capacitors around the desired values. In the literature, some studies have been proposed in order to limit the oscillations and the steady-state tracking errors on the load current and the internal voltages. Here, the decoupling control method is applied [5]. Three controllers (PI) are assigned to regulate each internal voltage, i.e., V_{c_1}, V_{c_2} and the load current I. The variable duty cycles, denoted $\alpha_1, \alpha_2, \alpha_3$ are computed for each commutation cell using a similar approach as in [5]. The switching signal is determined through the comparison between the duty cycles and triangular signals of period T_{MLI} and phase shift of $\frac{T_{\mathrm{MLI}}}{p}$ (see [6] for further details).

Example 12.3 Some examples of PWM are given in Fig. 12.2 for the three cells converter. To simplify the study, only constant duty cycles are considered (i.e., $\alpha_1 = \alpha_2 = \alpha_3 = Cste$). For each example, a duty cycle is considered (i.e., $\alpha_1 = \alpha_2 = \alpha_3 = Ref_i$). The three control inputs S_1, S_2 and S_3 are computed as follows: it equals to "1" if the duty cycle Ref_i is upper than the corresponding triangular signal otherwise it equals to zero. In Fig. 12.2, one can see the evolution of the control signals for the three cases (i.e., Ref_1, Ref_2 and Ref_3). It creates a cycle which depends on the load current reference. One can remark three possible cycles which depend on I_{ref}, i.e., "1, 2, 1, 3, 1, 5" for reference 1, "2, 4, 3, 7, 5, 6" for reference 2 and "8, 4, 8, 6, 8, 7" for reference 3.

12.3.2 Binary Control Strategy Based on Lyapunov Function

Let us consider the following common Lyapunov function candidate:

$$V = \frac{L(I - I_{\mathrm{ref}})^2}{2} + \sum_{j=1}^{p-1} \frac{c_j (V_{c_j} - V_{c_j,\mathrm{ref}})^2}{2} \tag{12.8}$$

Since there is no jump, V is positive, continuous and null only for $I = I_{\mathrm{ref}}$ and $V_{c_j} = V_{c_j,\mathrm{ref}}$, $j = 1, \ldots, p-1$. Its time derivative, which depends on the value of the control signals will be denoted by $\dot{V}$. Using Eq. (12.2), it becomes that:

$$\begin{aligned} \dot{V} &= (I - I_{\mathrm{ref}}) \left(-RI + ES_p\right) - (I - I_{\mathrm{ref}}) \left(\sum_{j=1}^{p-1} V_{c_j} (S_{j+1} - S_j) \right) \\ &+ \sum_{j=1}^{p-1} (V_{c_j} - V_{c_j,\mathrm{ref}}) I (S_{j+1} - S_j) \end{aligned} \tag{12.9}$$

Note that, Eq. (12.9) is continuous except at the switching instants. Let us denote

$$A_j = -(I - I_{\mathrm{ref}}) V_{c_j} + (V_{c_j} - V_{c_j,\mathrm{ref}}) I \tag{12.10}$$

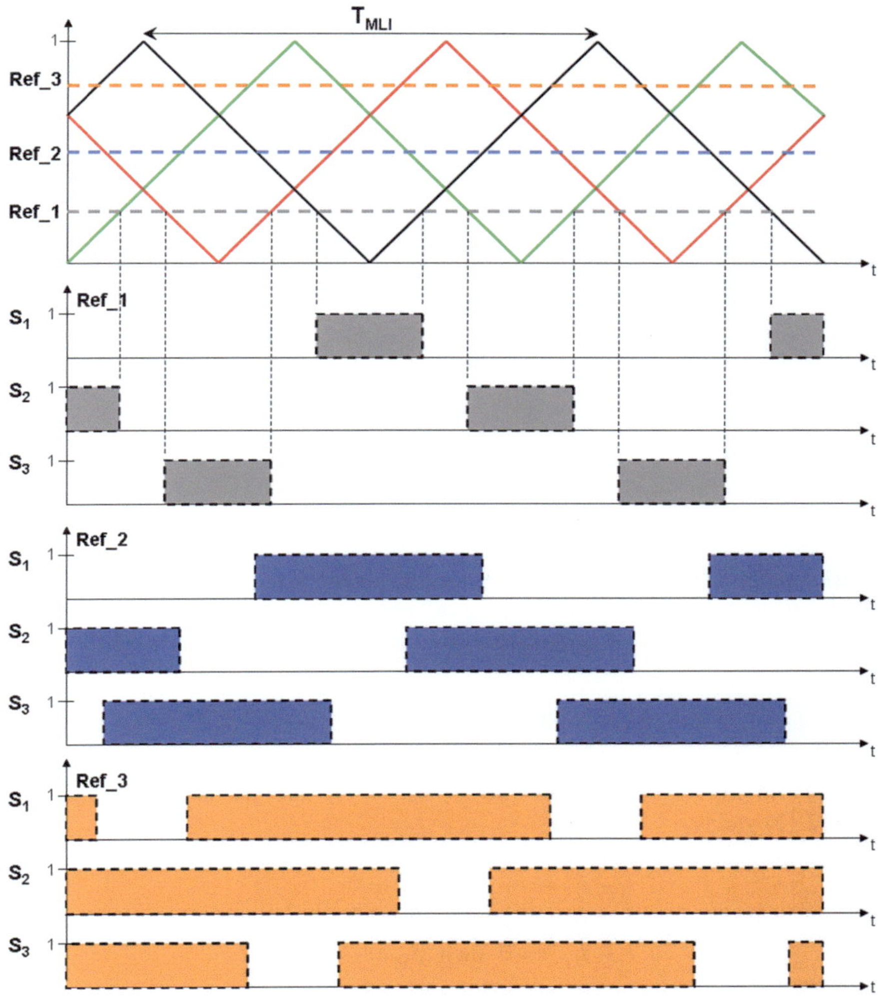

Fig. 12.2 Examples of PWM for the three cells converter

Hence (12.9) becomes:

$$\dot{V} = (I - I_{\text{ref}})\left(-RI + ES_p\right) - \sum_{j=1}^{p-1} A_j(S_j - S_{j+1}) \tag{12.11}$$

A state feedback control law, defining the switching sequence, can be deduced from this common Lyapunov function.

Theorem 12.4 *Under the control law:*

$$S_p = \frac{1 - f(I - I_{\text{ref}})}{2} \tag{12.12}$$

$$S_j = \frac{1 + g(A_j)}{2} \quad \forall j = 1, \ldots, p-1 \tag{12.13}$$

where:

$$f(I - I_{\text{ref}}) = \begin{cases} 1 & \textit{if } I - I_{ref} > 0 \\ -1 & \textit{if } I - I_{ref} < 0 \\ \text{Sign}(A_{p-1}) & \textit{if } I - I_{ref} = 0 \end{cases} \tag{12.14}$$

and

$$g(A_j) = \begin{cases} 1 & \textit{if } A_j > 0 \\ -1 & \textit{if } A_j < 0 \\ -\text{Sign}(A_{j-1}) & \textit{if } A_j = 0 \end{cases} \tag{12.15}$$

the multicellular converter supplies the load current I as close as possible to the desired load current I_{ref} while ensuring an appropriate distribution of the voltages across each cell. The signum function is described as:

$$\text{Sign}(x) = \begin{cases} 1 & \textit{if } x \geq 0 \\ -1 & \textit{if } x < 0 \end{cases} \tag{12.16}$$

Proof From Eq. (12.3), the quantities $(-RI + E)$ and RI are positive. Equation (12.12) yields:

$$\begin{array}{l} \text{if } I - I_{ref} > 0 \text{ then } S_p = 0 \\ \text{if } I - I_{ref} < 0 \text{ then } S_p = 1 \\ \text{if } I - I_{ref} = 0 \text{ then } S_p = \frac{1 - \text{Sign}(A_{p-1})}{2} \end{array} \tag{12.17}$$

Therefore, the term $(I - I_{\text{ref}})\left(-RI + ES_p\right)$ of Eq. (12.11) is negative or null.

From Eq. (12.13), one can note that:

$$\begin{array}{l} \text{if } A_j > 0 \text{ then } S_j = 1 \\ \text{if } A_j < 0 \text{ then } S_j = 0 \\ \text{if } A_j = 0 \text{ then } S_j = \frac{1 - \text{Sign}(A_{j-1})}{2} \end{array} \tag{12.18}$$

Since $S_{j+1} = 0$ or $S_{j+1} = 1$, the term $A_j(S_j - S_{j+1})$ of Eq. (12.11), is positive or null. Hence, one can conclude that $\dot{V} \leq 0$.

The time derivative of the Lyapunov function equals zero if $I - I_{\text{ref}} = 0$ and $V_{c_j} - V_{c_j,\text{ref}} = 0$ for all $j = 1, \ldots, p-1$. It can be proved using induction method.

The term $(I - I_{\text{ref}})\left(-RI + ES_p\right)$ of Eq. (12.11) equals zero if and only if $I = I_{\text{ref}}$. In this case, one gets

$$A_{p-1}(S_{p-1} - S_p) = A_{p-1}\left(S_{p-1} - \frac{1 - \text{Sign}(A_{p-1})}{2}\right) = 0 \Longleftrightarrow V_{c_{p-1}} = V_{c_{p-1},\text{ref}}$$

Assuming that $\forall j = k, \ldots, p-1\,, k > 1$, the following equivalence holds:

$$A_j(S_j - S_{j+1}) = 0 \quad \Longleftrightarrow \quad V_{c_j} = V_{c_j,\text{ref}}$$

At step $k - 1$, one gets:

$$A_{k-1}(S_{k-1} - S_k) = A_{k-1}\left(S_{k-1} - \frac{1 - \text{Sign}(A_{k-1})}{2}\right) = 0 \Longleftrightarrow V_{c_{k-1}} = V_{c_{k-1},\text{ref}}$$

Therefore, the control objective is fulfilled using the proposed controller.

12.4 Observer Design

12.4.1 Observability Analysis

This subsection is devoted to the observability analysis of the capacitor voltages V_{c_j} $(j = \{1, \ldots, p-1\})$ from the measurement of the current I and the knowledge of the control input sequence q.

12.4.1.1 Classical Approach

Considering system (12.7a–12.7c), one can see that there are several operating switching modes for which the state x is not observable. For instance, if V_{c_j} $(j = \{1, \ldots, p-1\})$, the voltages V_{c_j} $(\forall j \in \{1, \ldots, p-1\})$ become completely unobservable. Indeed, the observability matrix is not full rank.

12.4.1.2 Geometric Approach

The concept of $Z-$observability is introduced in [23] to consider the observability of some particular class of hybrid systems. Using a similar approach, Tanwani et al. gives a generalization of these concepts [20]. Let us recall some of these definitions.

Definition 12.5 The set $(\sigma^i, u^i, y^i, x^i)$ over an interval $\left[t_0, t_f\right)$ defines a trajectory for system (12.7a–12.7c). System (12.7a–12.7c) is said to be $\left[t_0, t_f\right)$—observable if the equality $(\sigma^1, u^1, y^1) = (\sigma^2, u^2, y^2)$ implies $x^1(t_0) = x^2(t_0)$.

Remark 12.6 For system (12.7a–12.7c), the switching signal $\sigma(t)$ changes its value, which correspond to the mode, at each switching time. In order to alleviate the writing, we rename the mode sequence $\{T(q)\}$ as increasing numbers $\{k\} \in \{1, 2, 3, \ldots\}$. The switching time t_k is the instant when transition from mode k to mode $k+1$ takes place.

Let $\mathcal{N}_k^m$, $(m \geq k)$ be the unobservable subspace on the interval $\left[t_{k-1}, t_{m-1}\right)$. On $[t_{k-1}, t_k)$, the observability matrix is given as:

$$G_k = \begin{bmatrix} C_k \\ C_k A_k \\ \vdots \\ C_k A_k^{p-1} \end{bmatrix} \tag{12.19}$$

Hence, one gets: $\mathcal{N}_k^k = ker(G_k)$. Using the measurement on $\left[t_{k-1}, t_{m-1}\right)$, more information about the state is obtained in general. Therefore, the unobservable subspace becomes smaller.

Proposition 12.7 [20] *System* (12.7a–12.7c) *with a given switching signal is* $[t_0, t_{m-1})$—*observable if and only if:*

$$\mathcal{N}_1^m = \{0\} \tag{12.20}$$

Remark 12.8 The application of Proposition 12.7 to the multicellular converter yields the following result. After $p-1$ time intervals (i.e., p switches) the different measurements of the load current I allows to obtain a set of $(p-1)$ linearly independent equations with respect to the voltages in the $(p-1)$ capacitors; which enables to estimate the capacitor voltages of the converter.

12.4.2 Hybrid Observer Design

Hereafter, a hybrid observer is designed in order to estimate the capacitor voltages of the multicellular converter by using only the measurement of the current I. The proposed observer gather partial information from individual modes of the switched system. The resetting procedure, applied at each switching time, is based on partial observers and some inversion formulae.

The system is persistent in the sense that there is $N \in \mathbb{N}$ such that:

$$N_{k-N}^k = \{0\} \quad \forall k \geq N+1 \tag{12.21}$$

Using the notations given in Remark 12.6, the hybrid observer is described by:

$$\dot{\hat{x}}(t) = A_k \hat{x}(t) + B_k u(t), \quad t \in [t_{k-1}, t_k) \tag{12.22a}$$

$$\hat{x}(t_k) = \hat{x}(t_k^-) - \xi_k(t_k^-) \tag{12.22b}$$

The initial state $\hat{x}(t_0)$ is chosen as zero.

The resetting procedure, applied at each switching time, is based on sliding mode partial observers. The correction vector ξ_k is computed using accumulated partial state information, obtained by a sliding mode observer, in order to ensure the convergence to zero of the error:

$$\tilde{x} = \hat{x} - x$$

The proposed procedure, similar to [20], to compute ξ_k is described hereafter.

A partial observer is designed in order to estimate the observable part of $\tilde{x}$ at time t_k^-. Let us choose a matrix Z^k(resp. W^k) such that its columns are an orthonormal basis of $Im(G_k^T)$ (resp. $ker(G_k)$). Therefore, on the time interval $[t_{k-1}, t_k)$, one can denote:

$$\begin{aligned} z^k &= (Z^k)^T \tilde{x} \\ w^k &= (W^k)^T \tilde{x} \\ S_k &= (Z^k)^T A_k (Z^k)^{-T} \\ R_k &= C_k (Z^k)^{-T} \end{aligned} \tag{12.23}$$

with $z^k \in \mathbb{R}^{l_k}$, $w^k \in \mathbb{R}^{p-l_k}$, $S_k \in \mathbb{R}^{l_k \times l_k}$ and $R_k \in \mathbb{R}^{1 \times l_k}$.

One should highlight that the pair (S_k, R_k) is observable. Hence, matrix $\bar{G}_k$ has a full rank, i.e.,

$$\text{rank}(\bar{G}_k) = \text{rank} \begin{bmatrix} R_k \\ R_k S_k \\ \vdots \\ R_k S_k^{l_k - 1} \end{bmatrix} = l_k$$

A sliding mode observer associated to the operation mode is proposed:

$$\begin{aligned} \dot{\hat{z}}^k(t) &= S_k \hat{z}^k(t) + \bar{G}_k^{-1} \nu_k(t), \quad \forall t \in [t_{k-1}, t_k) \\ \hat{z}^k(t_{k-1}) &= 0 \end{aligned} \tag{12.24}$$

where the correction terms are given as:

$$\nu_k = \begin{bmatrix} -\alpha_1 M_k^{\frac{1}{l_k}} |R_k \hat{z}^k - \tilde{y}|^{\frac{l_k - 1}{l_k}} \text{Sign}(R_k \hat{z}^k - \dot{\tilde{y}}) \\ -\alpha_2 M_k^{\frac{1}{l_k - 1}} |\nu_{2,k} - \dot{\nu}_{1,k}|^{\frac{l_k - 2}{l_k - 1}} \text{Sign}(\nu_{2,k} - \dot{\nu}_{1,k}) \\ \vdots \\ -\alpha_{l_k} M_k \text{Sign}(\nu_{l_k,k} - \dot{\nu}_{l_k - 1,k}) \end{bmatrix} \tag{12.25}$$

The high order sliding mode differentiator is used as an auxiliary dynamics. It has the following form:

$$\begin{cases} \dot{v}_{1,k} &= v_{2,k} - \alpha_1 M_k^{\frac{1}{l_k}} |R_k \hat{z}^k - \tilde{y}|^{\frac{l_k-1}{l_k}} \operatorname{Sign}(R_k \hat{z}^k - \dot{\tilde{y}}) \\ \dot{v}_{2,k} &= v_{3,k} - \alpha_2 M_k^{\frac{1}{l_k-1}} |v_{2,k} - \dot{v}_{1,k}|^{\frac{l_k-2}{l_k-1}} \operatorname{Sign}(v_{2,k} - \dot{v}_{1,k}) \\ &\vdots \\ \dot{v}_{l_k,k} &= -\alpha_{l_k} M_k \operatorname{Sign}(v_{l_k,k} - \dot{v}_{l_k-1,k}) \end{cases} \tag{12.26}$$

M_k are known Lipschitz constants for the function $R_k S_k^{l_k} z^k$. The constants α_l are chosen recursively and sufficiently large.

Theorem 12.9 *Using the sliding mode observer* (12.24) *with the correction terms* (12.25) *and providing that constants α_l and M_k are well chosen, the state estimation error $\tilde{z}^k(t) = \hat{z}^k(t) - z^k(t)$ converges to zero in finite time, i.e.,*

$$\tilde{z}^k(t) = O(\tau), \quad \forall t \in [t_{k-1} + T^*, t_k) \tag{12.27}$$

where τ is the sampling interval and $T^* > 0$.

Proof From (12.23), the observable state of error system is:

$$\dot{z}^k(t) = S_k z^k(t), \quad t \in [t_{k-1}, t_k) \tag{12.28a}$$

$$z^k(t_{k-1}) = (Z^k)^T \tilde{x}(t_{k-1}) \tag{12.28b}$$

$$\tilde{y} = R_k z^k \tag{12.28c}$$

Using the transformation $\varsigma^k = \bar{G}_k z^k$, the observable linear system (12.28a–12.28c) can be transformed into the following triangular observable form:

$$\begin{aligned} \dot{\varsigma}^k &= \bar{S}_k \varsigma^k + \bar{T}_k(z^k) \\ \tilde{y} &= \begin{bmatrix} 1 & 0 & \cdots & 0 \end{bmatrix} \varsigma^k \end{aligned} \tag{12.29}$$

with

$$\bar{S}_k = \begin{bmatrix} 0 & 1 & 0 & 0 & 0 \\ 0 & 0 & 1 & 0 & 0 \\ \vdots & \vdots & \vdots & \ddots & \vdots \\ 0 & 0 & 0 & 0 & 1 \\ 0 & 0 & 0 & 0 & 0 \end{bmatrix} \in \mathbb{R}^{l_k \times l_k}, \quad \bar{T}_k(z^k) = \begin{bmatrix} 0 \\ \vdots \\ 0 \\ R_k S_k^{l_k} z^k \end{bmatrix} \in \mathbb{R}^{l_k}$$

According to [19], the observer defined as follows:

$$\dot{\hat{\varsigma}}^k = \bar{S}_k \hat{\varsigma}^k + v_k \tag{12.30}$$

yields to the finite time stabilization of the observation error $\hat{\varsigma}^k - \varsigma^k$. Therefore, using the change of coordinate $\varsigma^k = \bar{G}_k z_k$, one can conclude that the sliding mode observer (12.24) with the correction terms (12.25) provides the convergence of the observation error $\tilde{z}^k$ in a finite time T^*. □

The partial state information is then accumulated to obtain an approximation of the estimation error. This approximation is transported to the current time using the state transition matrix:

$$\Psi_i^j = \prod_{l=j}^{i+1} (e^{A_l \tau_l}), \quad j > i \tag{12.31}$$

with $\Psi_i^i = I$. Following [20], let us define matrix Θ_i^k such that: $\forall i \in \{k-N, \ldots, k\}$,

$$Im(\Theta_i^k) = Im(\Psi_i^k W^i)^{\perp} \tag{12.32}$$

Matrix Θ_i^k in (12.32) is used to filter out the unobservable component in the state estimate obtained from mode i after being transported at time t_k^-. Then, one can use the terms Θ_i^k computed from the last N modes in order to get the following matrix:

$$\Theta_k = \left[\Theta_k^k \cdots \Theta_{k-N}^k\right] \tag{12.33}$$

which has rank p from Eq. (12.21). At last, let us introduce the following vector:

$$\Omega_k = \begin{bmatrix} \Theta_k^{kT} \Psi_k^k Z^k \hat{z}^k(t_k^-) \\ \vdots \\ (\Theta_{k-N}^k)^T \left(\Psi_{k-N}^k Z^{k-N} \hat{z}^{k-N}(t_{k-N}^-) - \sum\limits_{l=k-N}^{k-1} \Psi_l^k \xi_l(t_l^-) \right) \end{bmatrix} \tag{12.34}$$

Therefore, using information on $[t_{k-N-1}, t_k)$, one gets the resetting term defined as:

$$\xi_k(t_k^-) = \begin{cases} (\Theta_k^T)^{\dagger} \Omega_k & \text{if} \quad k > N \\ 0 & \text{else} \end{cases} \tag{12.35}$$

with $(\Theta_k^T)^{\dagger} = (\Theta_k \Theta_k^T)^{-1} \Theta_k$.

Theorem 12.10 *Let us consider system* (12.7a–12.7c) *satisfying (12.21). Using the hybrid observer* (12.22a, 12.22b)*, the estimation error $\tilde{x}$ converges toward zero in finite time, i.e.,*

$$\tilde{x}(t) = 0, \quad t_{k-1} + T^* \le t \le t_k, \ k > N \tag{12.36}$$

Proof Proof is similar to [20] and is omitted.

12.5 Numerical Study

In this section, we describe an application of the controller and observer schemes for a three cells converter.

12.5.1 Simulation Results Using the Control Schemes

In order to illustrate the performance of the control scheme, a comparative study between a PWM feedback control and the proposed binary control is presented. Simulations using a three-cells multicellular converter associated to a RL load highlight the characteristics for both strategies. The sampling period is 5×10^{-6} s for all the simulations.

The dynamics of the three-cells converter, with a load consisting in a resistance and an inductance is given in Eq. (12.2) with $p = 3$. The parameters are set as $c_1 = 40 \times 10^{-6}$ F, $c_2 = 40 \times 10^{-6}$ F, $R = 200\,\Omega$ and $L = 1$ H. The voltage of the source is $E = 60$ V. According to Eq. (12.1), the reference voltages of the floating capacitors V_{c_1} and V_{c_2} are defined as:

$$V_{c_1,\mathrm{ref}} = 20\,\mathrm{V}, \quad V_{c_2,\mathrm{ref}} = 40\,\mathrm{V}$$

In order to test the proposed strategy through different reference trajectories, the desired load current is chosen as a piecewise continuous and satisfies the following physical condition:

$$0 < I_{\mathrm{ref}} < E/R$$

with $\frac{E}{R}$ =0.3 A. The initial state of the converter is: $V_{c_1}(0)$ =0 V, $V_{c_2}(0)$ =0 V and $I(0)$ =0 A.

In the simulation results, the PWM controller uses a period T_{MLI} =0.001 s and the gains of PI controllers are chosen in order to obtain the best performances. Figure 12.3a–b show, respectively, the load current I and the evolution of the internal voltages V_{c_1} and V_{c_2}. The stabilization of the errors on the load current and on the floating capacitor voltages is ensured. The PWM feedback controller produces an average regulation around the reference. When the reference load current changes, one can note a small transient state.

The simulation using the binary controller is realized under the same condition as the previous case. Figure 12.4 shows the load current I and the internal voltages V_{c_1} and V_{c_2}. The stabilization of the errors on the load current and on the floating capacitor voltages is ensured. One can see that the proposed controller allows the converter to follow different values of the reference load current. Indeed, the transition phases for the load current and floating capacitor voltages are short. The proposed controller also provides better performance than the PWM strategy in terms of steady-state error. Therefore, the binary controller manages to accomplish the control objective with higher performance.

Table 12.1 summarizes the obtained simulation results, in steady state, for the PWM feedback and the binary controller.

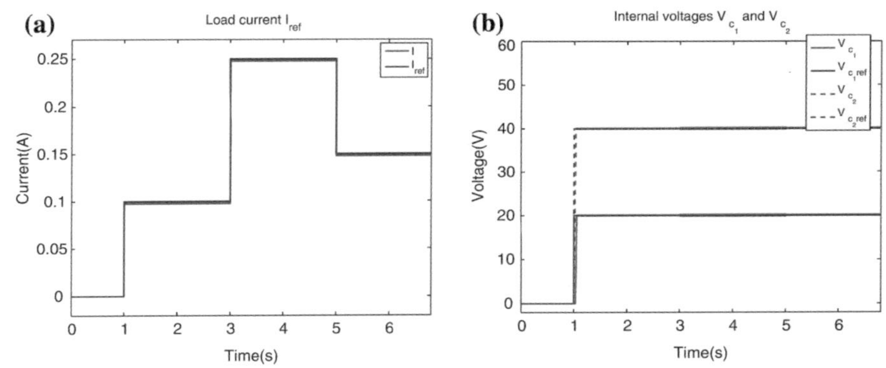

Fig. 12.3 a–b Simulation results using a PWM control strategy. **a** Load current I. **b** Internal voltages V_{c_1} and V_{c_2}

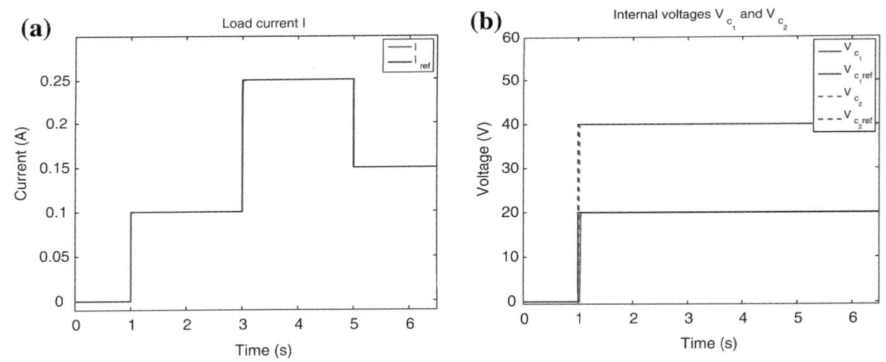

Fig. 12.4 a–b Simulation results using the proposed binary control. **a** Load current I. **b** Internal voltages V_{c_1} and V_{c_2}

Table 12.1 Comparison between the PWM and the binary control law

	PWM	Binary control
Error max on V_{c_1}	0.18 V	0.03 V
Error max on V_{c_2}	0.18 V	0.04 V
Error max on I	0.0026 A	0.00025 A

12.5.2 Simulation Results Using the Observer Scheme

Simulations using a three-cells multicellular converter associated to a RL load highlight the efficiency of the proposed hybrid observer. The dynamics of the three-cells converter, with a load consisting in a resistance and an inductance is given in Eq. (12.2) with $p = 3$. The parameters are set as E =1,500 (V), c_1 =40 (μF),

c_2 =40 (μF), $R = 10\,(\Omega)$ et L =0.0005 (H). It exists $2^3 = 8$ modes of working in the system. The matrices used in the hybrid representation (12.7a–12.7c) corresponding to the different modes are defined as follows:

$$A_1 = \begin{bmatrix} 0 & 0 & 0 \\ 0 & 0 & 0 \\ 0 & 0 & -2e4 \end{bmatrix}, \quad A_2 = \begin{bmatrix} 0 & 0 & -25e3 \\ 0 & 0 & 0 \\ 2e3 & 0 & -2e4 \end{bmatrix},$$

$$A_3 = \begin{bmatrix} 0 & 0 & 25e3 \\ 0 & 0 & -25e3 \\ -2e3 & 2e3 & -2e4 \end{bmatrix}, A_4 = \begin{bmatrix} 0 & 0 & 0 \\ 0 & 0 & -25e3 \\ 0 & 2e3 & -2e4 \end{bmatrix},$$

$$A_5 = \begin{bmatrix} 0 & 0 & 0 \\ 0 & 0 & 25e3 \\ 0 & -2e3 & -2e4 \end{bmatrix}, \quad A_6 = \begin{bmatrix} 0 & 0 & -25e3 \\ 0 & 0 & 25e3 \\ 2e3 & -2e3 & -2e4 \end{bmatrix},$$

$$A_7 = \begin{bmatrix} 0 & 0 & 25e3 \\ 0 & 0 & 0 \\ -2e3 & 0 & -2e4 \end{bmatrix}, \quad A_8 = \begin{bmatrix} 0 & 0 & 0 \\ 0 & 0 & 0 \\ 0 & 0 & -2e4 \end{bmatrix},$$

$$B_1 = \begin{bmatrix} 0 \\ 0 \\ 0 \end{bmatrix} B_2 = \begin{bmatrix} 0 \\ 0 \\ 0 \end{bmatrix} B_3 = \begin{bmatrix} 0 \\ 0 \\ 0 \end{bmatrix}, B_4 = \begin{bmatrix} 0 \\ 0 \\ 0 \end{bmatrix},$$

$$B_5 = \begin{bmatrix} 0 \\ 0 \\ 2e3 \end{bmatrix} B_6 = \begin{bmatrix} 0 \\ 0 \\ 2e3 \end{bmatrix} B_7 = \begin{bmatrix} 0 \\ 0 \\ 2e3 \end{bmatrix} B_8 = \begin{bmatrix} 0 \\ 0 \\ 2e3 \end{bmatrix}$$

The three cycles which can calculated by the PWM control are: "1, 2, 1, 3, 1, 5" or "2, 4, 3, 7, 5, 6" or "8, 4, 8, 6, 8, 7".

In order to analyse the observability of the system, we calculate the different matrices of observability:

$$G_1 = \begin{bmatrix} 0 & 0 & 1 \\ 0 & 0 & -2e4 \\ 0 & 0 & 4e8 \end{bmatrix}, \quad G_2 = \begin{bmatrix} 0 & 0 & 1 \\ 2e3 & 0 & -2e4 \\ -4e7 & 0 & 35e7 \end{bmatrix},$$

$$G_3 = \begin{bmatrix} 0 & 0 & 1 \\ -2e3 & 2e3 & -2e4 \\ 4e7 & -4e7 & 3e8 \end{bmatrix}, G_4 = \begin{bmatrix} 0 & 0 & 1 \\ 0 & 2e3 & -2e4 \\ 0 & -4e7 & 35e7 \end{bmatrix},$$

$$G_5 = \begin{bmatrix} 0 & 0 & 1 \\ 0 & -2e3 & -2e4 \\ 0 & 4e7 & 35e7 \end{bmatrix}, \quad G_6 = \begin{bmatrix} 0 & 0 & 1 \\ 2e3 & -2e3 & -2e4 \\ -4e7 & 4e7 & 3e8 \end{bmatrix},$$

$$G_7 = \begin{bmatrix} 0 & 0 & 1 \\ -2e3 & 0 & -2e4 \\ 4e7 & 0 & 35e7 \end{bmatrix}, \quad G_8 = \begin{bmatrix} 0 & 0 & 1 \\ 0 & 0 & -2e4 \\ 0 & 0 & 4e8 \end{bmatrix},$$

Remark 12.11 If we consider the three cycles of PWM control, we can note that the system is completely observable after three commutations.

We can determine for the first time the matrices noted Z^k associated to the orthonormal basis $Im(G_k^T)$:

$$\begin{array}{ll} Z^1 = \begin{bmatrix} 0 \\ 0 \\ -1 \end{bmatrix} & Z^2 = \begin{bmatrix} 0.11 & 0.99 \\ 0 & 0 \\ -0.99 & 0.11 \end{bmatrix} \\ Z^3 = \begin{bmatrix} -0.13 & 0.69 \\ 0.13 & -0.69 \\ -0.98 & -0.19 \end{bmatrix} & Z^4 = \begin{bmatrix} 0 & 0 \\ 0.11 & 0.99 \\ -0.99 & 0.11 \end{bmatrix} \\ Z^5 = \begin{bmatrix} 0 & 0 \\ -0.11 & 0.99 \\ -0.99 & -0.11 \end{bmatrix} & Z^6 = \begin{bmatrix} 0.13 & 0.69 \\ -0.13 & -0.69 \\ -0.98 & 0.18 \end{bmatrix} \\ Z^7 = \begin{bmatrix} -0.11 & -0.99 \\ 0 & 0 \\ -0.99 & 0.11 \end{bmatrix} & Z^8 = \begin{bmatrix} 0 \\ 0 \\ -1 \end{bmatrix} \end{array} \tag{12.37}$$

We can also determine the matrices W^k associated to the orthonormal basis of $ker(G_k^T)$.

Note that the modes "1" and "8" have a single element observable and the others modes have two elements observable. Therefore, we can conclude that the system is observable if $N = 3$. In other words, it is necessary that the observer uses three commutations to reconstruct all of the capacitor voltages of the converter. The Eq. (12.23) allows to calculate S_k and R_k for each modes ($k = 1, \ldots, 8$). The hybrid observer (12.22a, 12.22b) is used as follows:

Initialisation :

$Z^k, W^k, R^k, S^k, N = 1$ (12.37), (12.23)

$\hat{x}, \hat{z}, \tau, k$

Main program

Loop

Execution of the observer (12.22a, 12.22b) on the system (12.2):

Compute $\hat{x}$ (12.22a, 12.22b)

Compute $\hat{z}$ (12.28a–12.28c)

If Changing of mode **and** $t > \tau N$ **Then**
$\xi \leftarrow$ Compute_Xi
Update $\hat{x}(t_k)$ (12.22b), $\hat{z}^k(t_k) = 0$
End if
End loop
End main program

Function Compute_Xi
For $i = 1$ **to** N
Compute Ψ^k_{k-i} (12.31)
End for
For $i = 1$ **to** $N + 1$
Compute Θ^k_i (12.32)
End for
Compute Θ_k (12.33)
Compute Ω_k (12.34)
Compute $\xi_k(t_k^-) = (\Theta_k^T)^\dagger \Omega_k$ (12.35);
Return $\xi_k(t_k^-)$
End function

In order to illustrate the performance of the proposed observer, some MATLAB simulations are presented in this subsection. The PWM control is defined to ensure the distribution of the capacitor voltages around the reference values $V_{c1,\text{ref}} = E/3 = 500\,(\text{V})$, $V_{c2,\text{ref}} = 2E/3 = 1000\,(\text{V})$ and the reference of the current load is such that $I_{\text{ref}} = 60\,(\text{A})$ and changes at $t = 0.02\,(\text{s})$ with $I_{\text{ref}} = 30(A)$. For the simulations, the control PWM uses a sampling period of $T_{\text{MLI}} = 1\,(\text{ms})$. The Figs. 12.5 and 12.7 (Fig. 12.6 (ZOOM)) show the evolution of the internals voltages (V_{c1}, V_{c2}), the current load and their estimated. The Fig. 12.8 presents the estimation error for the both internal voltages.

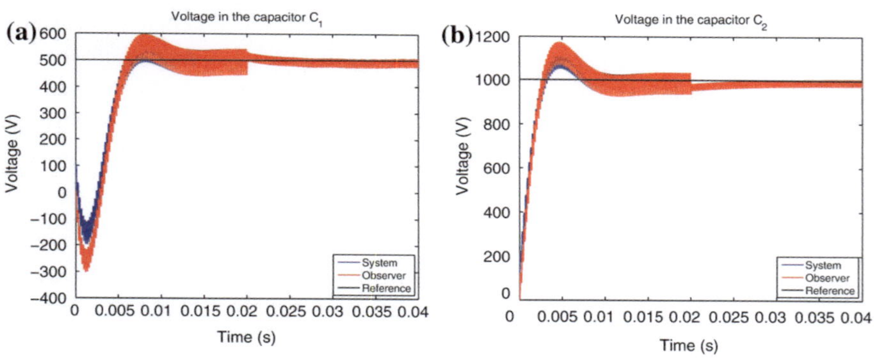

Fig. 12.5 **a–b** Simulation results using the proposed observer. **a** Evolution of the voltage V_{c1} (in *blue*), its estimation (in *red*) and the reference (in *black*) **b** Evolution of the voltage V_{c2} (in *blue*), its estimation (in *red*) and the reference (in *black*)

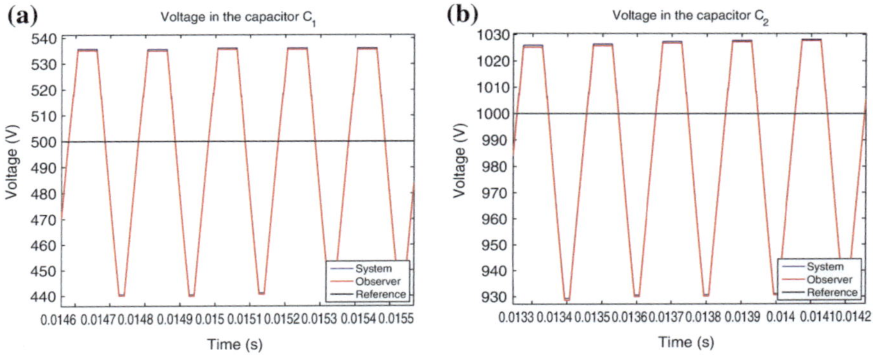

Fig. 12.6 a–b Simulation results using the proposed observer. **a** Evolution of the voltage V_{c1} (in *blue*), its estimation (in *red*) and the reference (in *black*) **b** Evolution of the voltage V_{c2} (in *blue*), its estimation (in *red*) and the reference (in *black*) (ZOOM)

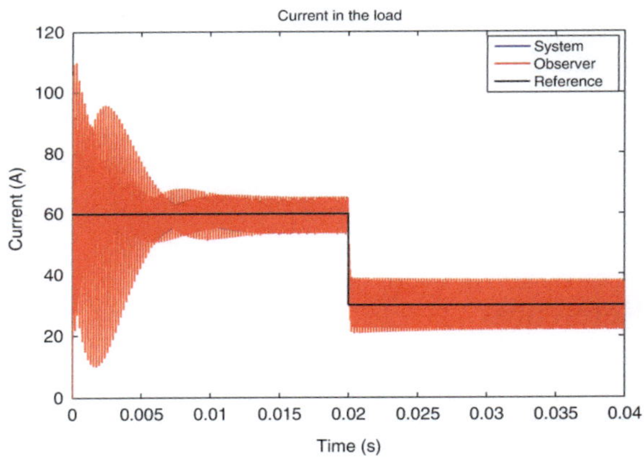

Fig. 12.7 Evolution of the current I (in *blue*), its estimation (in *red*) and the reference (in *black*)

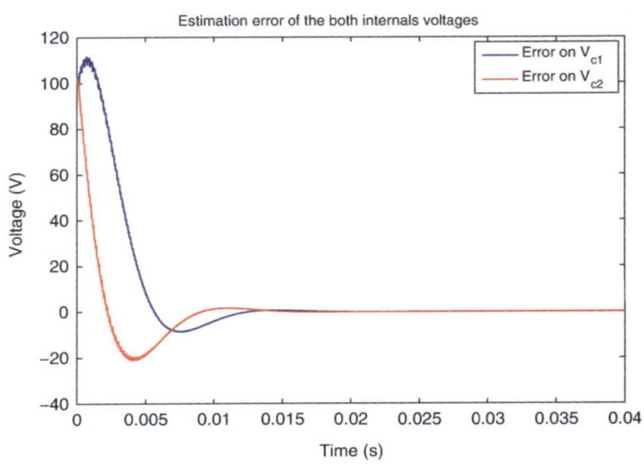

Fig. 12.8 Evolution of the estimation errors (voltages of the capacitors)

We can note that the estimation state converges and follows toward the state of the system. The evolution of the estimation errors show the efficiency of the hybrid observer. When the current desired changes, the estimation keeps to the objective.

12.6 Conclusion

The first part of this chapter was devoted to the controller design for switching power converters, which are a particular class of switched systems. Two controllers are introduced. The first one is based on the average model and exhibits good performances in the transient. The second one is based on a Lyapunov function to jointly control the load current and the capacitor voltages across each commutation cell while improving the steady state behavior. In the second part of this chapter, an observer has been designed to solve the capacitor voltages estimation while taking into account the hybrid behavior of the converter. Some simulations highlight the efficiency of the proposed control and observer schemes for the three cells converter.

Acknowledgments The research leading to these results has received funding from International Campus on Safety and Intermodality in Transportation, the Nord-Pas-de-Calais Region, the Regional Delegation for Research and Technology, the Ministry of Higher Education and Research, and the National Center for Scientific Research.

References

1. Erickson, R., Maksimovic, D.: Fundamentals of Power Electronics, 2nd edn. Kluwer Academic Publishers, Dordrecht (2001)
2. Meynard, T., Foch, H.: French patent N91,09582 du 25 juillet 1991, dépôt international PCT (Europe, Japon, USA, Canada), N92,00652 du 8 juillet 1992 (1991)
3. Meynard, T., Foch, H., Thomas, P., Courault, J., Jakob, R., Nahrstaedt, M.: Multi-cell converters: basic concepts and industry applications. IEEE Trans. Ind. Appl. **49**(5), 955–964 (2002)
4. Beaudesson, P.: Sûreté de fonctionnement, reconfiguration et marches dégradées des convertisseurs multiniveaux á IGBT, Ph.D. Thesis, INP Toulouse (2000)
5. Bethoux, O., Barbot, J.P.: Multi-cell chopper direct control law preserving optimal limit cycles. Proc. CCA Glasgow **2**, 1258–1263 (2002)
6. Djemaï, M., Busawon, K., Benmansour, K., Marouf, A.: High order sliding mode control of a DC motor via a switched controlled multi cellular converter. Int. J. Syst. Sci. **42**(11), 1869–1882 (2011)
7. Amjadi, Z., Williamson, S.S.: A novel control technique for a switched-capacitor-converter-based hybrid electric vehicle energy storage system. IEEE Trans. Ind. Electron. **57**(3), 926–934 (2010)
8. Lezana, P., Rodriguez, J., Perez, M.A., Espinoza, J.: Input current harmonics in a regenerative multicell inverter with single-phase PWM rectifiers. IEEE Trans. Ind. Electron. **56**(2), 408–417 (2009)
9. Bhagwat, P.M., Stefanovic, V.R.: Generalized structure of a multilevel PWM inverter. IEEE Trans. Ind. Appl. **32**, 509–517 (1996)
10. Silva, E.I., McGrath, B., Quevedo, D., Goodwin, G.: Predictivecontrol of a flying capacitor converter. In: IEEE American Control Conference, New York, USA (2007)

11. Defoort, M., Djemai, M., Van Gorp, J.: Sliding Mode Control of Switching Electrical Power Converters Associated to an Inductive Load. IFAC World Congress, Milan (2011)
12. Baja, M., Patino, D., Cormerais, H., Riedinger, P., Buisson, J.: Hybrid control of a three-level three-cell dc-dc converter. In: American Control Conference, pp. 5458–5463 (2007)
13. Dominguez-Garcia, A.D., Trenn, S.: Detection of impulsive effects in switched daes with applications to power electronics reliability analysis. In: IEEE Conference on Decision and Control, Atlanta, USA (2010)
14. Bejarano, F.J., Ghanes, M., Barbot, J.P.: Observability and observer design for hybrid multicell choppers. Int. J. Control **83**(3), 617–632 (2010)
15. Fliess, M., Join, C., Perruquetti, W.: Real-time estimation for switched linear systems. In: IEEE Conference on Decision and Control, Cancun, Mexico (2008)
16. Benmansour, K., De Leon, J., Djemai, M.: Adaptive observer for multi-cell chopper. In: International Symposium on Communications, Control and Signal Processing (2006)
17. Barbot, J.P., Saadaoui, H., Djemaï, M., Manamanni, N.: Nonlinear observer for autonomous switching systems with jumps. Nonlinear Analysis: Hybrid Systems, pp. 537–547 (2007)
18. Defoort, M., Djemaï, M., Floquet, T., Perruquetti, W.: Robust finite time observer design for multicellular converters. Int. J. Syst. Sci. **42**(11), 1859–1868 (2011)
19. Bejarano, F.J., Fridman, L.: State exact reconstruction for switched linear systems via a super-twisting algorithm. Int. J. Syst. Sci. **42**(5), 717–724 (2011)
20. Tanwani, A., Shim, H., Liberzon, D.: Observability of switched linear systems: characterization and observer design. IEEE Trans. Autom. Control **58**(4), 891–904 (2013)
21. Vu, L., Liberzon, D.: Invertibility of switched linear systems. Automatica **44**(4), 949–958 (2008)
22. Tanwani, A., Liberzon, D.: Invertibility of switched nonlinear systems. Automatica **46**(12), 1962–1973 (2010)
23. Kang, W., Barbot, J., Xu, L.: On the Observability of Nonlinear and Switched Systems. Lecture notes in control and information sciences. Springer, Berlin (2009)

Chapter 13
Application of Hybrid Dynamical Theory to the Cardiovascular System

Taous Meriem Laleg-Kirati, Zehor Belkhatir and Fernando Diaz Ledezma

Abstract In hybrid dynamical systems, the state evolves in continuous time as well as in discrete modes activated by internal conditions or by external events. In the recent years, hybrid systems modeling has been used to represent the dynamics of biological systems. In such systems, discrete behaviors might originate from unexpected changes in normal performance, e.g., a transition from a healthy to an abnormal condition. Simplifications, model assumptions, and/or modeled (and ignored) nonlinearities can be represented by sudden changes in the state. Modeling cardiovascular system (CVS), one of the most fascinating but most complex human physiological systems, with a hybrid approach, is the focus of this chapter. The hybrid property appears naturally in the CVS thanks to the presence of valves which, depending on their state (closed or open), divide the cardiac cycle into four phases. This chapter shows how hybrid models can be used for modeling the CVS. In addition, it describes a preliminary study on the detection of some cardiac anomalies based on the hybrid model and using the standard observer-based approach.

13.1 Introduction

The dynamic evolution of hybrid systems occurs over continuous time, but also includes discrete instants subject to some conditions [1].

The use of a hybrid approach is suitable in modeling biological systems as such an approach can provide the means to model behaviors with good approximations and simplifications. For instance, systems with multiscale dynamics can be further

T.M. Laleg-Kirati (✉) · Z. Belkhatir · F.D. Ledezma
Computer, Electrical and Mathematical Science and Engineering (CEMSE) Division,
King Abduallah University of Sciences and Technology (KAUST),
Thuwal, Kingdom of Saudi Arabia
e-mail: taousmeriem.laleg@kaust.edu.sa

Z. Belkhatir
e-mail: zehor.belkhatir@kaust.edu.sa

F.D. Ledezma
e-mail: fernando.diaz@kaust.edu.sa

M. Djemai and M. Defoort (eds.), *Hybrid Dynamical Systems*, Lecture Notes in Control and Information Sciences 457, DOI 10.1007/978-3-319-10795-0_13

simplified by introducing constant piecewise approximations of the slow changing variables [2]. In many biological models, the continuous part is described by differential equations. However, the dynamics of biological systems often include nonlinear functions and/or sudden changes in behavior, corresponding to changes in one or many variables of the model.

Recurrent nonlinearities in biological systems include sigmoidal functions that are used in continuous models. These functions, depending on how steep they are, can be approximated by discrete transitions between piecewise-linear regions. For example, a very steep sigmoidal curve could be replaced by a step function. Piecewise constant or piecewise linear approximations are derived in hybrid models using logic-based discrete transitions. For instance, in gene regulation modeling, steep sigmoidal functions are approximated assuming that a particular gene can be in either *on* or *off* state.

Discrete behaviors in biological systems originates from processes that can be treated as naturally discrete phenomena. That is, the physical laws involved in the derivation of the continuous state equations of the model do not apply to all possible operating conditions. For instance, when dealing with molecules, the law of mass action holds only for sets of large, homogeneously mixed molecules. However, this is not always the case because there are dynamics that result from the action of a few molecules within a cell. Discrete transitions are more suitable for describing the dynamics of small numbers of molecules. Both possibilities suggest that hybrid models could be used. An example can be found in the growth of cancerous cells. Because cancer starts with the mutation of a single cell, the initial stage of tumor growth is described differently than when the tumor has reached a certain size. Hybrid modeling allows us to describe the cancerous cells as discrete entities in some parts of the tumor and continuous entities in others [3].

In this chapter, we take advantage of the presence of valves in the human heart to discuss hybrid modeling of the cardiovascular system. The model is adapted to include some anomalies, then we seek to detect these anomalies using a classical observer fault detection approach.

13.2 Cardiovascular System Physiology

The human cardiovascular system (CVS) consists of the heart, which acts as a pump by pushing blood throughout the circulatory system, and arteries and veins, which distribute the blood to all the organs and tissues of the body (see Fig.13.1).

The heart is divided into four chambers: the right atrium, the right ventricle, the left atrium, and the left ventricle. Deoxygenated blood, which returns to the heart from the venous system, enters the right atrium in which it accumulates until it is pumped into the right ventricle. The ventricle contracts and pushes the blood through the pulmonary artery into the lungs where it is loaded with oxygen. Then, blood enters the left atrium through the pulmonary veins. The left atrium contracts and sends the blood to the left ventricle. Finally, the left ventricle contracts to pump

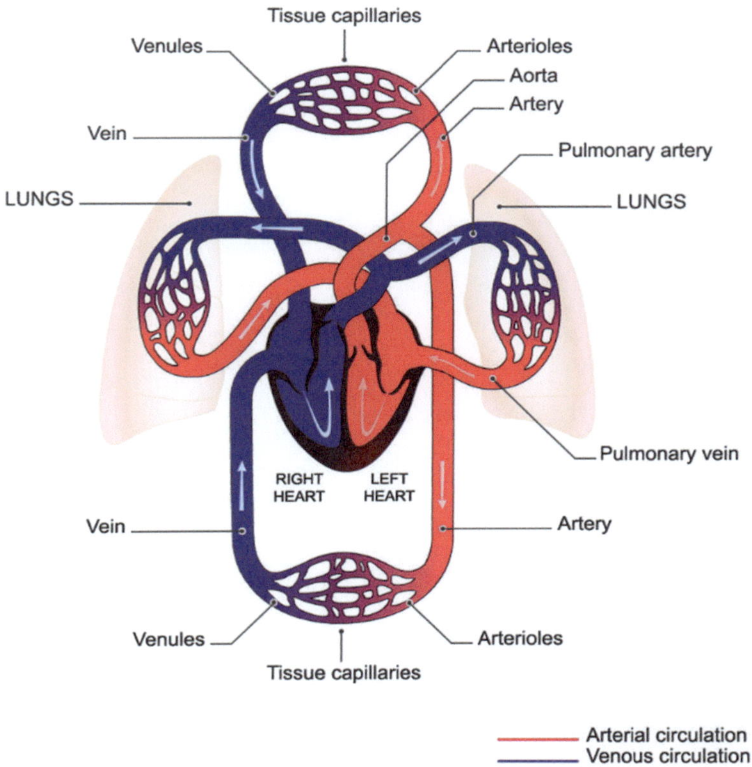

Fig. 13.1 Physiological components of the cardiovascular system

the oxygenated blood to the aortic artery, which in turn transfers the blood to the organs through smaller arteries and capillaries.

At the boundary of the heart's four chambers are four valves that control the transfer of blood and allow the blood to flow in a unique direction. These four valves, depicted in Fig. 13.2, are listed as follows:

- Tricuspid valve, divides the right atrium and right ventricle.
- Pulmonary valve, located between the right ventricle and the pulmonary artery.
- Mitral valve, controls the flow from the left atrium to the left ventricle.
- Aortic valve, is located between the left ventricle and the aorta.

These four valves transit from an open state to a closed state based on the gradient of pressure between left/right atrium and left/right ventricle for the tricuspid and mitral valves, and based on the gradient of pressure between the left/right ventricle and the aorta/pulmonary artery for pulmonary and aortic valves.

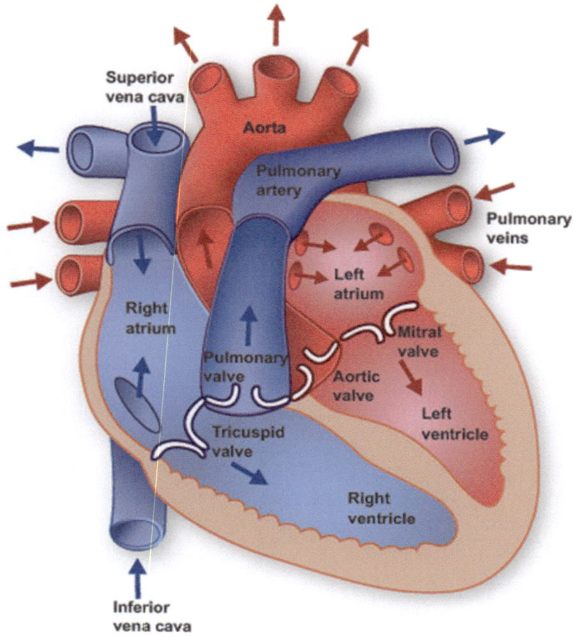

Fig. 13.2 Valves in the heart

13.3 A Hybrid Model of the Cardiovascular System

Two main types of models of the CVS have been proposed. They are lumped parameter models and distributed models.

- Lumped parameter models are based on an analogy with electrical circuits in which blood pressure and blood flow are described as voltage and current, respectively. In such models, resistors (R) represent the resistance of the arteries to the blood flow. Capacitors (C) describe the elastic (or the compliance) properties of the arterial walls. Inductors (L) account for the inertia of the blood mass. These models are simple to use and describe the behavior of CVS well [4, 5].
- Distributed models use fluid dynamics principles. They have the advantage of describing spatial and temporal properties. They are, therefore, more accurate than are lumped parameter models. However, they are, in general, complex, which limits their utilization in clinical practice [6, 7].

The lumped parameter model shown in Fig.13.3 has been proposed and studied in [8–10]. Pulmonary circulation is represented by the capacitor, C_{r}, the mitral valve is represented by the resistor, R_{m}, in series with ideal diode, D_{m}, the aortic valve is represented by the resistor, R_{a}, in series with another ideal diode, D_{a}, the aortic compliance is represented by C_{a}; the systemic circulation is given by the so-called

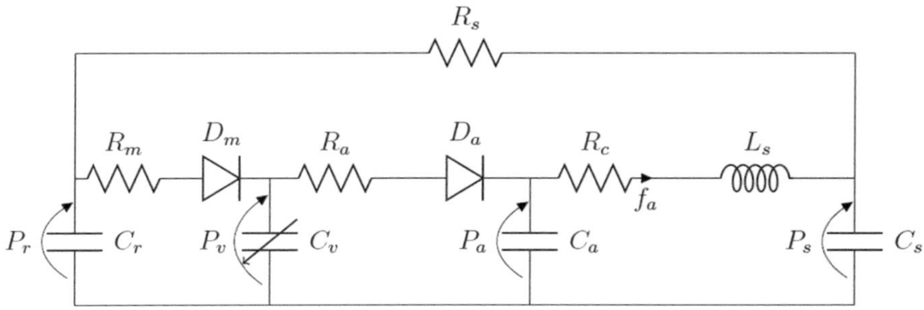

Fig. 13.3 A lumped parameter CVS model

Table 13.1 Physiological meaning of the model's components where resistances are in (mmHg.s /ml), compliances are in (ml/mmHg) and inductor is in (mmHg.s^2/ml)

Components	Physiological meaning
D_{m} and R_{m}	Mitral valve
D_{a} and R_{a}	Aortic valve
R_{c}	Characteristic resistance
R_{s}	Systemic resistance
$C(t)$	Left ventricular compliance
C_{a}	Aortic compliance
C_{s}	Systemic compliance
C_{r}	Pulmonary compliance
L_{s}	Inertia of blood in the aorta

four-element Windkessel model [11] consisting of R_{c}, L_{S}, C_{s} and R_{s} and the behavior of the left ventricle is modeled by means of time-varying compliance, $C_{\mathrm{v}}(t)$.

The physiological meaning of the model's components and variables is given in Tables 13.1 and 13.2.

The model of the CVS described above has continuous variables (pressure, flow) and discrete variables (the state of the diodes). It is a hybrid dynamical system. For the discrete variables, in the ideal case, each diode takes the value of 0 or 1, describing closed and open states of the valve, respectively. As the circuit has two

Table 13.2 Physiological meaning of cardiovascular variables, where pressures are in (mmHg) and flow is in (ml/s)

Variables	Physiological meaning
P_{v}	Ventricular pressure
P_{r}	Atrial pressure
P_{s}	Systemic pressure
P_{a}	Aortic pressure
f_{a}	Aortic flow

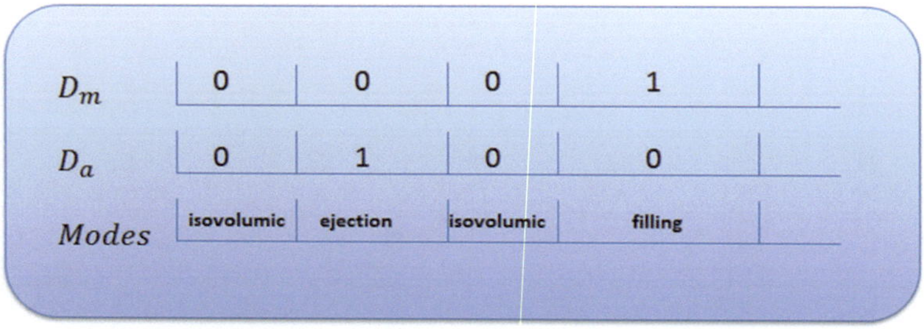

Fig. 13.4 State of the valves in the cardiac phases

valves, it results in four operating modes. However, physiologically, the mode in which the two diodes are passed simultaneously is impossible, which reduces the actual operating modes to three [12].

Figure 13.4 shows different possible modes of the model in terms of the state of the valves. When both mitral and aortic valves are closed, the mode is called isovolumic; this means that the volume of blood in the ventricle is constant, while it becomes full in the filling phase and empty in the ejection phase. The transition from one phase to another is purely physiological, and it appears according to the state of the valves.

The discrete modes are noted as $Q = (q_1, q_2, q_3)$ and each mode is defined on the space $X_{q_i} = \mathbb{R}^5$.

The form of the continuous dynamics in each cardiac phase is

$$\dot{X}(t) = A_i(t)X(t), \tag{13.1}$$

where X is the state vector given by:

$$X(t) = [x_1, x_2, x_3, x_4, x_5] = [P_\text{v}, P_\text{r}, P_\text{s}, P_\text{a}, f_\text{a}]. \tag{13.2}$$

The sets of transition between the continuous modes are defined as

$$E = (q_i, q_j), \forall i \neq j, \quad \text{for} \quad i, j = \overline{1, 3}. \tag{13.3}$$

Figure 13.5 illustrates the hybrid automata describing the CVS.
The conditions of transition, T_{ij}, are given by

$$\begin{cases} T_{12} = \{X \in \mathbb{R}^5 \backslash D_\text{a} = 0, D_\text{m} = 1\}, \\ T_{13} = \{X \in \mathbb{R}^5 \backslash D_\text{a} = 1, D_\text{m} = 0\}, \\ T_{21} = T_{31} = \{X \in \mathbb{R}^5 \backslash D_\text{m} = 0, D_\text{a} = 0\}. \end{cases} \tag{13.4}$$

The dynamic matrix, $A_i(t)$, in each cardiac phase is given by

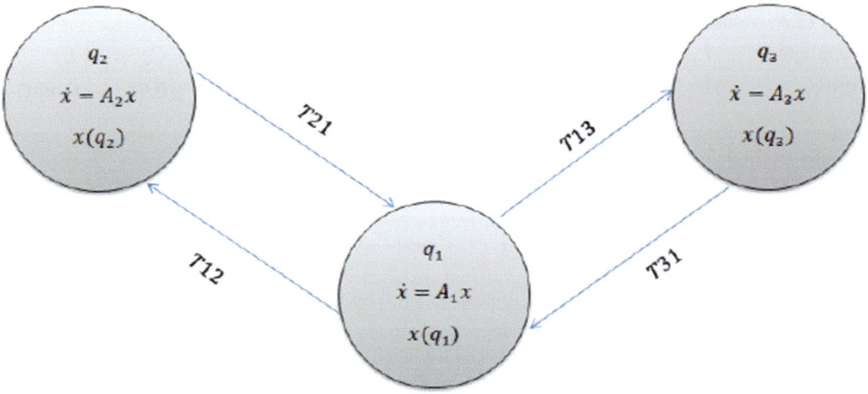

Fig. 13.5 Hybrid automata of the cardiovascular system

- Isovolumic phase:

$$A_1(t) = \begin{pmatrix} \frac{\dot{E}(t)}{E(t)} & 0 & 0 & 0 & 0 \\ 0 & -\frac{1}{C_r R_s} & \frac{1}{C_r R_s} & 0 & 0 \\ 0 & \frac{1}{R_s C_s} & -\frac{1}{R_s C_s} & 0 & \frac{1}{C_s} \\ 0 & 0 & 0 & 0 & -\frac{1}{C_a} \\ 0 & 0 & -\frac{1}{L_s} & \frac{1}{L_s} & -\frac{R_c}{L_s} \end{pmatrix}. \tag{13.5}$$

- Ejection phase:

$$A_2(t) = \begin{pmatrix} \frac{\dot{E}(t)}{E(t)} - \frac{E(t)}{R_a} & 0 & 0 & \frac{E}{R_a} & 0 \\ 0 & -\frac{1}{C_r R_s} & \frac{1}{C_r R_s} & 0 & 0 \\ 0 & \frac{1}{R_s C_s} & -\frac{1}{R_s C_s} & 0 & \frac{1}{C_s} \\ \frac{1}{R_a C_a} & 0 & 0 & -\frac{1}{R_a C_a} & -\frac{1}{C_a} \\ 0 & 0 & -\frac{1}{L_s} & \frac{1}{L_s} & -\frac{R_c}{L_s} \end{pmatrix}. \tag{13.6}$$

- Filling phase:

$$A_3(t) = \begin{pmatrix} \frac{\dot{E}(t)}{E(t)} - \frac{E(t)}{R_m} & \frac{E(t)}{R_m} & 0 & 0 & 0 \\ \frac{1}{C_r R_m} & -\frac{1}{C_r R_s} - \frac{1}{C_r R_m} & \frac{1}{C_r R_s} & 0 & 0 \\ 0 & \frac{1}{R_s C_s} & -\frac{1}{R_s C_s} & 0 & \frac{1}{C_s} \\ 0 & 0 & 0 & 0 & -\frac{1}{C_a} \\ 0 & 0 & -\frac{1}{L_s} & \frac{1}{L_s} & -\frac{R_c}{L_s} \end{pmatrix}. \tag{13.7}$$

where $E(t)$ is the varying time ventricle elastance given in [8] as follows

Table 13.3 Values of the parameters extracted from [8]

Parameters	R_s	R_m	R_a	R_c	C_r	C_s	C_a	L_s
Values	1	0.005	0.001	0.0398	4.4	1.33	0.08	0.0005

$$E(t) = (E_{\max} - E_{\min})E_n(t_n) + E_{\min}, \tag{13.8}$$

where $E_n(t_n)$ is the *double hill* function [8] defined as follows

$$E(t_n) = 1.55\left(\frac{\left(\frac{t_n}{0.7}\right)^{1.9}}{1+\left(\frac{t_n}{0.7}\right)^{1.9}}\right)\left(\frac{1}{1+\left(\frac{t_n}{1.17}\right)^{21.9}}\right), \tag{13.9}$$

and, $$\dot{E}(t_n) = 1.55\left[\left(\frac{2.714\left(\frac{t_n}{0.7}\right)^{2.8}}{\left(1+\left(\frac{t_n}{0.7}\right)^{1.9}\right)^2}\right)\left(\frac{1}{1+\left(\frac{t_n}{1.17}\right)^{21.9}}\right)\right] - 1.55\left[\left(\frac{\left(\frac{t_n}{0.7}\right)^{1.9}}{1+\left(\frac{t_n}{0.7}\right)^{1.9}}\right)\left(\frac{18.718\left(\frac{t_n}{1.17}\right)^{20.9}}{\left(1+\left(\frac{t_n}{1.17}\right)^{21.9}\right)^2}\right)\right],$$

with $t_n = \frac{t}{T_{\max}}$, $T_{\max} = 0.2+0.15t_c$ and t_c is the cardiac cycle interval ($t_c = 60/HR$), where HR is the heart rate. $E_{\max}$ and $E_{\min}$ are the end-systolic pressure volume relationship (ESPVR) and the end-diastolic pressure volume relationship (EDPVR), respectively.

The numerical implementation of the hybrid model of the CVS has been done on MATLAB using the parameters given in Table 13.3.

Figure 13.6 shows the results of the simulation. Note that the values are among those considered to be normal [8].

13.4 Detection and Isolation of Cardiovascular Anomalies

The parameters of the CVS model presented in Table 13.1 can be affected by the occurrence of some diseases, and can vary from their nominal values when an anomaly occurs [13]. Therefore, detecting the variations of these parameters can be thought of as a detection of the occurrence of an anomaly. The variations could be related to cardiovascular anomalies. For instance, consider these common cardiovascular diseases, atherosclerosis, aneurysm, high blood pressure. The first two refer, correspondingly, to the hardening of the arteries and the dilatation of the arterial walls. Both could be associated with a change in the systemic compliance, C_s, or a vari-

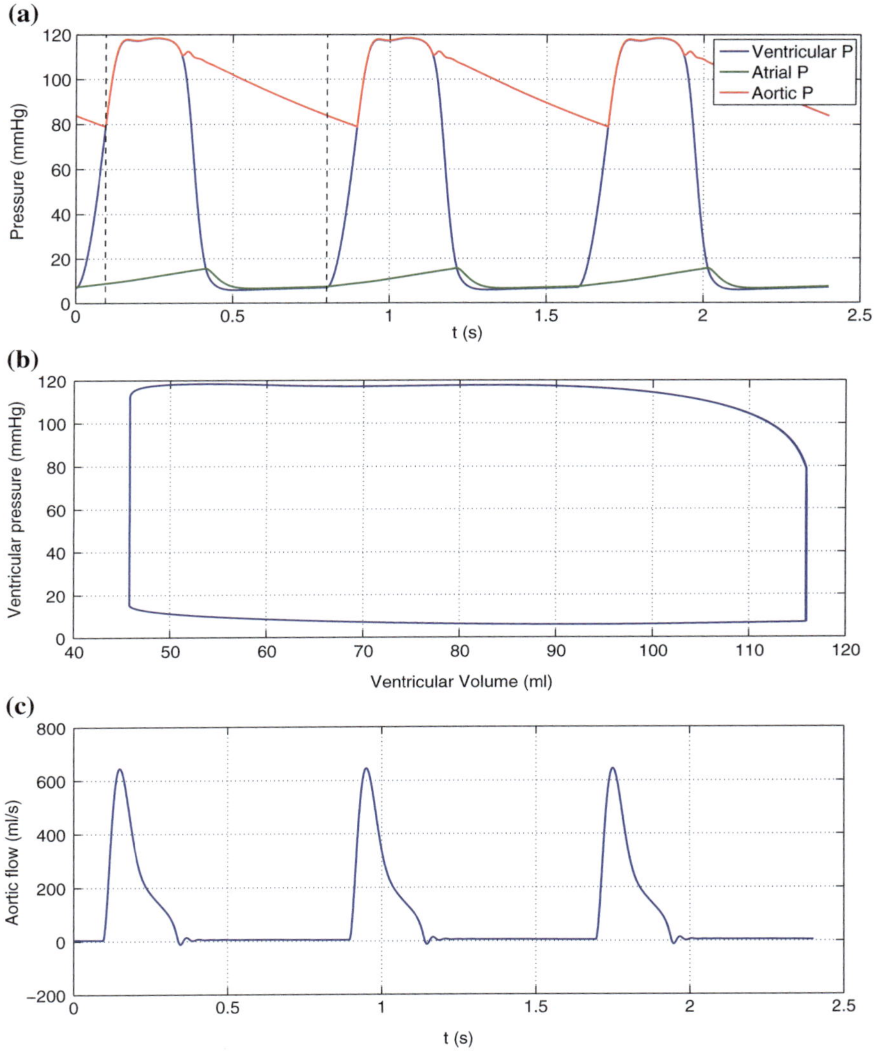

Fig. 13.6 Hemodynamic waveforms of the CVS. **a** Aortic, ventricular and atrial pressure. **b** Left ventricle pressure/volume loop. **c** Aortic flow

ation in the systemic vascular resistance, R_s. High blood pressure is a condition in which the pressure is abnormally high. The source of this condition could be related to deviations in the normal value of aortic compliance, C_a. It is worth noting that the variations in these parameters can also be due to physiological changes. Despite the importance of distinguishing between pathological variations and physiological variations, in this section, we focus on the detection of variations of a selected set of parameters (characteristic resistance, systemic resistance, aortic compliance and systemic compliance) from their nominal values without identifying whether the changes are physiological or pathological.

13.4.1 Output Vector

We first point out that not all states can be measured noninvasively. For instance, the three first state variables require invasive methods to be measured. However, aortic pressure and flow can be measured noninvasively using, for example, N-point moving average method or SphygmoCor or BPro device for aortic pressure measurements [14] and esophageal doppler, ultrasound monitoring, and impedance cardiography for aortic flow measurement [15]. Considering the above information, we assume that the aortic pressure (x_4) and the aortic flow (x_5) can be measured. The output equation for the system can be written as follows

$$y(t) = CX(t), \quad C = \begin{bmatrix} 0 & 0 & 0 & 1 & 0 \\ 0 & 0 & 0 & 0 & 1 \end{bmatrix}. \tag{13.10}$$

13.4.2 Detection of Changes in the Parameters

The model of Eq. (13.1) is adapted to include the anomalies related to variation of the parameters. Let us define a fault vector containing the difference, with respect to the nominal values, of the considered parameters, and which is given by the following equation

$$f_p = \left[\Delta R_\mathrm{c}\ \Delta C_\mathrm{a}\ \Delta R_\mathrm{s}\ \Delta C_\mathrm{s}\right]. \tag{13.11}$$

Let us also define θ as the vector containing the parameters subject to changes (i.e., $\theta = [R_\mathrm{c}, C_\mathrm{a}, R_\mathrm{s}, C_\mathrm{s}]^T$). Under these notations and taking the first-order Taylor expansion of Eq. (13.1), small variations in the vector of parameters, θ are considered in the hybrid model of CVS as follows

$$\dot{X}(t) = A_i(t)X(t) + E_{f,i}(t)f_p(t), \tag{13.12}$$

where $E_{f,i}(t)$ is the Jacobian matrix with respect to the considered parameters, which is given by

$$E_{f,i}(t) = \left.\frac{\mathrm{d}A_i(t)X(t)}{\mathrm{d}\theta}\right|_{\theta_N}, \tag{13.13}$$

where θ_N is the vector containing the nominal values of the parameters. The matrix $E_{f,i}(t)$, relating the effects of the anomalies (f_p) on the state, is given in each phase of the cardiac cycle as follows

- Isovolumic phase

$$E_{f,1}(t,X) = \begin{pmatrix} 0 & 0 & 0 & 0 \\ 0 & 0 & \frac{x_2}{C_r {R_s}^2} - \frac{x_3}{C_r {R_r}^2} & 0 \\ 0 & 0 & \frac{-x_2}{C_s {R_s}^2} + \frac{x_3}{C_s {R_s}^2} & \frac{-x_2}{{C_s}^2 R_s} + \frac{x_3}{{C_s}^2 R_s} - \frac{x_5}{{C_s}^2} \\ 0 & \frac{x_5}{C_a^2} & 0 & 0 \\ \frac{-x_5}{L_s} & 0 & 0 & 0 \end{pmatrix}. \tag{13.14}$$

- Ejection phase

$$E_{f,2}(t,X) = \begin{pmatrix} 0 & 0 & 0 & 0 \\ 0 & 0 & \frac{x_2}{C_r {R_s}^2} - \frac{x_3}{C_r {R_s}^2} & 0 \\ 0 & 0 & \frac{-x_2}{C_s {R_s}^2} + \frac{x_3}{C_s {R_s}^2} & \frac{-x_2}{{C_s}^2 R_s} + \frac{x_3}{{C_s}^2 R_s} - \frac{x_5}{{C_s}^2} \\ 0 & \frac{-x_1}{C_a^2 R_a} + \frac{x_4}{C_a^2 R_a} + \frac{x_5}{C_a^2} & 0 & 0 \\ \frac{-x_5}{L_s} & 0 & 0 & 0 \end{pmatrix}. \tag{13.15}$$

- Filling phase

$$E_{f,3}(t,X) = \begin{pmatrix} 0 & 0 & 0 & 0 \\ 0 & 0 & \frac{x_2}{C_r R_s^2} - \frac{x_3}{C_r {R_s}^2} & 0 \\ 0 & 0 & \frac{-x_2}{C_s {R_s}^2} + \frac{x_3}{C_s {R_s}^2} & \frac{-x_2}{{C_s}^2 R_s} + \frac{x_3}{{C_s}^2 R_s} - \frac{x_5}{{C_s}^2} \\ 0 & \frac{x_5}{C_a^2} & 0 & 0 \\ \frac{-x_5}{L_s} & 0 & 0 & 0 \end{pmatrix}. \tag{13.16}$$

We propose to detect anomalies related to variations in the selected parameters (f_p) using an extended kalman filter (EKF). To do so, the state vector, $X(t)$, is augmented with the vector, $f_p(t)$, such that

$$\dot{\tilde{X}} = \begin{bmatrix} \dot{X}(t) \\ \dot{f}_p(t) \end{bmatrix} = \begin{bmatrix} A_i(t) & E_{f,i}(t,X) \\ 0 & 0 \end{bmatrix} \begin{bmatrix} X(t) \\ f_p(t) \end{bmatrix}, \tag{13.17}$$

and

$$y(t) = \begin{bmatrix} C & 0 \end{bmatrix} \begin{bmatrix} X(t) \\ f(t) \end{bmatrix}, \tag{13.18}$$

The dynamic of the augmented state estimate vector is given by

$$\dot{\hat{\tilde{X}}}(t) = \tilde{A}_i(\hat{X},t)\hat{\tilde{X}}(t) + K_i(t)[y(t) - \tilde{C}\hat{\tilde{X}}(t)]. \tag{13.19}$$

The gain of EKF is

$$\dot{K}_i(t) = P_i(t)\tilde{C}^T R^{-1}, \tag{13.20}$$

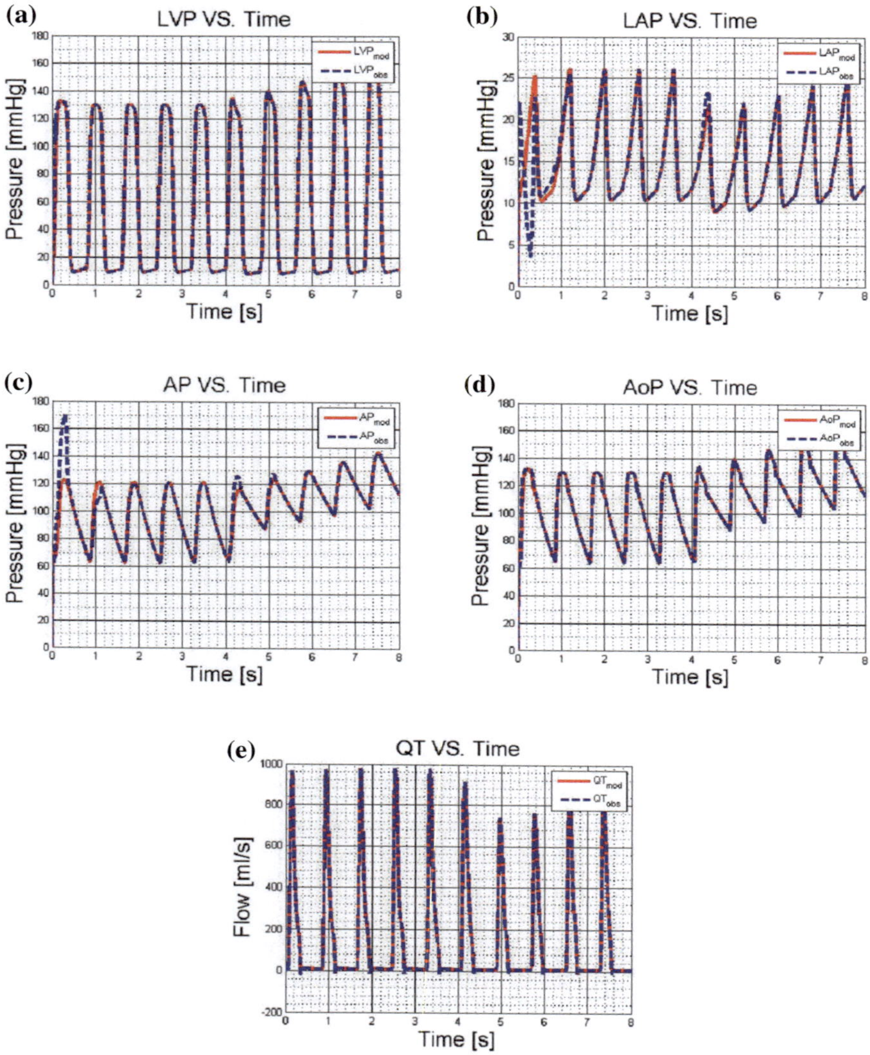

Fig. 13.7 Modeled and Estimated physiological variable, *mod* refers to the modeled vector state using the augmented system and *obs* refers to the observed state using an EKF. **a** LVP. **b** LAP. **c** AP. **d** AoP. **e** QT

where Q and R are gain matrices and $P_i(t)$ is the solution of the following differential Riccati equation:

$$\dot{P}_i(t) = -P_i(t)\tilde{C}^T R^{-1}\tilde{C} P_i(t) + B_i(t)P_i(t) + P(t)B_i(t)^T + Q. \tag{13.21}$$

The vector $\tilde{X}$ and the matrices $\tilde{A}_i(X,t)$, $B_i(t)$, $\tilde{C}$ are defined as follows:

$$\tilde{X} = [X(t),\ f_p(t)]^T, \tag{13.22a}$$

$$\tilde{A}_i(X,t) = \begin{bmatrix} A_i(t) & E_{f,i}(t,X) \\ 0 & 0 \end{bmatrix}, \tag{13.22b}$$

$$B_i = \frac{\partial \tilde{A}_i(X,t)\tilde{X}(t)}{\partial \tilde{X}(t)}\Big|_{\tilde{X}(t)=\hat{\tilde{X}}(t)}, \tag{13.22c}$$

$$\tilde{C} = \begin{bmatrix} C & 0 \end{bmatrix}. \tag{13.22d}$$

Equation (13.12) is used to introduce parameter anomalies in the CVS model. The abnormal vector (13.11) is composed of step changes that are equivalent to a percentage of the nominal value for the parameters. As an example, consider a simulation time of 8 s (corresponding to 10 cardiac cycles at 75 BPM). Changes in the parameters are taken as 30 % of their nominal value and appear at $t = 4$ s. Figure 13.7 shows the result of simulation using system (13.17). We observe that when the anomaly occurs, the systems deviate from the nominal behavior exhibited from time 0 to 4 s. Figure 13.8 shows the estimation of the anomalies using an EKF. We note that the estimates converge to the true values.

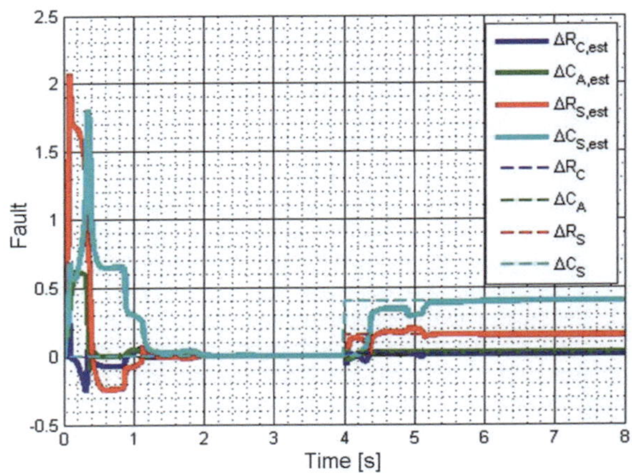

Fig. 13.8 Fault occurrence simulation

13.5 Conclusion

A hybrid model of the CVS has been discussed in this chapter. The model has been adapted to estimate cardiovascular anomalies related to variations in model parameters. We believe that the detection of some cardiovascular anomalies would be enhanced by taking advantage of the hybrid behavior of the system. Many hybrid tools are currently being developed for anomaly detection in hybrid systems, which can be used, in the future, to provide reliable techniques for detection of cardiovascular anomalies.

References

1. Lygeros, J.: Lecture Notes on Hybrid Systems. Department of Electrical and Computer Engineering, University of Patras, Patras (2004)
2. Lincoln, P., Tiwari, A.: Symbolic systems biology: hybrid modeling and analysis of biological networks. Syst. Biol. **2993**, 660–672 (2004)
3. Aihara, K., Suzuki, H.: Theory of hybrid dynamical systems and its applications to biological and medical systems. Philos. Trans. R. Soc. A **368**, 4893–4914 (2010)
4. Figueroa, A., Jansen, K.E., Hughes, T.J.R., Taylor, C.A.: A Coupled Momentum Method to Model Blood Flow in Deformable Arteries. Tsinghua University and Springer, Berlin (2004)
5. Olufsen, M., Nadim, A.: On deriving lumped models for blood flow and pressure in the systemic arteries. Math. Biosci. Eng. **1**(1), 61–80 (2004)
6. Frank, C.P., Zhaorong, L.: Estimating arterial resistance and compliance during transient conditions in humans. Am. J. Physiol. **257**, 190–197 (1989)
7. Westerhoff, N., Lankhaar, J., Westerhof, B.: The arterial windkessel, Medical and Biological Engineering and Computing, pp. 131–141 (2000)
8. Simaan, M.A., Ferreira, A., Chen, S., Antaki, J.F., Galati, D.G.: A dynamical state space representation and performance analysis of a feedback-controlled rotary left ventricular assist device. IEEE Trans. Control Syst. Technol. **17**(1), 1–6 (2009)
9. Faragallah, G., Wang, Y., Divo, E., Simaan, M.A.: A new control system for left ventricular assist devices based on patient-specific physiological demand. Inverse Prob. Sci. Eng. **53**(5), 721–734 (2012)
10. Wang, Y., Faragallah, G., Divo, E., Simaan, M.A.: Feedback control of a rotary left ventricular assist device supporting a failing cardiovascular system. In: Proceedings of the American Control Conference ACC, June 17–29, Fairmont Queen Elizabeth, Montréal, Canada (2012)
11. Stergiopulos, N., Westerhof, B.E., Westerhof, N.: Total arterial inertance as the fourth element of the windkessel model. Am. Physiol. Soc. **276**, 81–88 (1999)
12. Ledezma, F.D., Laleg-Kirati, T.M.: Detection of cardiovascular anomalies: hybrid systems approach. In: 4th IFAC Conference on Analysis and Design of Hybrid Systems (2012)
13. Belkhatir, Z., Laleg-Kirati, T.M., Tadjine, M.: Residual Generator for Cardiovascular Anomalies Detection, The 13th European Control Conference. Strasbourg, France (2014)
14. Williams, B.: Development of a novel method to device central aortic systolic from the radial pressure using an N-Point moving average method. J. Am. Coll. Cardiol. **57**, 951–961 (2011)
15. Ostrovsky, G.: USCOM non-invasive cardiac output monitor MedGadget,http://medgadget.com/2007/10/uscom_noninvasive_cardiac_output_monitor.html (2007)

Index

M. Djemai and M. Defoort (eds.), *Hybrid Dynamical Systems*, Lecture Notes in Control and Information Sciences 457, DOI 10.1007/978-3-319-10795-0

MIX
Papier aus verantwortungsvollen Quellen
Paper from responsible sources
FSC® C105338

If you have any concerns about our products,
you can contact us on
ProductSafety@springernature.com

In case Publisher is established outside the EU,
the EU authorized representative is:
Springer Nature Customer Service Center GmbH
Europaplatz 3, 69115 Heidelberg, Germany

Printed by Libri Plureos GmbH
in Hamburg, Germany